An Introduction to Fluorescence Correlation Spectroscopy

About the Series

The Biophysical Society and IOP Publishing have forged a new publishing partnership in biophysics, bringing the world-leading expertise and domain knowledge of the Biophysical Society into the rapidly developing IOP ebooks program.

The program publishes textbooks, monographs, reviews, and handbooks covering all areas of biophysics research, applications, education, methods, computational tools, and techniques. Subjects of the collection will include: bioenergetics; bioengineering; biological fluorescence; biopolymers *in vivo*; cryo-electron microscopy; exocytosis and endocytosis; intrinsically disordered proteins; mechanobiology; membrane biophysics; membrane structure and assembly; molecular biophysics; motility and cytoskeleton; nanoscale biophysics; and permeation and transport.

An Introduction to Fluorescence Correlation Spectroscopy

Thorsten Wohland
Departments of Biological Sciences and Chemistry,
National University of Singapore, Singapore

Sudipta Maiti
Department of Chemical Sciences, Tata Institute of Fundamental Research,
Mumbai, India

Radek Macháň
Singapore Centre for Environmental Life Sciences Engineering,
Nanyang Technological University in Singapore, Singapore

IOP Publishing, Bristol, UK

Multimedia and supplementary interactive exercises are available at http://iopscience.iop.org/book/978-0-7503-2080-1.

ISBN 978-0-7503-2080-1 (ebook)
ISBN 978-0-7503-2078-8 (print)
ISBN 978-0-7503-2081-8 (myPrint)
ISBN 978-0-7503-2079-5 (mobi)

DOI 10.1088/978-0-7503-2080-1

Version: 20201001

IOP ebooks

British Library Cataloguing-in-Publication Data: A catalogue record for this book is available from the British Library.

Published by IOP Publishing, wholly owned by The Institute of Physics, London

IOP Publishing, Temple Circus, Temple Way, Bristol, BS1 6HG, UK

US Office: IOP Publishing, Inc., 190 North Independence Mall West, Suite 601, Philadelphia, PA 19106, USA

Cover image adapted from artwork by Sonia Monti.

Contents

Preface

Fluorescence correlation spectroscopy (FCS) was developed in the 1970s based on the recognition early in the 20th century of the importance of fluctuations and how they relate to the underlying molecular processes. Since then it has developed and kept pace with technological advances and has branched nowadays into many different fluorescence fluctuation techniques. FCS has widely spread and is used in many different fields of research from materials to the life sciences. Just a search for FCS shows that in the ten years from 1972–81, 21 articles were published on this topic. In the ten year period from 2009 to 2018, there were 3099 publications on FCS.

Consequently, courses, workshops and seminars on fluorescence correlation techniques multiplied. And over the years, we have taught many of these courses on FCS and trained many students. During that time, we have all developed our own ways of providing insights into correlation functions and explaining the concepts of FCS. And in the end we realized that our approaches converged on similar teaching strategies and explanations. However, there was no book available that would have outlined these explanations and would have provided students with the reading material that should accompany such a course or workshop, or that would allow students to revise and study further after the inevitably short introductions provided by workshops. Therefore, we decided to come together and produce a book with the necessary reading material a beginner would need to start and then advance to an experienced user. And with the advent of e-books that allow including animations and interactive contents (available for this book at https://iopscience.iop.org/book/978-0-7503-2080-1), we feel that it is the right time to provide a compendium that can be used for self-study but can also support courses.

The book contains 10 chapters and is accompanied by an interactive computable document file. The chapters can be read in sequence or individually, depending on the reader's level of confidence in FCS and the need to study particular aspects of FCS. The first chapter is a general overview of FCS, explaining some background on how it developed and how it is linked to other techniques, and provides first ideas of the concepts involved. Chapter 2 introduces in detail the two main concepts of FCS, namely fluctuations and correlations in a stepwise fashion and discusses general properties of correlation functions. Chapter 3 provides an overview of the different detection schemes used in FCS before chapter 4 explains different data recording and processing strategies. Chapters 5 and 6 derive the theoretical models, which are required for data fitting, for various fluorescence correlation and cross-correlation spectroscopy (FCS and FCCS) methods. Chapter 6 explains in particular how FCCS can be used to obtain quantitative data on molecular interactions. Chapter 7 describes general artefacts that researchers encounter during FCS measurements and which need to be corrected for proper qualitative and quantitative interpretations of measurements. Chapter 8 describes noise, bias and FCS data fitting and model selection, providing the necessary background to understand how FCS data can be evaluated to obtain quantitative interpretations. Chapter 9 describes various experimental strategies for how the reader can proceed to plan FCS experiments. And chapter 10 provides a short summary and gives an outlook on current or expected future developments in FCS. To help the user visualize FCS concepts we include animated or interactive figures, where appropriate, to help explain concepts in FCS measurement strategies and FCS data evaluation. Finally, we include an interactive computable document file, that allows the reader to experiment with various FCS setup and fit parameters to develop a better understanding of how each one of them changes the correlation functions. These interactive graphics are all based on the theoretical models developed in the book and provide not only a good tool to test one's understanding but also to simulate experimental outcomes.

This book is meant as an introduction to the basics of FCS, including suggestions on where to look for more advanced information for the interested reader. It is in many ways incomplete. It is not a review and thus we focused on citing literature we deemed appropriate for helping in book's aim of teaching FCS concepts and articles we think are good gateways into more advanced topics. We tried to include original articles where possible but apologize for the references not representing a full overview of the field. Furthermore, it is not possible to capture all the different branches of fluctuation techniques in a small volume that is aimed at teaching the basics of FCS. We try to point towards extra information on these wherever possible. Nevertheless, we hope that this is a first introduction for the beginner into the world of fluctuations, a voyage we ourselves undertook many years ago and which has been full of rewarding insights and surprises.

Acknowledgements

This book would not have been possible without the help of many people not least the publishers who have patiently accompanied us on this journey. Many if not most of the animations in this book have been designed by Sonia Monti, who also made sure we used a color-deficient friendly palette. The confocal and total internal reflection microscope figures in chapter 3 are based on earlier versions drawn by Xioaxiao Ma and Ping Liu, both former graduate students in Thorsten's lab. Some of the figures and data shown in chapters 3 and 7 are from Vicky Visvakarma, Barun Maity and Arpan Dey in Sudipta's lab. The accompanying computable document format (CDF) files were largely written by us; the FCCS CDF is based on an early version written by Yong Hwee Foo, when he was a graduate student in Thorsten's lab. And finally, we would like to thank our families who had to suffer through the periods when we had to spend time writing the book but who were always supportive and understanding.

About the authors

Thorsten Wohland

Departments of Biological Sciences and Chemistry, National University of Singapore, Singapore

Thorsten studied Physics at the Technical University Darmstadt and the University of Heidelberg. He came into contact with biophysics while working on his diploma thesis in the lab of Ernst Stelzer on optical tweezers at the European Molecular Biology Laboratory (EMBL), Heidelberg. He then pursued his PhD at the École Polytechnique Fédéral de Lausanne (EPFL) in the group of Horst Vogel, where he encountered FCS for the first time and where he also met and collaborated with Rudolf Rigler. After a postdoctoral stay at Stanford University in the group of Richard Zare, where he worked on single molecule fluorescence spectroscopy, he took up a position at the National University of Singapore (NUS) where he is now Professor in the Departments of Biological Sciences and Chemistry. His interests lie in the development of fluorescence techniques and their application to biological systems. He has continued to further develop FCS and has organized and participated in multiple workshops on this topic which motivated the writing of this book.

Sudipta Maiti

Departments of Chemical Sciences, Tata Institute of Fundamental Research, Mumbai, India

Sudipta pursued his undergraduate studies in Physics at IIT Kanpur (India), and then obtained a PhD in Biophysics from the University of Pennsylvania, working with Robin Hochstrasser and Leslie Dutton on femtosecond spectroscopy of proteins. He subsequently joined Watt Webb at Cornell University, the lab that invented fluorescence correlation spectroscopy, and investigated protein fluctuations. There he also worked on label-free multiphoton imaging of neurotransmitters. He subsequently joined the Tata Institute of Fundamental Research (India), where he is currently a Professor of Chemical Sciences. His current research interest spans several different problems, namely, protein aggregation, monoaminergic neurotransmission, membrane interactions, and the development of optical techniques for studying biological phenomena. He is on the editorial board of several journals such as ACS Chemical Neuroscience, Frontiers of Physiology and Membrane Biophysics, and Biomedical Physics and Engineering Express. He has been the recipient of several prizes and fellowships, such as the SICO award from the National Academy of Sciences, TIFR Alumni Association Patent Award, and the Wellcome Trust Senior Overseas Research Fellowship. He is the co-founder and President of the Fluorescence Society (India) and President of the Society of Optics Within Life Sciences (Germany). He is a Fellow of the Indian Academy of Sciences.

Radek Macháň

Singapore Centre for Environmental Life Sciences Engineering, Nanyang Technological University in Singapore, Singapore

Radek Macháň obtained his PhD in Biophysics from Charles University in Prague in 2012. His doctoral thesis was done in the laboratory of Martin Hof and involved FCS as one of the main techniques. During his postdoctoral work in the laboratory of Thorsten Wohland, Radek used imaging FCS and FCCS to study protein dynamics in live-cell membranes. Since 2016, he has worked as imaging scientist in microscopy core facilities; first, at Charles University in Prague in a facility headed by Aleš Benda and since 2019 at Singapore Centre for Environmental Life Sciences Engineering located at Nanyang Technological University in Singapore. Throughout his work at the microscopy core facilities, Radek has continued to promote FCS and related techniques by assisting facility users with FCS experiments and by teaching FCS at courses and practical workshops.

IOP Publishing

An Introduction to Fluorescence Correlation Spectroscopy

Thorsten Wohland, Sudipta Maiti and Radek Macháň

Chapter 1

Introduction

This chapter provides an overview of the main principles and the historic development of fluorescence correlation spectroscopy (FCS) before more detailed discussions in the following chapters. FCS is based on the fact that noise carries information. Properties of a molecular system are inherently noisy since the molecules incessantly and randomly sample all the available physical and chemical states, driven by their thermal energy. The timescale of this wandering is reflected in the noise, and bears a signature of the molecular properties. FCS is one of the most exquisite methods to extract those molecular properties from the noise. It can measure the molecular size and the kinetics of physico-chemical transformations of fluorescent molecules in an equilibrium solution. This chapter introduces the basic concepts of FCS. It also describes its development in the initial years, introduces the basic design of an FCS instrument, and discusses measurements amenable to this technique.

1.1 What is fluorescence correlation spectroscopy all about?

FCS measures the rate of change in a chemical system. Of course, so do many other techniques. After all, measuring chemical kinetics has been a goal of scientists for centuries. However, there is something special about FCS that makes it different from any other technique. While other techniques typically measure kinetics by following reactants turning over to products, and finish the measurement process when the system ultimately attains equilibrium, FCS measurements begin at that point. They follow the tiny departures from equilibrium that spontaneously occur all the time in a system, even when it is perfectly at equilibrium. Due to the inherent entropy of the system, the system moves from microstates to microstates, randomly. An open system which can exchange energy and particles with the outside can fluctuate within the accessible states, with a timescale determined by the temperature and by the free energy barriers between these states. Any signal that is proportional (or at least deterministically related) to the population in a particular state also fluctuates.

doi:10.1088/978-0-7503-2080-1ch1

The timescales of these fluctuations, when analyzed, measures the same rate constants that other macroscopic techniques measure by following a departure from equilibrium. In doing so, FCS separately measures both the forward and backward rate constants, conformational dynamics, sizes of molecules (to Angstrom scale resolutions), and many other parameters that a single bulk kinetic measurement cannot access. And it accomplishes all this even at picomolar concentrations, just by using light, and does it very well in a test tube, in a living cell, or even in a living organism.

1.2 What do 'fluorescence', 'correlation' and 'spectroscopy' have to do with measuring change?

In FCS, measure of concentrations fluctuations is provided by fluorescence. Fluorescence is typically proportional to the population of the fluorescent state, and can be sensitively determined with relative ease. If the fluorescent molecules are reactive, and they take part in a reversible chemical reaction which changes their fluorescence properties, then the concentration fluctuations of individual species will result in fluctuations in fluorescence. Even without any reaction, if they randomly move from one place to another, the concentration at any given place will keep fluctuating, providing the basic signal for FCS.

But how do we get information about the actual parameters, e.g. forward and backward rate constants or diffusion constants, from these seemingly random fluctuations? FCS puts this time trace of fluorescence fluctuations through a mathematical funnel to distil out the relevant numbers. The function of 'correlation' (explained in chapter 2) acts as that funnel. It gives a measure of how long a time varying function remains self-similar, thereby measuring the characteristic decay time of the fluctuations. So in FCS the actual raw fluctuation data is almost never looked at. It is the auto (or sometimes 'cross') correlation of the fluorescence fluctuations that is considered to be the primary output of the experiment. The decay of the correlation function can be related to the underlying chemical kinetics or diffusion. Depending on the model of the chemical system, a function can be derived (though not always in a closed analytical form) whose fit to the data yields the underlying parameters.

So we understand the 'F' (fluorescence) and the 'C' (correlation) parts. What about 'spectroscopy'? Of course, fluctuations can have a spectrum, in the sense that a Fourier transform of the fluctuations in the frequency space will reveal its spectral components. An 'autocorrelation' is an inverse Fourier transform of the power spectrum. However, 'spectroscopy' for fluorescence conventionally refers to separating different colours of the emitted light. There is not much spectral determination that needs to be performed in FCS, beyond the selection of the appropriate spectral window for fluorescence collection, so the term 'spectroscopy' can be misleading to a newcomer to the field. In fact, Elliot Elson, one of the co-inventors of FCS, jokingly admitted in a teaching workshop on FCS (Fluorescence 2009, Mumbai, India) that 'we started with that name, and it stuck'.

1.3 What can FCS do for you?

FCS, depending on the characteristics of the chemical system under study, can measure a large number of properties. These include size, viscosity, ligand binding (through auto- or cross-correlation), absolute concentration of the solutes, velocity, chemical kinetics, intracellular characteristics, intramolecular dynamics, fluorescence lifetimes, triplet state lifetimes, anisotropy (rotational correlation time), number of fluorophores in a molecular complex, absolute diffusion constants, domain sizes for constrained diffusion, non-equilibrium steady state parameters, correlation between lifetime and size, and many others. Here we briefly describe a few of these applications.

The most obvious use of FCS is in measuring the hydrodynamic radius through diffusion. For solutes in a solution, diffusion is always present, leading to the fluctuation in the concentration of solutes at any given location. The time scale of these fluctuations is dictated by diffusion, and thus FCS can provide the diffusion constant. An accurate measure of the diffusion constant can provide the size of the macromolecule, provided the temperature and the viscosity of the medium is known. Here 'size' actually means the 'hydrodynamic radius', which is the effective radius of a molecule calculated from the Stokes–Einstein relation (equation (5.4)) when the diffusion coefficient is known. The hydrodynamic radius depends not only on the size of the molecule but also on its hydration shell. This size can be accurate to nearly an Angstrom, despite using only visible light microscopy to perform the experiment [1]. This is one of the most appealing features of FCS and is the reason why FCS is widely used in the field of protein aggregation. Alternatively, if the hydrodynamic radius of the particle is already known, then the diffusion constant can provide information about the viscosity of the medium. This can be a major source of information about the hydrodynamic properties of the cellular interior. Of course, the cellular interior can be a complex environment lacking a homogeneous viscosity. For example, a meshwork of microtubules may show that the diffusion constant is dependent on the length scale over which it is measured. This then provides valuable information about the internal structure of a cell.

The same measurement also yields the absolute concentration of a species in solution. The amplitude of the concentration fluctuations is related to the concentration itself. If a single species is present, the amplitude of the autocorrelation function at time $t = 0$ is the inverse of the number of molecules present in the probe volume. This does not depend on the brightness of the fluorophore. This inverse relationship makes FCS a curious technique where the amplitude of the measured signal actually increases as the concentration of the probe decreases (of course to a limit, as explained in chapters 5 and 7). The power of this absolute measurement has been exploited in estimating actual copy numbers of sparsely expressed proteins in the cell. Very few other techniques can give such an absolute measure of the concentration, especially in a living cell.

FCS also yields a wealth of other information, completely unrelated to the diffusive process. In principle, any process that can modulate the fluorescence properties of a molecule can give rise to fluctuations, provided they are faster than the diffusive process, and can be measured by FCS. In general, such processes can involve reversible physical or chemical reactions. The simplest to think of is an isomerization reaction which turns a bright fluorescent molecule into a dark one or vice versa. The fluctuation timescale can be followed by FCS. The chemical kinetic rate constants, together with the relative population of the two species, can be recovered from the data. More complex reaction schemes, involving multiple reactants, can also be followed using FCS. The major difficulty posed by complex reaction schemes lies in obtaining an analytical expression that relates the chemical kinetics to the decay of the autocorrelation function. Treatments starting with a generalized master equation can usually do the job, but it is not always possible to obtain a simplified description of the rate constants. However, the remarkable fact is that chemical kinetics can be measured from a system completely in equilibrium. This, therefore, marks a watershed in decades of progress in measuring chemical kinetics. It is interesting that this progress comes in the form of a technique which does not involve any externally imposed departure from equilibrium.

1.4 What does an FCS measurement involve?

In its most common implementation, an FCS spectrometer is nothing but a modified fluorescence microscope. It observes a sample on the stage, which may be just a drop of liquid for an *in vitro* measurement, or possibly a cell culture dish. It uses a laser as an excitation source, with a wavelength which is tuned to the absorption wavelength of the molecule being studied. In its most typical implementation, the laser is focused at a spot through an objective lens. The fluorescence is collected by the microscope and focused onto a small-aperture detector, which has high time resolution and sensitivity. The instrument keeps observing the same spot in the sample with time. Any changes in fluorescence emission are recorded by the detector. The raw data is therefore just a time trace of the fluorescence that emanates from a single spot in the sample. This time trace is mathematically transformed to a correlation function by the computer, which is typically fitted with special data processor electronics or efficient software algorithms for speedy computation of the correlation function. The correlation curve is fitted with different models to obtain the kinetic parameters, such as the timescales of diffusion or chemical reaction. The data fitting requires some knowledge of the system, but when performed with care, it can usually provide robust quantitative measures of the underlying parameters.

1.5 A brief history of FCS

1.5.1 Early work

Sometime in the late sixties, Magde, Elson and Webb at Cornell University understood the importance of measuring fluorescence fluctuations. They did not have to start from scratch. As in most things in science, FCS did not just fall from the sky one fine day. It had a precursor technique which prepared the ground for it. The technique that preceded FCS was dynamic light scattering (DLS), which is still a major biophysical method used in many labs. Concentration fluctuations of solutes in a solution cause local changes in its refractive index. This is what DLS measures.

When light falls on the liquid, these random local changes in refractive index cause light to be scattered. The scattering is time dependent, since the local fluctuations are continuously modulated by random diffusive motion. The scattered light is (at least partially) coherent, so light scattered from different locations interferes constructively or destructively, depending on their path length to the detector. The amplitude of the scattered light depends on the refractive index variations at the particular location, which in turn depends on the solute concentration. So, as the concentration fluctuates continuously about a mean value, the interference pattern at the detector also changes continuously. An analysis of these scattering signals in time, usually using an 'auto-correlation function' (a concept that we will talk about in chapter 2) reports how quickly the scattering changes. However, since 'how quickly it changes' is determined by diffusion, this is a way to measure the parameters that control diffusion. That, in turn, is determined by the size of the solutes in a solution of a given viscosity, so finally it measures molecular size (i.e. the hydrodynamic radius). This is how DLS works (*for a detailed exposition, see* [2]). This useful technique suffers from several major constraints which limit its effectiveness in many situations. First, the minimum concentration at which one can get appropriate signals from typical proteins is in the micromolar range. Second, the scattering is mostly dependent on the size and shape of the scatterer, and is insensitive to other properties or modifications. Third, it is not possible to distinguish between biomolecules in a solution using their scattering properties (other than the fact that a larger molecule will scatter more). Fourth, the scattering properties of a biomolecule, such as a protein, changes little upon reaction, so it is practically impossible to measure chemical kinetics from scattering measurements. Finally, DLS is not a microscopic technique, so it is not possible to measure diffusion inside a living cell with it.

Magde, Elson and Webb (MEW) decided that the limitations of DLS were quite crippling for many biophysical measurements. They were looking for a technique to measure intracellular local conformational changes of the genomic DNA. DLS was not appropriate to perform such a measurement. However, the concepts of probing equilibrium fluctuations, and using the autocorrelation function to characterize the fluctuations, served as guiding lights for MEW to circumvent the limitations of DLS. They moved from scattering to fluorescence, which is a property that can be detected at the single molecule level (as was to be discovered later [3, 4]). It is also extremely sensitive to local changes, such as folding or ligand binding, but insensitive to size. In the process, all the major limitations of DLS were avoided, and the palette of measurable properties expanded from simple diffusion to chemical kinetics. The penalty to be paid was the requirement that the molecule of interest had to be labelled with a reasonably stable fluorescent moiety. MEW accepted that limitation, and a revolution in biophysics ensued [5]. It started slowly, due to the extremely difficult signal-to-noise conditions offered by the instrumentation of that era. At around the same time, Rudolf Rigler's group in Europe was separately pushing for higher sensitivity and time resolution [6], using newer optical geometries and electronic hardware. Modern semiconductor photodetectors and computing hardware became widely available by the nineties, and FCS served as an easily implementable method in the lab, and indeed, commercial FCS machines started appearing on the market. MEW and Rigler created a new field that is still developing, enabling new measurements which have been enriching biophysics and chemical biology for nearly half a century.

1.5.2 The year of the invention

MEW had to struggle for a few years before the experiments and the theory fell in place. It was not an easy experiment to perform half a century ago. The detectors were insensitive, the laser had just been invented, and the data processor (the 'correlator card') was slow. The first paper was truly a tour de force [5], as it detailed all the important aspects of the theory and the experiment. They determined the diffusion constant of the DNA binding dye ethidium bromide, and also its binding constant to DNA. The initial wave of excitement in the Cornell labs mellowed down in a few years. However, the elegant method was separately pushed by the group of Rudolf Rigler at the Karolinska Institute in Sweden [6]. They had described an elegant FCS method to measure rotational diffusion. Rigler's group was responsible for pioneering technical improvements of FCS, and they continuously improved the size resolution and the signal-to-noise ratio of the experiment. They also provided a detailed study of the photophysics of dye molecules, which laid the basis for much of the later work in this area [7]. Their group pushed the sensitivity past the single molecule limit, which was an important milestone [6]. Their experiments caught the attention of the well-known chemical physicist Manfred Eigen. Eigen had earlier received a Nobel Prize on measuring chemical kinetics using a perturbative approach that disturbs the equilibrium [8]. The idea of spontaneous equilibrium fluctuations reporting many parameters of a chemical reaction was perhaps naturally appealing to him. Europe soon became the hotbed of FCS, with multiple centres developing their own instruments, and more importantly, coming up with beautiful experiments based on FCS. Some of these progresses were encapsulated in a review paper by Eigen and Rigler [9], which provided a brief overview of the technique. This was received enthusiastically by researchers in the area, and it helped fuel an exponential growth of the field in the subsequent years.

1.5.3 The initial progress

Some of the key initial successes of the technique helped it achieve the position of a predominant technique in biophysics. Initial forefronts were measurement of the intracellular viscosity, rates of biomolecular and photophysical reactions, analysis of multiple fluorophores, analysis of molecular aggregation, and measurement of the conformational dynamics of biomolecules. Experimental progress went hand-in-hand with theoretical developments. First comprehensive theoretical signal-to-noise analysis of FCS was done by Dennis Koppel [10]. Nancy Thompson showed that higher order correlation functions are very powerful in studying molecular aggregates (reviewed in [11]). Nils Petersen demonstrated 'scanning FCS', which circumvented the limitations of FCS in studying molecular species which diffuse too slowly for convenient recording, and also opened up FCS measurements over an extended area [12]. Dual colour fluorescence cross correlation spectroscopy was demonstrated by the Rigler group [13], and provided an elegant method to study molecular interactions between different species. Kinetics of chemical and physical transformations of biomolecules were demonstrated by the nineties. Libchaber and coworkers measured the conformational dynamics of a hairpin DNA molecule, using a FRET pair as a molecular beacon for DNA hairpin formation [14]. The rate constant of protonation of GFP was measured in the Webb lab [15]. Gratton *et al* brought two-photon excitation to FCS and measured intracellular diffusion [16].

Measurement of protein conformational fluctuations and folding needed to wait for a little more time. In principle, FCS can measure fluctuations of proteins around the midpoint of folding, which yield the folding and the unfolding rates of a protein. The protein folding problem was enticing for FCS, because the typical approaches for measuring the kinetics of folding involved a perturbation at time = 0. Mixing was an easy way to define $t = 0$, but its time resolution was limited to about a millisecond. Much faster approaches involved the change of conditions by a laser flash, but this could work only in special cases involving light absorbing proteins, or under specialized solution conditions. The emergence of FCS removed the requirement for a $t = 0$ trigger in the folding kinetics experiments. Moreover, it yielded the folding equilibrium constant, and the forward and backward rate constants all together in one single experiment. If there were multiple transitions at multiple time scales, the technique was capable of measuring that too. It is still one of the foremost techniques for probing the protein folding landscape. Thus, some of the early experiments using FCS showed the way to solve some of the key biophysical problems of the era.

Intracellular FCS opened another new era for chemical kinetics and diffusion measurements in a living system. There were very few techniques available which could measure chemical kinetics in a living cell. Diffusion measurements were usually performed using FRAP, which is a strongly perturbative technique. The initial intracellular FCS measurements were performed by the group of Enrico Gratton [16]. It also embodied another first, that of applying multiphoton excitation in FCS. This technique used femtosecond lasers to cause a simultaneous two (or higher order)-photon excitation of fluorophores at the focus of a laser beam. The excitation usually decays quickly to a single-photon excited state and the fluorescence occurs from that state. Its nonlinear dependence on the excitation power automatically provides a small probe volume, which is appropriate for FCS. It had another major advantage. The wavelength of excitation is far removed from the wavelength of emission in two-photon excitation, so it is easy to remove the scattering background. Yet another advantage is that the toxicity of the two-photon excitation for most types of cells is lower than that of the single-photon excitation of the same fluorophore, since the excitation is confined to the excitation volume only. Gratton *et al* used fluorescently labelled beads, which were bright and slow moving, and therefore easier to probe. Of course, single-photon FCS of biomolecules inside cells soon became a reality. Employing GFP as a reporter (solution state studies: [17]) opened up the most fruitful avenues for intracellular measurements. Politz *et al* [18] measured the diffusion of fluorescently labelled DNA probes inside the nucleus. By the end of the nineties, all this progress had made FCS a well-known name, at least in the fluorescence and biophysics community.

From that early progress, FCS made further strides in the next two decades. While there have been further improvements in the hardware (e.g. higher time resolution detectors, better single-photon avalanche photodiodes (SPAD), SPAD detector arrays, higher time resolution correlator cards, etc), more significant advances have happened in the development of different types of methodologies based on FCS, or in allied techniques. These are described in detail in later sections of this book.

1.6 Critical technical steps of the revolution

1.6.1 Fluorescence: towards single molecule sensitivity

In fact, FCS is the first 'single molecule level' fluorescence measurement technique. It is the forerunner of all the true single molecule techniques that arose later, which were rewarded with a Nobel Prize. The big difference from scattering to fluorescence as the measurable quantity was that a single molecule scatters very little, but it can fluoresce brightly enough to be observed (though true single-molecule sensitivity in FCS emerged mostly in the nineties with the advent of the single-photon counting avalanche photodiodes).

1.6.2 Microscopic volume

In FCS it was no longer necessary to collect scattering from many neighbouring volumes for interference and detection, as was required in DLS. Collecting the fluorescence from a very small volume was enough for the correlation data to yield the underlying physical properties. This gave rise to the ability to measure the fluctuations in the smallest optically definable volume, which was diffraction limited and had sub-femtoliter size, inside a specific location of a living cell. This completely changed the variety of systems accessible for measurement. It went from a carefully filtered solution in a test tube, as was used in DLS, to a sub-femtoliter volume *in vitro* containing sub-nanomolar concentrations of a fluorescently labelled protein, and finally to proteins inside the cell.

1.6.3 The confocal technique

The confocal technique revolutionized FCS as it provided a new way of optically defining a small volume in three dimensions. This technique of course became widely known only in the eighties with the advent of the confocal microscopes, though the confocal principal was known in the fifties (Minsky, US Patent US 3013467, 1961). The initial FCS measurements used a quasi-two dimensional volume. The confocal geometry allowed the probe volume to be limited to sub-femtoliter volumes, and also offered the ability to perform it within an imaging set-up, thereby allowing the probing of particular sites in a sample, e.g. in a living cell. The signal from FCS is inversely proportional to the number of fluorescent particles in the volume, so a smaller probe volume increased the signal amplitude. It helped improve another limitation of the technique, which was the low concentrations required for the experiments. At higher concentrations, the number of particles in a volume would be higher, so a smaller volume would be needed to keep the amplitude of the FCS signal reasonable. It also helped in keeping the signal to background levels high. The relationship between the sample concentration, the probe volume size and the signal-to-noise is somewhat complex (as described in chapter 7). However, within some limits, a smaller probe volume helps generate better data. All this translated to a larger signal-to-noise ratio for FCS experiment when a confocal geometry was employed in the experiment.

1.6.4 Modern detectors

It was not just the confocal geometry, but the advent of the SPADs (single-photon counting avalanche photodiodes, also often simply called avalanche photodiode—APD) which made FCS signals easily observable. The APDs boosted the quantum efficiency of the detection process by almost an order of magnitude compared to the photomultiplier tube (PMT), which was the detector of choice until the eighties. This was especially true in the ~600 nm region, where many photostable organic dyes fluoresced. There have been continuous developments in this field, in improving the quantum efficiency, dead time, multiple pixel nature, and time resolution of the detectors. These have bread new types of methods which have surpassed what was perhaps the original aim of the inventors of this method.

1.6.5 The data processors

The next challenge in FCS was to process the data. A correlation is a computationally expensive operation, as it requires a multiplication of two waveforms (i.e. time traces). The amount of calculation depends on the number of data points in the waveforms. Assuming a 2 min (i.e. ~100 s) long data record, which contains a data point at every 10 ns (a fairly common time resolution for FCS), the waveform will have about 10 billion points. This would require about 10 billion calculations for every 'delay time' point in the correlation curve. The number of delay time points in the correlation curve could also be 10 billion. This would require 10^{20} calculations for the 2-min data. If we want the correlation curve to be calculated 'online' (i.e. without delay), then this would need 10^{18} multiplications per second. This is beyond the capability of a typical personal computer. The original correlators had few time bins, and were rather slow in calculating the correlation function. This problem was solved by a combined algorithmic approach (called the 'multi-tau' algorithm) and by employing specialized data processors [19]. Fast processors emerged in the nineties, just as the optical and detection methods were coming of age, and FCS started being recognized for the power it offered.

1.7 Where is FCS now?

From the early invention, FCS made further strides in the next two decades. The major advancement has happened in the development of different types of methodologies based on FCS, or in allied techniques. The initial problem of slow acquisition times for slowly diffusing species, e.g. those diffusing on the membrane, has been turned into an advantage. Scanning FCS [20, 21] was a powerful approach to tackle this problem, but initially diffusion was not a parameter that could be measured in scanning FCS. Later approaches which measured diffusion for slowly diffusing species involved implementing FCS in a laser scanning confocal microscope. Using the confocal scanning hardware, Gratton *et al* scanned a sample in a circular pattern. This in effect performed FCS sequentially at multiple probe volumes, each pixel being one such. The scanner came back to the same pixel every ~ms or so, and therefore one could autocorrelate the intensity versus time trace for each pixel separately [22]. Later, Maiti *et al* demonstrated a somewhat simpler arrangement of scanning a single line to obtain similar data [23]. If the diffusion is slower than the repetition rate, then one could average these autocorrelations to obtain an FCS curve. The advantage of this method lies in the multiplexing. This gave acceptable signal-to-noise even for the inferior collection efficiency of the PMT based confocal detectors, compared to the more expensive APD. Since the region of interest for scanning could be chosen from the image of a sample, it also automatically allowed measurements of local viscosity in different regions of a heterogeneous sample.

In a parallel development, detectors with multiple detection elements were used to obtain various fluorescence correlation functions. Petersen introduced image correlation spectroscopy that exploited spatial correlations to determine membrane organization of receptor proteins [12]. This was extended by the Weismann group to the temporal domain in spatiotemporal image correlation spectroscopy (STICS) [24] that analysed spatial correlations in time. Brinkmeier *et al* had shown earlier how two-beam cross-correlation provides additional information for temporal FCS measurements [25]. Goesch *et al* demonstrated then the first parallel FCS measurements on a 2 × 2 detector [26]. And this approach was brought to full cameras with spinning disk confocal [27], TIRF [28, 29] and light sheet microscopy [30]. In these latter approaches, the advantage as that FCS was performed at all the pixels of the image, and one could actually provide a diffusivity map of the whole sample (chapter 5).

Another major advancement has been in shrinking the probe volume of FCS. In a typical microscope, the probe volume is limited to the diffraction limit. Webb *et al* used evanescent field excitation through a sub-resolution aperture to achieve volumes which are about 10 000 times smaller than the diffraction limit [31]. Such a small probe volume allowed analyte concentrations as high as ~mM. Later, with the advent of stimulated emission depletion (STED) microscopy, sub-resolution FCS was achieved inside a solution without any physical aperture [32]. Reactions which happen only at high concentrations could be accessed through such small volume measurements. Probe volumes in FCS will be further discussed in chapter 5.

The decay of the autocorrelation function can yield diffusion constants, provided one knows the size of the probe volume. This is frequently not possible to determine in a very accurate manner. So typical FCS based size measurements compare the diffusion time of an unknown analyte with that of a known one, resulting in a relative measurement. However, for absolute determination of diffusion constants, one needs to know the exact distance parameters. This problem has been elegantly solved by Enderlein *et al*, by performing two-focus FCS [33] (chapter 5). The distance between the foci is separately determined and known, and the cross correlation between the two foci shows how the fluorescence molecules diffuse through the two spots. This allows the determination of an absolute diffusion constant.

Cross correlation FCS of course does not need to be confined to two signals only. One can define cross correlation between any number of signals, and this would confirm the co-diffusion of more than two types of molecules together. This was demonstrated recently in the context of multivalent ligand binding of a protein [34].

One major disadvantage of conventional FCS is the need to label the molecule of interest with a fluorescent moiety. However, proteins containing aromatic amino acids display their own fluorescence, and in principle, FCS can be performed using that native fluorescence [35, 36]. In practice, though, this has proven to be extremely difficult. The aromatic amino acids such as tryptophan are only weakly fluorescent and are very easily photo-bleached. Obtaining multiple photons from each molecule (a requirement for FCS data) during their transit through the probe volume becomes difficult. Moreover, they are fluorescent only when excited in the ultraviolet, and this can destroy the sample. One possibility that was explored was to probe proteins with many Trp residues, which together can provide adequate fluorescence. Though such proteins are not common, aggregation of proteins can provide the same large number of Trp residues in an aggregated particle being investigated. This was performed by the group of Maiti *et al*, when they pursued the aggregation of the protein barstar [37].

The size of an FCS instrument is, at the minimum, equivalent to the size of a fluorescence microscope, which is not exactly an easily portable device. The use of FCS may go up substantially if a small portable device can be made. This was demonstrated by the group of Maiti *et al* when they made a 'palmtop' FCS device [38]. This makes use of fiber optics for both light delivery and light collection from the objective, and the objective and the sample are held in a small device that can fit on one's palm. In a further miniaturized version, the objective lens is done away with, and the tip of a single mode fiber acts as the collection lens. The collection NA is only around 0.1, and so the sensitivity is low, but this opens up tremendous opportunities to perform FCS in a remote or hostile/toxic environment, or through very small orifices, or even in a biological cavity.

A major advancement in FCS involves its combination with fluorescence lifetime measurements. The lifetime of a fluorophore is a sensitive reporter of its local environment, and provides complementary information to what is available from FCS. The lifetime of a fluorophore is reliably determined by time correlated single-photon counting (TCSPC). This is perfectly compatible with FCS, the only modification needed is that the laser source should be pulsed (ideally with a sub-nanosecond pulse width), and the detector should have similarly high time resolution. Of course, the data processor cards then would also have to keep track of both types of photon arrival times, i.e. the time delay between the last excitation pulse and the fluorescence photon (sometimes called the 'microtime', which should be accurate to ~100 ps or better for reasonable TCSPC), and the arrival time according to a lab clock (which should be better than ~1 μs for FCS autocorrelation measurements, sometimes called 'macrotime'). All of these became available early this century, and the combined power of these two techniques has made new types of measurements possible. For example, it is now possible to determine the diffusion time of a short lifetime species separately from that of the long lifetime species present in the same solution, and marked by the same dye. This technique, christened 'fluorescence lifetime correlation spectroscopy' (FLCS), has become a very powerful method to probe ligand interactions and conformational heterogeneity of a molecule in solution [39]. In a recent development, the laboratory of Tahei Tahara has demonstrated a two-dimensional FCS method, which provides a 2D map of the population of interconverting species in a solution [40]. Though it requires very high signal-to-noise, and therefore long collection times, it has the potential to become the most effective utilization of these two types of temporal information.

As we will see in the rest of this book, there have been major further developments which have enriched the field of FCS and enabled it to solve many biophysical and biological mysteries. Given all the advantages of FCS, especially its high sensitivity and its non-perturbative nature, one can expect the development of many more allied methods in the future. In the remaining chapters of the book, we aim to clarify the main concepts of FCS that are necessary for the reader to dive deeper into the subject, and equip them to explore even these advanced methods.

References

[1] Goossens K, Prior M, Pacheco V, Willbold D, Müllen K, Enderlein J, Hofkens J and Gregor I 2015 Accurate diffusion coefficients of organosoluble reference dyes in organic media measured by dual-focus fluorescence correlation spectroscopy *ACS Nano.* **9** 7360–73

[2] Berne B and Pecora R 1976 Dynamic light scattering with applications to biology *Chemistry and Physics* vol 1 (New York: Wiley)

[3] Orrit M and Bernard J 1990 Single pentacene molecules detected by fluorescence excitation in a p-terphenyl crystal *Phys. Rev. Lett.* **65** 2716–9

[4] Brooks Shera E, Seitzinger N, Davis L, Keller R and Soper S 1990 Detection of single fluorescent molecules *Chem. Phys. Lett.* **174** 553–7

[5] Magde D, Elson E and Webb W 1972 Thermodynamic fluctuations in a reacting system—measurement by fluorescence correlation spectroscopy *Phys. Rev. Lett.* **29** 705–8

[6] Ehrenberg M and Rigler R 1974 Rotational Brownian motion and fluorescence intensify fluctuations *Chem. Phys.* **4** 390–401

[7] Widengren J, Mets U and Rigler R 1995 Fluorescence correlation spectroscopy of triplet states in solution: a theoretical and experimental study *J. Phys. Chem.* **99** 13368–79

[8] Eigen M 1967 Immeasurably fast reactions. Nobel lecture https://www.nobelprize.org/prizes/chemistry/1967/eigen/lecture/

[9] Eigen M and Rigler R 1994 Sorting single molecules: application to diagnostics and evolutionary biotechnology *Proc. Natl Acad. Sci. USA* **91** 5740–7

[10] Koppel D 1974 Statistical accuracy in fluorescence correlation spectroscopy *Phys. Rev.* A **10** 1938

[11] Thompson N L 1991 Fluorescence correlation spectroscopy *Topics in Fluorescence Spectroscopy vol 1: Techniques* ed J R Lakowicz (New York: Plenum Press) pp 337–78
[12] Petersen N O, Höddelius P L, Wiseman P W, Seger O and Magnusson K E 1993 Quantitation of membrane receptor distributions by image correlation spectroscopy: concept and application *Biophys. J.* **65** 1135–46
[13] Schwille P, Meyer-Almes F J and Rigler R 1997 Dual-color fluorescence cross-correlation spectroscopy for multicomponent diffusional analysis in solution *Biophys. J.* **72** 1878–86
[14] Bonnet G, Krichevsky O and Libchaber A 1998 Kinetics of conformational fluctuations in DNA hairpin-loops *Proc. Natl. Acad. Sci. USA* **95** 8602–6
[15] Haupts U, Maiti S, Schwille P and Webb W W 1998 Dynamics of fluorescence fluctuations in green fluorescent protein observed by fluorescence correlation spectroscopy *Proc. Natl. Acad. Sci. USA* **95** 13573–8
[16] Berland K M, So P T and Gratton E 1995 Two-photon fluorescence correlation spectroscopy: method and application to the intracellular environment *Biophys. J.* **68** 694–701
[17] Terry B R, Matthews E K and Haseloff J 1995 Molecular characterisation of recombinant green fluorescent protein by fluorescence correlation microscopy *Biochem. Biophys. Res. Commun.* **217** 21–7
[18] Politz J C, Browne E S, Wolf D E and Pederson T 1998 Intranuclear diffusion and hybridization state of oligonucleotides measured by fluorescence correlation spectroscopy in living cells *Proc. Natl. Acad. Sci. USA* **95** 6043–8
[19] Schaetzel K 1985 New concepts in correlator design *Inst. Phys. Conf. Ser.* **77** 175–84
[20] Petersen N O, Johnson D C and Schlesinger M J 1986 Scanning fluorescence correlation spectroscopy. II. Application to virus glycoprotein aggregation *Biophys. J.* **49** 817–20
[21] Petersen N O 1986 Scanning fluorescence correlation spectroscopy. I. Theory and simulation of aggregation measurements *Biophys. J.* **49** 809–15
[22] Levi V, Ruan Q, Kis-Petikova K and Gratton E 2003 Scanning FCS, a novel method for three-dimensional particle tracking *Biochem. Soc. Trans.* **31** 997–1000
[23] Balaji J and Maiti S 2005 Quantitative measurement of the resolution and sensitivity of confocal microscopes using line-scanning fluorescence correlation spectroscopy *Microsc. Res. Tech.* **66** 198–202
[24] Hebert B, Costantino S and Wiseman P W 2005 Spatiotemporal image correlation spectroscopy (STICS) theory, verification, and application to protein velocity mapping in living CHO cells *Biophys. J.* **88** 3601–14
[25] Brinkmeier M, Dörre K, Stephan J and Eigen M 1999 Two-beam cross-correlation: a method to characterize transport phenomena in micrometer-sized structures *Anal. Chem.* **71** 609–16
[26] Gösch M, Serov A, Anhut T, Lasser T, Rochas A, Besse P-A, Popovic R S, Blom H and Rigler R 2004 Parallel single molecule detection with a fully integrated single-photon 2 × 2 CMOS detector array *J. Biomed. Opt.* **9** 913
[27] Sisan D R, Arevalo R, Graves C, Mcallister R and Urbach J S 2006 Spatially resolved fluorescence correlation spectroscopy using a spinning disk confocal microscope *Biophys. J.* **91** 4241–52
[28] Kannan B, Guo L, Sudhaharan T, Ahmed S, Maruyama I and Wohland T 2007 Spatially resolved total internal reflection fluorescence correlation microscopy using an electron multiplying charge-coupled device camera *Anal. Chem.* **79** 4463–70

[29] Kannan B, Har J, Liu P, Maruyama I, Ding J and Wohland T 2006 Electron multiplying charge-coupled device camera based fluorescence correlation spectroscopy *Anal. Chem.* **78** 3444–51

[30] Wohland T, Shi X, Sankaran J and Stelzer E 2010 Single plane illumination fluorescence correlation spectroscopy (SPIM-FCS) probes inhomogeneous three-dimensional environments *Opt. Express* **18** 10627–41

[31] Levene M J 2003 Zero-mode waveguides for single-molecule analysis at high concentrations *Science* **299** 682–6

[32] Kastrup L, Blom H, Eggeling C and Hell S 2005 Fluorescence fluctuation spectroscopy in subdiffraction focal volumes *Phys. Rev. Lett.* **94** 178104

[33] Dertinger T, Pacheco V, Von Der Hocht I, Hartmann R, Gregor I and Enderlein J 2007 Two-focus fluorescence correlation spectroscopy: a new tool for accurate and absolute diffusion measurements *ChemPhysChem* **8** 433–43

[34] Ridgeway W K, Millar D P and Williamson J R 2012 Quantitation of ten 30 s ribosomal assembly intermediates using fluorescence triple correlation spectroscopy *Proc. Natl Acad. Sci.* **109** 13614–9

[35] Wennmalm S, Blom H, Wallerman L and Rigler R 2001 UV-Fluorescence correlation spectroscopy of 2-aminopurine *Biol. Chem.* **382** 393–7

[36] Lippitz M, Erker W, Decker H, van Holde K E and Basché T 2002 Two-photon excitation microscopy of tryptophan-containing proteins *Proc. Natl. Acad. Sci. USA* **99** 2772–7

[37] Sahoo B, Balaji J, Nag S, Kaushalya S K and Maiti S 2008 Protein aggregation probed by two-photon fluorescence correlation spectroscopy of native tryptophan *J. Chem. Phys.* **129** 075103

[38] Garai K, Muralidhar M and Maiti S 2006 Fiber-optic fluorescence correlation spectrometer *Appl. Opt.* **45** 7538–42

[39] Böhmer M, Wahl M, Rahn H J, Erdmann R and Enderlein J 2002 Time-resolved fluorescence correlation spectroscopy *Chem. Phys. Lett.* **353** 439–45

[40] Ishii K and Tahara T 2013 Two-dimensional fluorescence lifetime correlation spectroscopy 1. principle *J. Phys. Chem.* B **117** 11414–22

Thorsten Wohland, Sudipta Maiti and Radek Macháň

Chapter 2

Correlation functions

Fluorescence correlation spectroscopy (FCS) is based on two concepts, fluctuations and correlations. This chapter introduces these two concepts and explains the calculation of a correlation coefficient for the case of discrete variables (viz. the toss of two coins). The correlation coefficient is then generalized to a correlation function for the correlation between two continuous variables (viz. a signal in time). We deal with both cases, the correlation between two distinct variables, called a cross-correlation function (CCF), and the correlation of a variable with itself, called the autocorrelation function (ACF), which describes the self-similarity of a signal in time. We end with the derivation of a number of properties of the ACF and CCF that are important for the understanding of FCS.

2.1 Introduction

FCS aims at gaining information about molecular processes occurring in a system. The system under investigation can be simple solutions, complex mixtures and environments, cells, tissue or organisms. Any molecular process in such a system is constituted by molecular events that lead to changes in an observable parameter. This can be simply a change in the position of molecules, e.g. as happens in diffusion or transport, it can be a change in the fluorescent state of a molecule, e.g. a transition between a singlet and a triplet state, it can be the change of the identity of the molecule as happens in a chemical reaction. The time course of these changes provides information about the molecular processes.

To understand how FCS measures characteristics of these molecular processes based on their time course, two concepts are required. The first is the idea of fluctuations and why they carry information about molecular processes. An excellent discussion of fluctuations can be found in the book by MacDonald [1]. The second is the concept of correlations and how they can extract information from a signal and in particular from fluctuations. Correlations have been treated in many

doi:10.1088/978-0-7503-2080-1ch2

textbooks on statistics and random variables; for an introduction see, e.g. chapter 8 in the book *Probability and Random Processes* [2].

2.2 Fluctuations

So let's start with fluctuations. As just explained, a molecular process changes some, in principle, observable parameter. We have to differentiate between two cases for a process, the case of an equilibrium and the case of kinetics where a process continuously changes a parameter, i.e. the period before equilibrium is reached.

In this first part we will use chemical reactions as an example as the idea of fluctuations and their information content is easily explained using chemical concepts. Once we have clarified the ideas of fluctuations and correlations we can then extend it to any process.

Let's assume we have a chemical reaction of compounds A and B, whose concentrations we will denote by square brackets [...]. Compound A reacts with compound B to give a compound AB:

$$\mathrm{A} + \mathrm{B} \underset{k_b}{\overset{k_f}{\rightleftarrows}} \mathrm{AB} \tag{2.1}$$

where k_f and k_b represent the forward and backward reaction rate constants. The change in concentration of each of the three compounds is given by the following three equations

$$\frac{d[\mathrm{AB}]}{dt} = k_f[\mathrm{A}][\mathrm{B}] - k_b[\mathrm{AB}] \tag{2.2a}$$

$$\frac{d[\mathrm{A}]}{dt} = -k_f[\mathrm{A}][\mathrm{B}] + k_b[\mathrm{AB}] \tag{2.2b}$$

$$\frac{d[\mathrm{B}]}{dt} = -k_f[\mathrm{A}][\mathrm{B}] + k_b[\mathrm{AB}] \tag{2.2c}$$

The first equation describes how the concentration [AB] increases with the forward reaction of compounds A and B, depending on their concentration [A] and [B], and how it decreases with the backward reaction of the dissociation of AB into A and B, depending on the concentration [AB]. The next two equations describe how the concentrations [A] and [B] decrease when they react to form AB and how they increase by the dissociation of AB.

In an equilibrium the change for all three compounds must be 0 and therefore

$$k_f[\mathrm{A}][\mathrm{B}] = k_b[\mathrm{AB}] \tag{2.3}$$

Using this condition we can define the association constant K_a which describes the equilibrium.

$$K_a = \frac{k_f}{k_b} = \frac{[\mathrm{AB}]}{[\mathrm{A}][\mathrm{B}]} \tag{2.4}$$

Let's take a closer look at what happens in the reaction. Each reaction of A and B will lead to one more compound AB thus increasing the concentration [AB]. In the case of the backward reaction, when a compound AB falls apart into its components A and B, the concentration [AB] decreases. If we add at a time $t = 0$ two amounts of A and B into a solution they will start producing the compounds AB, and [AB] will increase on average with time, providing the reaction kinetics (figure 2.1). But of course the forward and backward reactions will take place at any time, it is only on average that we will see an increase of the concentration [AB]. Thus if we focus in on the kinetic rise of the curve for the concentration of [AB] we will see forward reactions and backward reactions, i.e. we see fluctuations of the concentration [AB]. In this case the forward reactions will dominate and [AB] will rise on average. However, with time [A] and [B] deplete and the number of forward reactions decreases. At the same time the concentration [AB] increases and thus the number of backward reactions increases. At one point the forward and backward reactions will be the same and we have reached equilibrium. However, we can still observe fluctuations around the equilibrium due to the continuously ongoing forward and

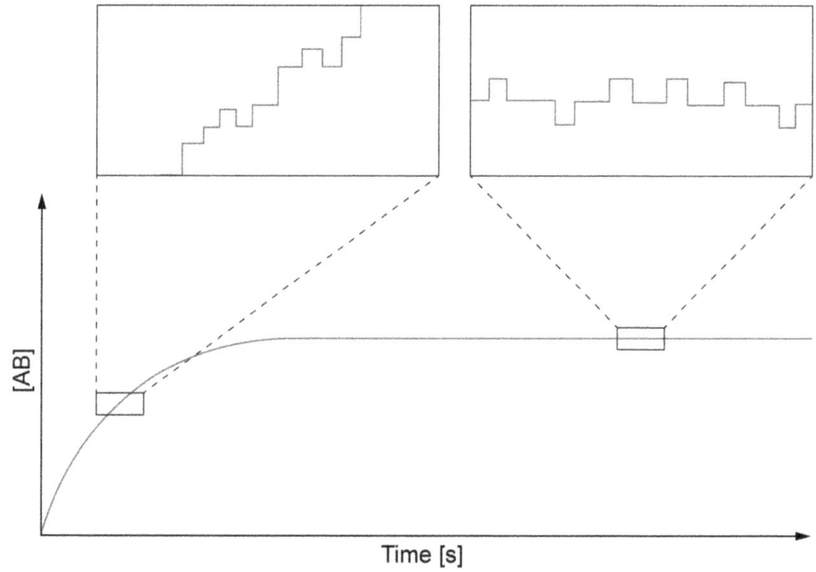

Figure 2.1. In a chemical reaction between two compounds A and B that yield a new compound AB, we have a competition between the forward reaction A + B → AB and the backward reaction AB → A + B. If we start a reaction by mixing A and B, we will obtain a rise in the amount of AB. This is reflected by the rising kinetics. As shown in the insert (a), during the rising kinetics the forward reaction dominates. When the reaction reaches equilibrium, the forward and backward reactions happen at the same rate and we reach a plateau in the amount of AB, as depicted in (b). The fluctuations are the same in both cases. Therefore, it is enough to measure the fluctuations in a system, even in equilibrium, to obtain all necessary information on its dynamic processes.

backward reactions. Thus, it is easy to see that the fluctuations that cause the kinetics of the reaction are the same as the ones that keep the equilibrium stable. Therefore, one can either record the reaction kinetics or record the fluctuations around the equilibrium, both contain the same information and thus characterize the reaction process.

The difference is that reaction kinetics can be measured with less sensitivity and at lower time resolution to characterize the reaction because in kinetics we observe the change in the average value. But a change in the average value is the sum of all fluctuations happening in a defined period of time and thus can be large and easy to detect. However, to obtain information on the fluctuations in an equilibrium, one needs to be able to detect the fluctuations of single reactions, so needing high sensitivity, and the ability to measure faster than the reactions happen. This is the reason why the study of fluctuations had to await the development of highly sensitive and fast detectors before it came of age even if the theory had long been worked out.

Furthermore, the observation of kinetics requires the ability to start a process at a particular time so that the reaction kinetics can be faithfully observed. But fluctuations can be recorded in a system that is in equilibrium at any time. And as we will see later, in principle the fluctuations can even be analysed in non-equilibrium systems if the change in kinetics is sufficiently different from the characteristic time of the fluctuations and the two can be separated.

2.3 Correlations

Now with the knowledge that fluctuations carry the same information about molecular processes as kinetics do, we can proceed to look at the analysis of fluctuations and how we can extract that information. But before we address the issue of how fluctuations are analysed let's first clarify the concept of correlations, starting with the definition of some statistical rules that are needed for their calculation.

For any process averages can be taken over time or over any other variable on which the process depends. For an experiment with N trials each of which with an outcome a_n this means

$$\langle a \rangle = \frac{1}{N}\sum_{n=1}^{N} a_n \tag{2.5}$$

We introduce here the angular brackets $\langle \ldots \rangle$, which we use as a shorthand version to denote the average or arithmetic mean.

If there is a finite number M of different unique possible outcomes a_m, and the number n_m how often they occur is known, then we can calculate the probability for each outcome a_m as

$$p_m = n_m / N \tag{2.6}$$

with

$$N = \sum_{m=1}^{M} n_m \tag{2.7}$$

When throwing a standard six-sided die, there are only $M = 6$ outcomes, and a_m can take on the numbers 1–6. When each outcome is equally likely, then $p_m = 1/6$. So when the die is rolled 100 times, we expect that $n_m \approx 17$. For this to be accurate N needs to be sufficiently large. For FCS we will discuss this condition in chapter 8, when we discuss the bias in FCS. Using the probabilities for each possible outcome we can calculate the average

$$\langle a \rangle = \frac{1}{N}\sum_{n=1}^{N} a_n = \sum_{m=1}^{M} \frac{n_m}{N} a_m = \sum_{m=1}^{M} p_m a_m \tag{2.8}$$

If we look at the outcome of two independent variables a and b, each of which have a finite number of possible outcomes, then the probability of each common result is just the product of the probabilities of each single outcome. So if we roll two six-sided dice, each can take a value from 1 to 6 with a probability $p_{am} = p_{bm} = 1/6$. Therefore, the probability of getting any pairs of values {1,1}, {1,2} … {2,1} … {6,6} is just $p_{am}p_{bm} = 1/36$.

We will start now our discussion of correlations with the mathematical definition of a correlation and then show that this definition yields sensible results using various examples. In general, when two variables are correlated it means that the value of one variable is to some extent dependent on the value of the other variable. The extent to which this is true will be the level of the correlation. For example, in a perfect correlation one can accurately predict the value of one parameter from the other, as for instance the volume and mass of a solid at standard conditions. If mass is added or taken away the volume increases or decreases in the same manner. It should be noted here that correlation does not imply causation. The relation between volume and mass has to be related back to the structure of matter and it is not the volume that causes the mass or vice versa.

In a partial correlation one can predict with some success the second parameter from the first but not perfectly. If you, for instance, stand at the luggage belt at an airport and look at the suitcases passing by. Can you predict the weight of the suitcase from the size? The answer is yes … to a certain extent. When comparing suitcases with the same items inside then the correlation holds. Most people pack only clothes, for instance. In that case a larger suitcase will be heavier. But some people may pack heavy items of small volume like books. So sometimes a smaller suitcase, containing books, can be heavier than a larger one, containing clothes. Therefore, the size and weight of luggage is only partially correlated. And other parameters might not be correlated at all. For example, you cannot predict the size of luggage on the belt from the hair colour of the owner of that luggage, so hair colour and luggage size are uncorrelated. I should state that we have not tested this particular case but a correlation between hair colour of the owner and luggage size seems to be sufficiently unlikely to us so that this serves as a good example. A more scientific example is the radioactive decay of two different substances. It is not possible to predict from a decay happening in one substance a decay in the other substance.

In order to discuss this in a more quantifiable way, we will take a look at the mathematical definition of a correlation. Let's assume we have two variables a and b that depend on some general parameter u. Here, a and b can be any observable property, and u is a general coordinate on which that property depends. The mathematical definition of a correlation states that two variables are correlated if the following is true:

$$\langle a(u)b(u)\rangle \neq \langle a(u)\rangle\langle b(u)\rangle \tag{2.9}$$

In words, two variables are correlated if the average of their product is different from the product of their averages. As a simple example of two variables, let's look at the outcome of a toss of two coins, where H denotes heads and T tails. The variables a and b are the outcomes for the first and second coin, and u will be the number of the toss. Here is a table of outcomes of a set of coin tosses:

u	$a(u)$	$b(u)$
1	H	T
2	H	T
3	T	H
4	H	H
5	T	T
6	…	…

If we assign a value of $H = 1$ and $T = 0$ to the outcomes, we get

u	$a(u)$	$b(u)$
1	1	0
2	1	0
3	0	1
4	1	1
5	0	0
6	…	…

From this table we can calculate the average of the product of $a(u)$ and $b(u)$ over all coin tosses u.

$$\langle a(u)b(u)\rangle = \frac{1}{n}\sum_{u=1}^{n} a(u)b(u) \tag{2.10}$$

We can simplify this calculation as we know that there are only four possible outcomes: HH, HT, TH, TT. If the coins are randomly showing heads (H) and tails (T), then H and T have a probability of occurring of 50%, then all possible outcomes for the coin toss will have the same probability of occurring, namely 25%. Again replacing H and T with 1 and 0, respectively, the average is given in the following table:

m	a_m	b_m	p_{am}	p_{bm}	p_{abm}	$p_{abm}a_m b_m$
1	1	1	0.5	0.5	0.25	0.25
2	1	0	0.5	0.5	0.25	0
3	0	1	0.5	0.5	0.25	0
4	0	0	0.5	0.5	0.25	0

We have used here the fact that if two independent events have probability p_a and p_b the probability of both events happening is just the product of each event happening, $p_{ab} = p_a p_b$.

We can now calculate the average of the product of $a(u)$ and $b(u)$ as well as the product of their averages, with values of $H = 1$ and $T = 0$.

$$\begin{aligned}\langle a(u)b(u)\rangle &= \frac{1}{n}\sum_{u=1}^{n} a(u)b(u) = \sum_{m=1}^{4} p_{abm}a_m b_m \\ &= 0.25(HH + HT + TH + TT) \\ &= 0.25(1 + 0 + 0 + 0) = 0.25\end{aligned}$$

$$\langle a(u)\rangle = \frac{1}{n}\sum_{u=1}^{n} a(u) = H \times p_1 + T \times p_2 = 1 \times 0.5 + 0 \times 0.5 = 0.5$$

$$\langle b(u)\rangle = \frac{1}{n}\sum_{u=1}^{n} b(u) = H \times p_1 + T \times p_2 = 1 \times 0.5 + 0 \times 0.5 = 0.5$$

So finally we have

$$\langle a(u)b(u)\rangle = 0.25 = \langle a(u)\rangle\langle b(u)\rangle$$

from which we can see that the equality holds and the two variables are not correlated and any occurrence of an outcome can be explained by chance. Please note, the lack of correlation does not mean that the two coins never show the same face. It just means that they do not show the same face more or less often than is expected in a random trial.

Let's assume that one of the coins is manipulated and shows 75% of the time H. In that case

m	a_m	b_m	p_{am}	p_{bm}	p_{abm}	$p_{abm}a_m b_m$
1	1	1	0.5	0.75	0.375	0.375
2	1	0	0.5	0.25	0.125	0
3	0	1	0.5	0.75	0.375	0
4	0	0	0.5	0.25	0.125	0

$$\langle a(u)b(u)\rangle = 0.375\ HH + 0.125\ HT + 0.375\ TH + 0.125\ TT = 0.375$$

$$\langle a(u)\rangle = \sum_{m=1}^{4} p_{am} a_m = 0.5$$

$$\langle b(u)\rangle = \sum_{m=1}^{4} p_{bm} b_m = 0.75$$

And overall,

$$\langle a(u)b(u)\rangle = 0.375 = \langle a(u)\rangle\langle b(u)\rangle$$

So again a and b are uncorrelated. They are not correlated because it is not the first coin that influences the second or vice versa but it just so happens that the second coin has a different behaviour independent of the first coin (it still means, though, that betting on outcomes where the second coin is H has a better chance of winning). But now let's assume that whenever $a(u) = H$, then in 75% of the cases $b(u) = H$, but if $a(u) = T$, then $b(u)$ is still 50% of the time H and 50% of the time T. In that case we have now the following situation.

m	a_m	b_m	p_{am}	p_{bm}	p_{abm}	$p_{abm} a_m b_m$
1	1	1	0.5	0.75	0.375	0.375
2	1	0	0.5	0.25	0.125	0
3	0	1	0.5	0.5	0.25	0
4	0	0	0.5	0.5	0.25	0

From the table we can calculate the overall averages

$$\langle a(u)b(u)\rangle = \frac{1}{n}\sum_{u=1}^{n} a(u)b(u) = 0.375\ HH + \ 0.125\ HT + 0.25\ TH$$
$$+0.25\ TT = 0.375$$

$$\langle a(u)\rangle = \frac{1}{n}\sum_{u=1}^{n} a(u) = 0.5$$

$$\langle b(u)\rangle = \frac{1}{n}\sum_{u=1}^{n} b(u) = 0.375 + 0.25 = 0.625$$

and finally arrive at

$$\langle a(u)b(u)\rangle = 0.375 > 0.3125 = \langle a(u)\rangle\langle b(u)\rangle$$

So in this case $a(u)$ and $b(u)$ are correlated. If you know that $a(u)$ is H you can predict with a probability of 75%, i.e. better than pure chance, that $b(u)$ will be H as well. Take a careful look at the last two tables and note the subtle differences in the probability that decide whether there are correlations or not.

So equation (2.9) above, defining correlations, means that one can predict the value of the variable $b(u)$ from the variable $a(u)$ with a probability that is better than pure chance.

However, if for some reason, once we have H for the first coin, we never find H for the second coin but only T, then we arrive at the following table

m	a_m	b_m	p_{am}	p_{bm}	p_{abm}	$p_{abm} a_m b_m$
1	1	1	0.5	0.0	0	0
2	1	0	0.5	1	0.5	0
3	0	1	0.5	0.5	0.25	0
4	0	0	0.5	0.5	0.25	0

Calculating the averages

$$\langle a(u)b(u)\rangle = \frac{1}{n}\sum_{u=1}^{n} a(u)b(u) = 0\ HH + 0.5\ HT$$

$$+0.25\ TH + 0.25\ TT = 0$$

$$\langle a(u)\rangle = \frac{1}{n}\sum_{u=1}^{n} a(u) = 0.5$$

$$\langle b(u)\rangle = \frac{1}{n}\sum_{u=1}^{n} b(u) = 0.25$$

yields

$$\langle a(u)b(u)\rangle = 0 < 0.125 = \langle a(u)\rangle\langle b(u)\rangle$$

In this case $a(u)$ and $b(u)$ are anti-correlated. This means coincidences happen less than what would have been expected by pure chance. Correlations and anti-correlations both allow predicting the outcome of one coin toss from the other. If the outcomes are correlated, one can predict that the second outcome will be the same as the first outcome with a probability better than pure chance. If the outcomes are anti-correlated, one can predict that the second outcome will be the opposite of the first outcome with a probability better than pure chance. In both cases one has a certain predictive power. If that were to happen in a casino, you would get rich very fast!

Of course many of the situations depicted here as instances of coin tosses serve only as explanatory examples. However, the different types of correlations can be found in nature. For example, particles carrying equal charge repulse each other. Therefore they will not be found in the same place. When one particle is at a particular position, a second particle would not be found in the immediate vicinity. Therefore, the location of these particles would be anticorrelated. For a derivation that the independence of two outcomes implies uncorrelatedness see math box 2.1.

Math Box 2.1

This can be also shown in general for any pair of discrete two-state random processes. Let us follow the notation introduced above and call the two random variables $a(u)$ and $b(u)$ and the two values they can assume H and T. We will denote the probabilities with which they assume these values as $p_a(H)$, $p_a(T)$, $p_b(H)$ and $p_b(T)$, respectively. We can then write for the averages in the case that $a(u)$ and $b(u)$ are independent:

$$\begin{aligned}
\langle a(u)\rangle &= p_a(H)H + p_a(T)T \\
\langle b(u)\rangle &= p_b(H)H + p_b(T)T \\
\langle a(u)b(u)\rangle &= p_a(H)p_b(H)H^2 + p_a(T)p_b(T)T^2 \\
&\quad + (p_a(H)p_b(T) + p_a(T)p_b(H))HT \\
&= (p_a(H)H + p_a(T)T)(p_b(H)H + p_b(T)T)
\end{aligned}$$

Thus, we have shown that the independence of the outcomes of $a(u)$ and $b(u)$ is equivalent to the fact that they are not correlated.

In the case where the two variables are not independent, when for example the value of $a(u)$ influences the value of $b(u)$, we need to introduce conditional probabilities of $b(u)$ assuming value X (where X stands for H or T) under the condition that $a(u)$ is equal to H or T; we will denote them $p_b(X)|_{a=H}$ and $p_b(X)|_{a=T}$, respectively. The influence of $a(u)$ over $b(u)$ is expressed in the fact that for at least one possible value of X:

$$p_b(X)|_{a=H} \neq p_b(X)|_{a=T}$$

For the average values, we can write:

$$\begin{aligned}
\langle a(u)\rangle &= p_a(H)H + p_a(T)T \\
\langle b(u)\rangle &= (p_b(H)|_{a=H}\ p_a(H) + p_b(H)|_{a=T}\ p_a(T))H \\
&\quad + (p_b(T)|_{a=H}\ p_a(H) + p_b(T)|_{a=T}\ p_a(T))T \\
\langle a(u)b(u)\rangle &= p_a(H)p_b(H)|_{a=H}\ H^2 + p_a(T)p_b(T)|_{a=T}\ T^2 \\
&\quad + (p_a(H)p_b(T)|_{a=T} + p_a(T)p_b(H)|_{a=T})HT
\end{aligned}$$

It can be easily shown that in this case the equality between the product of the average values and the average value of the product does not hold.

2.4 From correlation coefficient to correlation function

With equation (2.9) and our discussion we can now define a correlation coefficient in the following way:

$$g = \frac{\langle a(u)b(u)\rangle}{\langle a(u)\rangle\langle b(u)\rangle} \tag{2.11}$$

This coefficient can take three value ranges:

$$g = \begin{cases} > 1: & \text{the data is } \textbf{correlated} \\ = 1: & \text{the data is } \textbf{not correlated} \\ < 1: & \text{the data is } \textbf{anti-correlated} \end{cases}$$

With the definition of correlations, we can go ahead and apply this to more complicated data. Up to now we assumed u to be discrete, namely the number of the coin toss. So let's rewrite the correlation coefficient for a continuous parameter, e.g. the time t:

$$G = \frac{\langle a(t)b(t)\rangle}{\langle a(t)\rangle\langle b(t)\rangle} \tag{2.12}$$

The individual averages, for continuous functions of time, are now given by

$$\langle a(t)b(t)\rangle = \frac{1}{T}\int_0^T a(t)b(t)dt \tag{2.13a}$$

$$\langle a(t)\rangle = \frac{1}{T}\int_0^T a(t)dt \tag{2.13b}$$

$$\langle b(t)\rangle = \frac{1}{T}\int_0^T b(t)dt \tag{2.13c}$$

This coefficient has still the same properties as the discrete one. Let's look at an example. Remember the reaction of compounds A and B with concentrations [A] and [B]. The concentrations can be measured at any point in time and thus are a continuous function of time. Now at each step they form a product

$$[\mathrm{A}] + [\mathrm{B}] \xrightarrow{\text{yields}} [\mathrm{AB}] \tag{2.14a}$$

or are produced when the product dissociates

$$[\mathrm{AB}] \xrightarrow{\text{yields}} [\mathrm{A}] + [\mathrm{B}] \tag{2.14b}$$

It is clear that [A] and [B] always increase or decrease together and thus are correlated in time. And [A] or [B] always change in the opposite direction to [AB], and thus both pairs {[A], [AB]} and {[B], [AB]} are anti-correlated in time.

2.5 The autocorrelation function (ACF) and its properties

At this point we will leave cross-correlations, i.e. the correlations between two different variables, to which we will come back later, and turn to autocorrelations, i.e. we look at how a signal is correlated to itself:

$$G = \frac{\langle a(t)a(t)\rangle}{\langle a(t)\rangle\langle a(t)\rangle} = \frac{\langle a(t)^2\rangle}{\langle a(t)\rangle^2} \tag{2.15}$$

At first sight this does not seem to be a particularly interesting result, as any signal has to be correlated with itself. To analyse this we have to go back to the concepts of mean and variance. The mean, $\langle a(t)\rangle$, was already defined. The variance is given by the mean square deviation of $a(t)$ from the mean $\langle a(t)\rangle$:

$$\begin{aligned}\sigma^2 &= \langle (a(t) - \langle a(t)\rangle)^2\rangle \\ &= \langle a(t)^2 - 2a(t)\langle a(t)\rangle + \langle a(t)\rangle^2\rangle \\ &= \langle a(t)^2\rangle - 2\langle a(t)\rangle^2 + \langle a(t)\rangle^2 \\ &= \langle a(t)^2\rangle - \langle a(t)\rangle^2\end{aligned} \tag{2.16}$$

With this result we can now rewrite the correlation coefficient

$$G = \frac{\langle a(t)^2\rangle}{\langle a(t)\rangle^2} = \frac{\sigma^2 + \langle a(t)\rangle^2}{\langle a(t)\rangle^2} = \frac{\sigma^2}{\langle a(t)\rangle^2} + 1 \tag{2.17}$$

This shows that for the autocorrelation—but not the cross-correlation, as we have seen before—the correlation coefficient must be larger than or equal to 1 as both, σ^2 and $\langle a(t)\rangle^2$, are positive. And the value of the correlation coefficient is determined by the variance and mean of the signal. Knowing this allows us to make some more general remarks. In many processes the variance σ^2 is a function of the mean μ, which therefore allows us to say something about the magnitude σ^2/μ^2 of the autocorrelation coefficient. Because of the importance of the Poisson distribution we will use it here, but similar arguments could be made about other distributions whose variance depends on the mean. For a Poisson distribution $\sigma^2 = \mu$.

$$G = \frac{\sigma^2}{\langle a(t)\rangle^2} + 1 = \frac{\sigma^2}{\mu^2} + 1 = \frac{\mu}{\mu^2} + 1 = \frac{1}{\mu} + 1 \tag{2.18}$$

So, we see that the magnitude of the autocorrelation coefficient depends inversely on the mean for a Poisson distributed measurement.

But the autocorrelation does not have to be taken at the same time, as we have already said this is trivial. But we can now ask how the signal $a(t)$ is related to itself at a time τ later, i.e. $a(t + \tau)$. As we can ask that for any value of τ, the correlation coefficient becomes an ACF which depends on τ:

$$G(\tau) = \frac{\langle a(t)a(t + \tau)\rangle}{\langle a(t)\rangle\langle a(t + \tau)\rangle} \tag{2.19}$$

$G(\tau)$ contains the information how a signal depends on its history, or from another perspective, how well we can predict the future value from the present value of the signal.

Finally we can simplify the equation if we assume that we measure only stationary processes, i.e. processes in which the time point of measurement does not influence the average value and $\langle a(t)\rangle = \langle a(t + \tau)\rangle$

$$G(\tau) = \frac{\langle a(t)a(t+\tau)\rangle}{\langle a(t)\rangle^2} \tag{2.20}$$

Stationarity: Often we encounter non-stationary processes, e.g. when the overall signal increases or decreases on average. This can happen in cases when one measures in non-equilibrium conditions, e.g. when photobleaching slowly destroys fluorophores leading to an overall decrease in fluorescence intensity, or when measuring on a sample that moves and the instantaneous fluorescence intensity is a function of the position of the sample with respect to the observation volume, as can happen when measuring cell membranes. In this last case, cell membranes can fluctuate in their position moving the membrane to different parts of the observation volume and thus providing different amounts of total fluorescence. However, as long as the change of the average value is sufficiently slow, i.e. slower than the molecular processes to be observed, one can correct the signal. For this purpose Ries *et al* [3] have introduced a variance conserving correction in which the signal $a(t)$ is corrected by a function $f(t)$ describing the slowly changing average values of the signal. The function $f(t)$ can be of any form but is often an exponential or a polynomial to empirically fit the fluorescence decay or slow changes. The corrected signal $a_c(t)$ is then given by

$$a_c(t) = \frac{a(t)}{\sqrt{f(t)/f(0)}} + f(0)\left(1 - \sqrt{f(t)/f(0)}\right) \tag{2.21}$$

So in the rest of the book we assume stationarity, if not otherwise mentioned, as stationarity is either fulfilled or the signal can often be corrected with equation (2.21) to establish stationarity. For more discussions on this topic, see chapter 7.

From the definition of the ACF and our assumption of stationary processes, it follows that the ACF is an even function, i.e. $G(\tau) = G(-\tau)$. This can be easily shown.

$$G(-\tau) = \frac{\langle a(t)a(t-\tau)\rangle}{\langle a(t)\rangle^2} \tag{2.22}$$

If we shift the correlation function in time, which we can do as we deal with a stationary process, we replace t with $t + \tau$, i.e. we shift the whole function by τ. Therefore,

$$G(-\tau) = \frac{\langle a(t+\tau)a(t)\rangle}{\langle a(t)\rangle^2} = G(\tau) \tag{2.23}$$

showing that the ACF is symmetric (figure 2.2).

Another important property regards the value over which the correlation function is calculated. At the moment, the correlation function is defined over the signal not over the fluctuations. This makes sense in so far as the signal is the actual measurement that can be directly correlated. To calculate the fluctuations, in principle, first the average signal value has to be calculated before the autocorrelation of the fluctuations can be calculated. Therefore, we derive here the relation

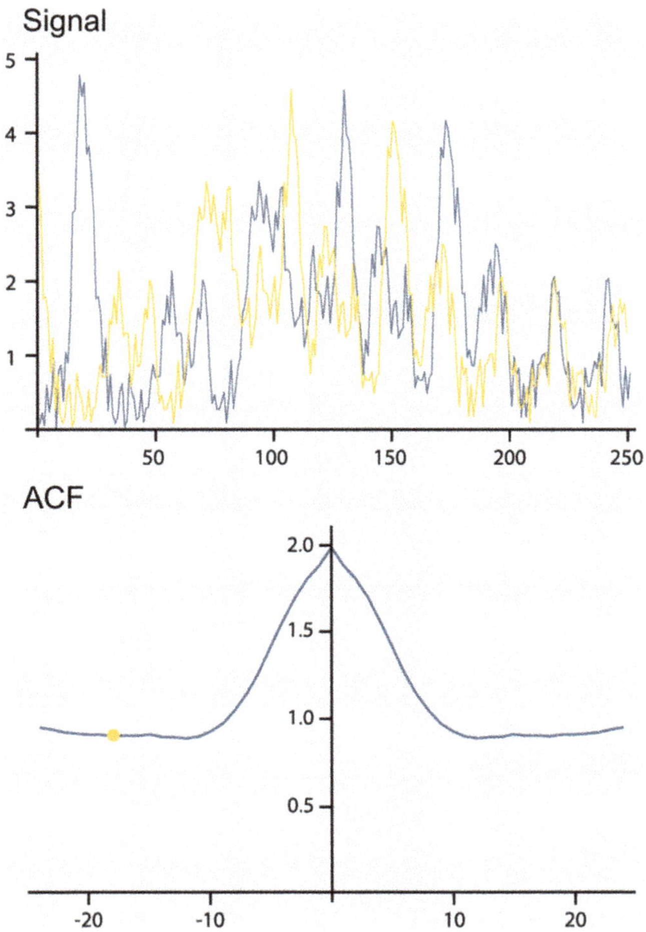

Figure 2.2. This figure shows the principle how ACFs are calculated. In the upper panel we show the original signal $F(t)$ in blue. In yellow we show the same signal shifted by a lagtime τ, $F(t+\tau)$. In the above example, τ can vary from negative to positive values. For each time shift τ we can calculate the ACF value of $G(\tau)$, shown in the lower panel. As shown in the text, the calculated ACF is symmetric in time. Note that the ACF reaches its peak when the signal perfectly overlaps with itself and decreases with decreasing overlap of the peaks. Animation available at https://iopscience.iop.org/book/978-0-7503-2080-1.

between the ACFs of the signal and the fluctuations to show that they contain the same information.

With the definition of a fluctuation, $\delta a(t)$, as the deviation from the mean

$$\delta a(t) = a(t) - \langle a(t) \rangle \tag{2.24}$$

We can show that the average of the fluctuation is 0:

$$\langle \delta a(t) \rangle = \langle a(t) - \langle a(t) \rangle \rangle = \langle a(t) \rangle - \langle a(t) \rangle = 0 \tag{2.25}$$

We can now express the signal as the addition of the average signal plus the fluctuations

$$a(t) = \delta a(t) + \langle a(t)\rangle \tag{2.26}$$

and can substitute this in the equation for the correlation function:

$$\begin{aligned} G(\tau) &= \frac{\langle a(t)a(t+\tau)\rangle}{\langle a(t)\rangle^2} \\ &= \frac{\langle(\delta a(t) + \langle a(t)\rangle)(\delta a(t+\tau) + \langle a(t+\tau)\rangle)\rangle}{\langle a(t)\rangle^2} \\ &= \frac{\langle \delta a(t)\delta a(t+\tau) + \delta a(t)\langle a(t+\tau)\rangle + \langle a(t)\rangle\delta a(t+\tau) + \langle a(t)\rangle\langle a(t+\tau)\rangle\rangle}{\langle a(t)\rangle^2} \end{aligned} \tag{2.27}$$

The average over the fluctuations is 0 as shown above, i.e. $\langle\delta a(t)\rangle = 0$. Furthermore, as we work with stationary processes, we know that $\langle a(t)\rangle = \langle a(t+\tau)\rangle$. Therefore, it follows that the second and third term in the numerator are 0:

$$\langle\delta a(t)\langle a(t+\tau)\rangle\rangle = \langle\delta a(t)\rangle\langle a(t+\tau)\rangle = 0 \tag{2.28a}$$

$$\langle\langle a(t)\rangle\delta a(t+\tau)\rangle = \langle a(t)\rangle\langle\delta a(t+\tau)\rangle = 0 \tag{2.28b}$$

The transition from the left hand side to the middle terms is possible as both $\langle a(t)\rangle$, and $\langle a(t+\tau)\rangle$ are constants.

Thus the ACF is given by

$$\begin{aligned} G(\tau) &= \frac{\langle\delta a(t)\delta a(t+\tau) + \langle a(t)\rangle^2\rangle}{\langle a(t)\rangle^2} \\ &= \frac{\langle\delta a(t)\delta a(t+\tau)\rangle + \langle a(t)\rangle^2}{\langle a(t)\rangle^2} \\ &= \frac{\langle\delta a(t)\delta a(t+\tau)\rangle}{\langle a(t)\rangle^2} + 1 \end{aligned} \tag{2.29}$$

Therefore, we have shown that

$$G(\tau) = \frac{\langle a(t)a(t+\tau)\rangle}{\langle a(t)\rangle^2} = \frac{\langle\delta a(t)\delta a(t+\tau)\rangle}{\langle a(t)\rangle^2} + 1 \tag{2.30}$$

This means that the ACFs for the signal and for the fluctuations have the same functional character and they only differ by a value of 1. Thus, except for an offset, the two ACFs contain the same information. In the rest of the book, we will refer to the signal correlation function as $G(\tau)$ and the fluctuation ACF as $g(\tau)$ and will point out when the offset makes a difference.

The property that the ACF for the signal and the fluctuations are the same except for an offset of 1, allows us to determine also where the maximum of the ACF can be found. To answer this we can compare the values of the ACF at different times.

For $\tau = 0$,

$$g(0) = \frac{\langle \delta a(t)^2 \rangle}{\langle a(t) \rangle^2} \tag{2.31a}$$

For $\tau \neq 0$,

$$g(\tau) = \frac{\langle \delta a(t) \delta a(t + \tau) \rangle}{\langle a(t) \rangle^2} \tag{2.31b}$$

The square of a value is always positive, so $\delta a(t)^2 > 0$. This is not the case for a product of different values, so $\delta a(t)\delta a(t + \tau)$ can be both larger and smaller than 0, as is expected from fluctuations. So when taking the average $\langle \delta a(t)^2 \rangle$, and $\langle \delta a(t)\delta a(t + \tau) \rangle$, the second but not the first term will have negative contributions that will reduce the value of the average, and therefore

$$\langle \delta a(t)^2 \rangle \geqslant \langle \delta a(t) \delta a(t + \tau) \rangle \tag{2.32}$$

This is of course true, independent of whether the ACF is defined over the signal or the fluctuations, and therefore the ACF will be maximum for $\tau = 0$. This means that, for any stochastic process, $g(\tau)$ will proceed from a value of $\frac{\langle a(t)^2 \rangle}{\langle a(t) \rangle^2}$ at the origin ($\tau = 0$) to 0 at long times ($\tau \rightarrow \infty$).

The equal sign in the inequality above holds for $\tau \neq 0$, i.e. $\langle \delta a(t)^2 \rangle = \langle \delta a(t)\delta a(t + \tau) \rangle$, for periodic processes with a period τ, in that case $\delta a(t) = \delta a(t + \tau)$ and the ACF will possess the same periodicity because every period it will take on the same maximum value $\langle \delta a(t)^2 \rangle$ (figure 2.3).

2.6 The cross-correlation function (CCF) and its properties

The definition of the cross-correlation function (CCF) is analogous to the definition of the ACF

$$G_X(\tau) = \frac{\langle a(t) b(t + \tau) \rangle}{\langle a(t) \rangle \langle b(t + \tau) \rangle} \tag{2.33}$$

Here the two signals $a(t)$ and $b(t)$ are both signals in time but they will differ in the observed parameter. In fluorescence spectroscopy this could be the fluorescence wavelength to determine how the signal of two differently labelled particles are correlated, and whether two particles interact. Or signals could be taken at different locations to test whether the signals at two locations are correlated and thus whether they are connected and particles can move from one location to the other. But other parameters can also be used as has been shown for fluorescence lifetime correlation spectroscopy (see section 4.4).

The CCF, however, is not symmetric and in general

$$\frac{\langle a(t) b(t + \tau) \rangle}{\langle a(t) \rangle \langle b(t + \tau) \rangle} \neq \frac{\langle b(t) a(t + \tau) \rangle}{\langle b(t) \rangle \langle a(t + \tau) \rangle} \tag{2.34}$$

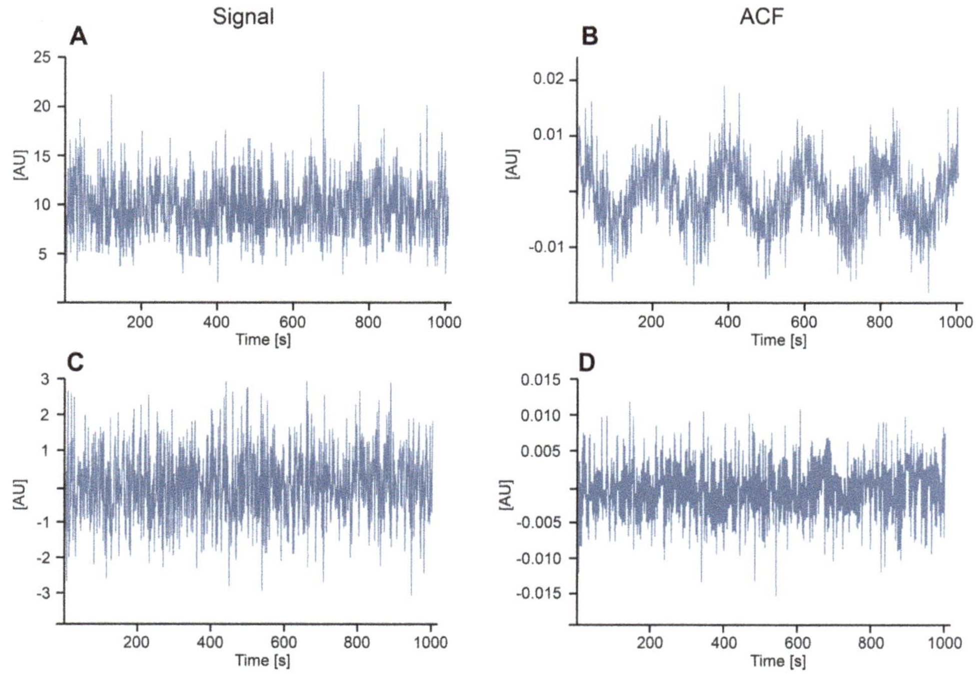

Figure 2.3. Signal and ACF for (A, B) a periodic process buried in noise and (C, D) a signal containing only noise. The periodic process (A) leads to an equally periodic ACF (B). For signal containing only noise (C), the ACF (D) is flat as there are no correlations. This figure demonstrates the power of autocorrelation analysis, namely to find signals even in cases where the noise is large compared to the signal.

Unlike in the ACF, it is possible in a cross-correlation that, e.g. a signal in $a(t)$ always precedes a signal in $b(t)$ with a certain lag time τ, while this is not the case the other way around; i.e. the time a signal in $a(t)$ follows after a signal in $b(t)$ is random. An example would be the relationship between the scoring of a goal in a soccer match and the fans cheering. The time between a goal being scored and the fans cheering will be measured in seconds or even parts of a second (depending how awake the fans are). However, one cannot predict when the next goal will be scored once the fans cheered. That time will be random.

This means cross-correlations can be non-symmetric and they, unlike autocorrelations, do not necessarily peak at the origin. The peak of the cross-correlation will be located at the time delay τ between the signal $a(t)$ and the following signal $b(t + \tau)$ (figure 2.4).

2.7 Fluctuations and correlations

The results of the previous sections provide us with all the basics to understand how fluctuations encode properties of molecular processes and how correlations help us extracting quantitative information from these fluctuations.

Processes in physical systems originate from the action of single molecules. These processes include anything from simple molecular movements to chemical reactions.

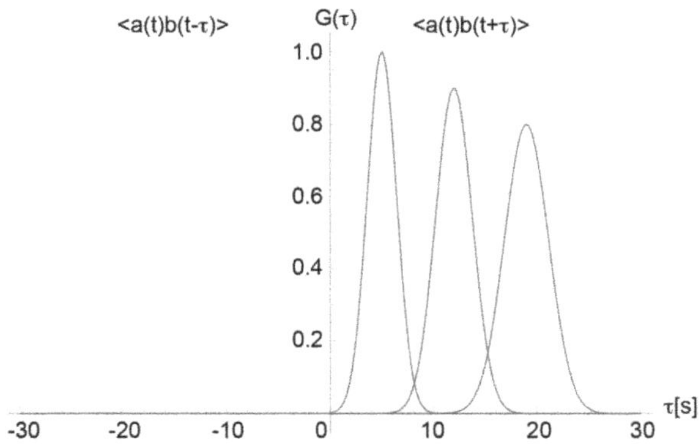

Figure 2.4. CCF for two signals $a(t)$ and $b(t)$, assuming that $a(t)$ precedes $b(t)$ at a certain time but $b(t)$ does not precede $a(t)$. This means the time difference between events $a(t)$ and $b(t)$ is within a defined range. However, the time difference between events $b(t)$ and $a(t)$ are random. With increasing time that $a(t)$ precedes $b(t)$, the peak of the CCF shifts to the right. It is also possible that $a(t)$ does not always precede $b(t)$, in which case the amplitude can change. And the time that $a(t)$ precedes $b(t)$ can have an uncertainty in which case the peak of the CCF widens. Note that the CCF is asymmetric as we assume that the time shift that $b(t)$ precedes $a(t)$ is random and thus there is no correlation for $\langle b(t)a(t+\tau)\rangle = \langle a(t)b(t-\tau)\rangle$.

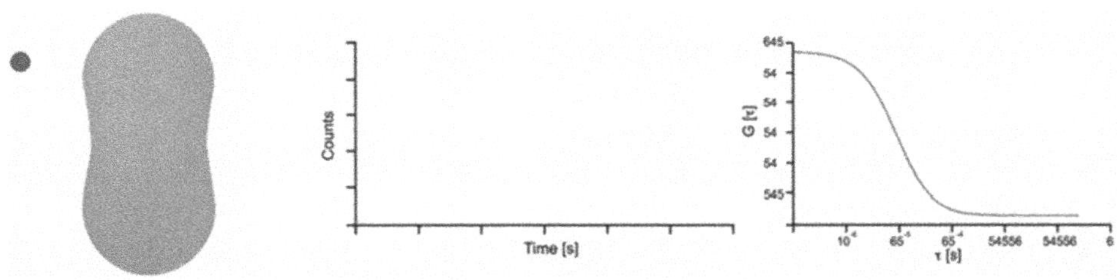

Figure 2.5. Fluorescent particles diffusing through an illuminated observation volume, depicted in blue, lead to fluctuations in the fluorescence intensity. The correlation of the signal that contains many of these transitions leads to the ACF, as shown in figure 2.2. Animation available at https://iopscience.iop.org/book/978-0-7503-2080-1.

Each molecular action results in a change in an observable parameter. If one can measure the action of these single molecules by recording the subtle changes in an observable, then one can in principle determine everything there is to know about the system. As an example we have seen that the fluctuations in concentrations of a chemical reaction in equilibrium are determined by exactly the same molecular processes as the kinetics of the same reaction.

In FCS, we measure the fluctuations in the fluorescence intensity that can be caused by a range of different processes. Mostly, FCS measures the transition of particles through a defined observation volume (figure 2.5). But as we will see in chapter 5 we can also measure molecular rotations, molecular blinking or isomerization reactions. And the first application of FCS was conducted to measure chemical kinetic constants.

These fluctuations are often difficult to quantify and one needs to observe many events to obtain sufficiently good statistics. To be able to efficiently extract information from these single molecule events, we use correlation functions, that are straightforward to calculate and rely mainly on multiplication and addition of data points. Qualitatively, the correlation function will immediately indicate time scales of the correlations, i.e. the order of magnitude of the characteristic time of the process under investigation, concentrations and frequency of occurrence of the process, and possibly interactions in cases of cross-correlations. However, the quantitative interpretation of the correlation functions is not as straightforward and one has to know the properties of the molecular process and has to take account of the particular experimental implementation to be able to calculate the expected form of the correlation function. We will show examples for the derivations of FCS fit models in chapter 5. But once that has been determined, the model function can be used to fit a wide range of experiments. And as we will discuss in chapter 8, the selection which model fits the data best is an important step in the analysis of FCS experiments.

2.8 Synopsis

- The understanding of FCS relies on two main concepts, fluctuations and correlations
- Any process creates a change in an observable. In equilibrium the average of the observable value stays constant but the value itself will fluctuate around the mean depending on the molecular process investigated.
- These fluctuations carry information about molecular processes (characteristic time of the process; frequency of occurrence).
- Correlation functions are a means to statistically treat a signal and extract its average properties.
- The correlation of a signal with itself in time is called an autocorrelation function (ACF). The ACF extracts information from the fluctuating signal.
- The ACF can be calculated either over the fluctuations ($\delta a(t)$) or over the signal ($a(t)$) itself. The two correlation functions are equivalent and differ only by an offset of 1.
- ACFs are symmetric in time and proceed from a maximum value at $\tau = 0$ to a minimum value at long times (1 for the signal ACF, 0 for the fluctuation ACF). Exceptions are periodic signals, in which case the ACF will also be periodic (section 4.1.2 and 'Introduction to FCS—interactive exercises.cdf'—available at https://iopscience.iop.org/book/978-0-7503-2080-1).
- The maximum of the correlation function is inversely related to the number of occurrences of the fluctuations. How fast the correlation function converges to the uncorrelated value of 0 (or 1) is a measure of the persistence in time of the molecular process that causes the fluctuations.
- A CCF measures the interdependence of two variables or signals
- CCFs are neither necessarily symmetric in time nor do they necessarily peak at $\tau = 0$.
- By analysing fluctuations with correlation functions we can obtain information on molecular processes

2.9 Exercises

1. Transfer the argument of section 2.2 about the fluctuations to diffusion. For this you need to assume that you have unequal concentrations within at least two connected volumes. How do the concentrations in the two volumes change and what is the role of concentration fluctuations? Are the concentration fluctuations in the two volumes correlated or anticorrelated? Describe the processes in the two volumes until concentrations equilibrate.

2. For the coin tossing examples in section 2.3, show that this is not only true when H and T are replaced by 1 and 0 but for any values for H and T as long as they are different.

3. Find examples of fully, partially and anti-correlated variables.

4. How does the addition of a constant a_0 to $a(t)$ influence the correlation function as outlined in section 2.5? First calculate this for the correlation coefficient assuming that $a(t)$ is Poisson distributed. Then calculate it for $g(\tau)$.

5. Following the same arguments as in section 2.5 (equations (2.16)–(2.18)) show the following relations:

$$g(0) = \frac{\langle \delta a(t) \delta a(t) \rangle}{\langle a(t) \rangle^2} = \frac{\sigma^2}{\langle a(t) \rangle^2} = \frac{\sigma^2}{\mu^2}$$

   What is $g(0)$ for a Poisson distributed variable?

References

[1] MacDonald D K 1962 *Noise and Fluctuations: An Introduction* (New York: Wiley)

[2] Miller S and Childers D 2012 *Probability and Random Processes: With Applications to Signal Processing and Communications* 2nd edn (Amsterdam: Academic)

[3] Ries J, Petrášek Z, García-Sáez A J and Schwille P 2010 A comprehensive framework for fluorescence cross-correlation spectroscopy *New J. Phys.* **12** 113009

Thorsten Wohland, Sudipta Maiti and Radek Macháň

Chapter 3

Fluorescence excitation and detection

For FCS, it is necessary to optically define an open observation volume. The number of fluorescent particles within this volume is counted by monitoring their fluorescence. Understanding the nature of the observation volume, and tuning it to one's needs, is important both for optimizing the signal-to-noise, and for applying the right function in fitting the data. In FCS, the probe volume is defined by the optics of excitation and collection. As we will see in this section, there are different types of probe volumes that are created in different types of FCS experiments. We also discuss the different types of detectors, which are integral parts of defining the probe volumes. Chapter 5 will deal with the question of deriving the right fitting functions for some of these probe volumes. Chapter 7 will deal with some of the serious artefacts that can arise in analysing data if the probe volumes are not tailored correctly, or if the limitations of the detectors are ignored.

3.1 The probe volume in FCS

3.1.1 Introduction

As we noted in chapter 1 (section 1.4), the raw data in FCS is just a time trace of the fluorescence that emanates from a single spot (which is called the 'probe volume') in the sample. Before we describe the different optical arrangements which create different types of probe volumes appropriate for FCS, we need to understand the concept of the probe volume a little better. The probe volume is a microscopic sub-volume of the sample which allows unfettered entry and exit of particles from the rest of the bulk solution. There are several important points to understand: if the probe volume has a diffuse boundary (as is the case in reality), what do we exactly mean by 'entry' and 'exit' of a particle? For a given molecular system, what geometry of excitation can give us the maximum amplitude of fluctuation? What shape and size are optimal? Should we have a large probe volume or a small one? We provide the answers in the following sections.

doi:10.1088/978-0-7503-2080-1ch3

3.1.2 The significance of the size of the probe volume

The size question is easy to answer, and it is related to the statistics of fluctuations. The molecules in a solution are randomly distributed everywhere, and their number within any defined small open volume obeys Poissonian statistics (see chapter 2, equation (2.18); for a more detailed discussion, see math box 3.1). The characteristics of Poissonian statistics is that the mean (μ) is equal to the variance (σ^2). So the fluctuations of the number (the standard deviation, σ) is proportional to the square root of the number itself. If we have 100 particles on an average, the variance, σ^2, is also 100, and so the σ is 10. So the ratio of fluctuation to the fluorescence signal is 0.1. On the other hand, if we have ten molecules, the average amplitude of fluctuations is approximately 3.1, and so the ratio of fluctuations to fluorescence is ~0.3. This is directly reflected in the initial amplitude of the autocorrelation function near time zero. The numerator of the autocorrelation function at time zero, barring the shot noise autocorrelation, is proportional to the variance of the number of particles in the probe volume, which follows from the definition of the autocorrelation function itself (see math box 3.1). The denominator is proportional to the square of the average number, and hence $G(0)$ is proportional to $1/N$, where N is the average number of particles.

So the smaller the number inside the volume is, the larger the relative fluctuations are. Does that mean we should go to the smallest number? Is there any limit? Well, it is the signal-to-noise (S/N) ratio, and not the signal itself, that is typically of concern to an experimenter. It turns out that at least about 1 molecule per volume (or larger) gives the best S/N ratio (math box 3.2). As long as the number is somewhat larger than one, the S/N ratio does not change much with the number. Of course, this assumes that the background remains low. The background is typically proportional to the probe volume, because it arises from scattering from the solvent, or from fluorescence of contaminants. So a smaller probe volume helps in lowering the background. Also, at large $\langle N \rangle$, the total photon rate may be too high for the detector, which may begin to get saturated, distorting the data. This would require one to lower the photon rate per molecule, which in turn would reduce the S/N ratio. So at high concentrations, it would be beneficial to have a smaller volume. This limits the largest concentration that can be probed by FCS. These factors imply that we should have a small probe volume (as long as $\langle N \rangle > 1$) for the best S/N in FCS. For a 10 nM solution, to have 1 molecule inside the volume, we need a probe volume of about 0.17 fL. How do we make an open probe volume so small?

This brings us to the optics of photo-excitation in FCS, and the need for lasers. In the most commonly used optical geometry, the excitation and the detection geometries are confocal in nature. This involves focusing of laser excitation through a high power objective lens, collection of the emitted fluorescence by the same objective lens, and finally detection through a pinhole. For some other types of excitations, such as multiphoton or TIRF or Light Sheet, the smallness of the volume is achieved without a pinhole, as we will see later. However, before discussing these geometries in detail, we start with an introduction to the process of generating the fluorescence signal itself.

Math box 3.1

Poissonian statistics
Diffusing particles obey Poissonian statistics. 'Statistics' in this context means an equation that predicts the probability of a particular event occurring. Imagine an ant randomly moving about on a chess board. What is the probability that it will be at a particular square at a particular time? The average probability is 1/64, a pretty small number. Now imagine that there are 64 ants on the board (imagine that they don't notice each other's presence). What is the probability that there will be one ant in a particular square? Surely the probability is not one (though the average number per box is one), since there is some chance of having zero or more than one ant to be there in a square at a particular time. This is the problem whose answer is provided by Poissonian statistics. When the probability of occurrence of a particular event (say the presence of a given ant in a given square) is very small, but there are a large number of possible events, then the probability of occurrence of a particular number of events (say n) is predicted by Poissonian statistics. It says that the probability of n occurrences, when the average occurrence is μ, is given by

$$P(n) = e^{-\mu}\frac{\mu^n}{n!}$$

For our ant example, $\mu = 64/64 = 1$, so the probability of finding two ants in a square at a given time is $P(2) = (e^{-1})/2 = 0.184$. One of the useful results obtained from the Poissonian statistics is that the variance of the number n equals the mean μ.

In a dilute solution, the diffusing molecules behave similarly and would be governed by the same Poissonian statistics. This is important in the context of FCS, as there we are monitoring the occupancy number in a small sub-volume of the solution. As we shall see now, the relationship between the variance (arising out of the occupation number fluctuations) and the mean allows FCS to determine the absolute number of molecules in the probe volume.

By definition, the fluorescence autocorrelations function at time zero (actually a small time after time zero, since the autocorrelation function at time 0 is dominated by the photon shot noise) is

$$G(0) = \frac{\langle \delta F(t)^2 \rangle}{\langle F(t) \rangle^2}$$

Here $\delta F(t) = F(t) - \langle F(t) \rangle$ is the deviation of the fluorescence signal from the mean signal at time t, and $< >$ denotes average over the measurement time. What is this shot noise at zero delay time? The time trace records the actual arrival time of the photons. Even if two time intervals have the same probability of recording photons, the actual number of photons recorded by them will have a Poisson distribution. This so-called 'photon shot noise' contributes to the correlation function only when the delay time is exactly zero. This zero time correlation gives a strong peak, and will be ignored in our subsequent discussions. Now, since $F(t)$ is proportional to $N(t)$, the number of molecules in the probe volume at time t,

$$G(0) = \frac{\langle \delta N(t)^2 \rangle}{\langle N(t) \rangle^2} = \frac{\sigma^2(N(t))}{\langle N(t) \rangle^2} = \frac{\langle N(t) \rangle}{\langle N(t) \rangle^2} = \frac{1}{\langle N(t) \rangle}$$

Here $\delta N(t) = N(t) - \langle N(t) \rangle$ is the deviation of the number of molecules in the probe volume from the mean number at time t, and $\langle \; \rangle$ denotes average over the measurement time. So the amplitude of the correlation function at time 0 provides a direct measure of the absolute number of molecules that is in the probe volume on an average. Of course, Poissonian statistics by itself does not say anything about the timescale of the fluctuations, which is the most frequently used information that FCS provides.

Math box 3.2

The S/N ratio in FCS

The S/N ratio in FCS depends on the observed brightness or 'detectivity' of each molecule, the background, the number of molecules in the probe volume $<N>$, the duration of each 'time bin' $\Delta\tau$, and of course on the total time of recording T_{tot}. The duration of the 'time bin' has to be suitably small to detect the fastest fluctuation processes of interest. The S/N in FCS was first derived by Dennis Koppel [1], in the limit of $<N> \gg 1$. He derived the important result that the S/N is proportional to the detectivity, defined as the number of photons detected per molecule in the probe volume per time bin (usually called counts per particle and second or cps, but also abbreviated as per particle brightness, ppb, or count rate per molecule, cpm). He also showed that increasing the cps (per time bin) beyond 1 does not benefit the S/N much. Later, Qian [2], Wohland *et al* [3] and Wenger *et al* [4] derived S/N including the possibility of $\langle N \rangle < 1$. They showed that at short times:

$$S/N = CPS \cdot \sqrt{\frac{T_{tot}\Delta\tau}{1 + 1/<N>}}$$

This shows that the S/N decreases as $\langle N \rangle$ approaches 1 or takes values below 1.

3.1.3 A brief introduction to the generation of the fluorescence signal

When photons interact with matter, they may get dispersed, absorbed, scattered and re-emitted. The re-emitted photon may have lower or higher energy (i.e. it can have a different colour), different direction, or a different polarization state compared to the original photon. Spectroscopy is the measurement of properties of photons which are interacting with matter, and in doing so, provides information about that material substance. Some of these processes are depicted in a simple energy level diagram (figure 3.1).

Of the many processes that can occur when a photon interacts with matter, the most important one for our purposes is fluorescence (process 3 in figure 3.1). The topic of fluorescence is well covered by many spectroscopy books, e.g. [5, 6], and by online texts/tutorials. Here we will briefly recount a few of the properties that will be useful for our discussion on fluorescence correlation spectroscopy. First, we consider

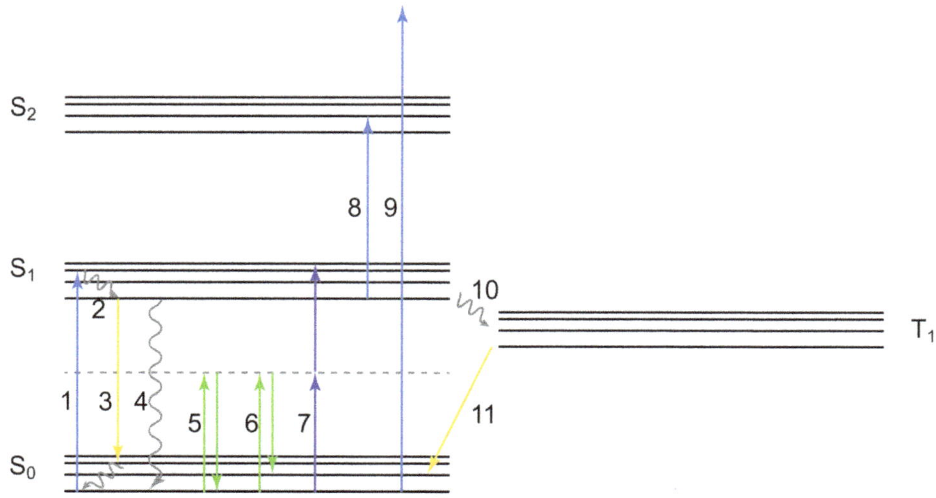

Figure 3.1. Photo-induced processes. S_0, ground state; S_1, excited state; S_2, higher excited state; T_1, lowest triplet state. Energy is plotted in the vertical axis. Horizontal axis has no particular significance. The processes are, 1: ground state absorption, 2: vibrational relaxation, 3: fluorescence, 4: internal conversion, 5: Rayleigh scattering, 6: Raman scattering, 7: two-photon absorption, 8: excited state absorption, 9: photoionization, 10: intersystem crossing, 11: phosphorescence.

the energy level diagram (so-called Jabłoński diagram after the Polish physicist Aleksander Jabłoński, figure 3.1). The ordinate here represents energy, while the abscissa has no particular significance. The horizontal lines depict the energy levels of the molecule, which can be computed from quantum mechanics, and can also be deciphered from experiments. Optical transitions between some of the levels are allowed, as has been depicted by the straight arrows shown in the diagram. An upward arrow depicts the absorption of a photon, while a downward one depicts an emission. Of course, for an upward transition to happen, photons of that energy must be impinging on the molecule. Also, a downward transition will eventually follow an upward transition, since the molecule will eventually return to the ground state (unless it is photo-converted to a different molecule). The squiggly arrows depict transitions between levels which occur through processes not involving photon absorption or emission (they are called 'non-radiative' transitions). Symmetry and conservation laws make some of the radiative transitions disallowed (or weak). The energy involved with a transition (i.e. the wavelength or the 'colour' of the transition), and its intensity, is determined by the wavefunctions of the two participating states, and the intensity, wavelength and polarization of the light impinging on the molecule. The strength of an absorptive transition occurring between levels 'a' and 'b' (assuming suitable approximations typically valid in an FCS experiment) is proportional to several factors, as given by the Fermi golden rule [7]:

$$\text{Rate} \sim |\langle\psi_b|\mu E(\nu)|\psi_a\rangle|^2\rho(\nu) = |\mu_{ab}|^2|E|^2\rho(\nu) \sim \varepsilon(\nu)I(\nu)\rho(\nu) \tag{3.1}$$

where E = electric field amplitude of the light ($|E|^2 \sim I$, intensity), ψ_a and ψ_b are the wavefunctions of the two levels between which the transition is occurring, and μ is the dipole moment operator ($\mu = e\boldsymbol{r}$, where e = the charge of the electron and $\boldsymbol{r}$ is its position, here assumed parallel to E), ρ is the density of states as a function of energy (or frequency) around the resonance (transition) energy, and $\varepsilon(\nu)$ is the molar absorptivity of the molecule at frequency ν. We note these frequency dependent quantities can also be expressed as wavelength (λ) dependent quantities. A notable exception to this relationship is an n-photon (e.g. two-photon) excitation, where the intensity term is replaced by I^n and $\varepsilon(\nu)$ is replaced by its n-photon counterpart. Therefore, the rate of absorption depends on the intensity of light at a given energy (frequency) around the resonance energy, and molecular properties.

Fluorescence is denoted by arrow number 3 (yellow) in figure 3.1. Of course, for fluorescence to occur, an absorption must have taken place before, indicated by arrow number 1. The rate of emission of fluorescence is proportional to the rate of absorption given by equation (3.1) (true for spontaneous fluorescence emission, not for stimulated emission), multiplied by a constant φ known as the quantum efficiency (QE). Also, we should remember that vibrational relaxation (a non-radiative transition) in the excited state typically happens before the fluorescence occurs, so the starting state is generally a lower lying state compared to the state to which the molecule was first excited. So, the rate of fluorescence combining all wavelengths λ_{em} (where λ_{em} = wavelength of emission) under steady uniform illumination $I(\lambda_{ex})$ (where λ_{ex} = wavelength of excitation) of a homogeneous solution of concentration C is given by:

$$F(\lambda, t) \sim I(\lambda_{ex})\varepsilon(\lambda_{ex})C(t)\varphi \tag{3.2}$$

The signal detected is therefore:

$$F(\lambda, t) \sim I(\lambda_{ex})\varepsilon(\lambda_{ex})C(t)\delta(\lambda_{em})\varphi \tag{3.3}$$

where $\delta(\lambda_{em})$ is the probability of detecting an emitted photon at wavelength λ_{em} (which is determined by the collection efficiency of the instrument, the QE of the detector, the filter characteristics, etc).

For FCS, the important thing to note in this equation is that I, δ, and C can all be a function of position $\vec{r}$ and time t in principle. We note that while the average concentration $\langle C \rangle$ in the microscopic probe volume in FCS is assumed to remain constant in time and independent of position, the instantaneous concentration at any given position $C(\vec{r}, t)$ is a function of both $\vec{r}$ and t (due to diffusion and photo-physics), and this has to be taken into account. The probability of detection $\delta(\lambda, \vec{r})$ is also a function of position. With these quantities defined, now let's calculate the $F(t)$ for FCS. In an FCS probe volume, e.g. in the confocal geometry, $I(\vec{r})$ and $\delta(\vec{r})$ are strongly varying functions of $\vec{r}$, and their product is localized to a very small volume (which we call the probe volume) near the focus (see appendix 3.1 for details). In most FCS experiments, $I(t)$ is not a function of time. Therefore the concentration in each sub-volume of the probe volume contributes differently to the signal. The instantaneous fluctuations of the concentration in each such small sub-volumes is non-negligible, and has to be taken into account. This fluctuation in $C(\vec{r}, t)$ about its time-averaged value is at the heart of FCS measurements.

We denote the probe volume (also called the detectivity profile) by $W(\vec{r})$, which is a product of the excitation probability profile $I(\vec{r})$ and the collection efficiency function $\delta(\vec{r})$, and is normalized to unity at the center where it is the highest (at a given wavelength). $F(t)$ can then be described by the following equation (which is equivalent to equation (5.1), described in chapter 5):

$$F(t) = \kappa \int W(\vec{r})C(\vec{r}, t)d\vec{r} \tag{3.4}$$

Here, κ is a constant related to the overall detection efficiency of the system, and to the molecular parameters ε and ϕ. This will be the starting point of our discussions on the theoretical equations for FCS in chapter 5.

For more sophisticated FCS experiments, other parameters may come into play. For example, in fluorescence lifetime correlation spectroscopy (FLCS), the time between excitation and fluorescence, which is related to the fluorescence lifetime, becomes important. For some other FCS experiments, such as those measuring rotational diffusion, the polarization of the emitted photon becomes an important quantity. Each molecule has one (or more) absorption dipole with a specific orientation, and emission dipole with the same or different orientation. The time between the absorption and emission allows the orientation to change due to rotational diffusion. Thus if the detector is polarization sensitive, then the detected photons will also carry information about rotational diffusion.

3.1.4 Optical designs applied to obtain an appropriate probe volume for FCS

3.1.4.1 Ideal volume versus real optical volumes

The optimal shape which gives rise to the largest fluctuations for a given volume is one that has sharp edges, where the detectivity (probability that a photon emitted from the volume will be detected by the detector) of a molecule is 1 inside (in some relative scale), and zero outside (math box 3.3).

Such a shape with the maximum contrast is difficult to create with light, as this would mean very sharp gradients in light intensity. The light intensity gradient (as determined by the distance between 0% of the peak intensity level to the 100% intensity level) is limited by diffraction to about half of the wavelength and the diameter of the diffraction limited volume is limited to about a wavelength, e.g. for visible light, the diameter is about 500 nm (appendix 3.1). The method of creating an optical volume depends both on the excitation and detection of fluorescence. Some of the common geometries are confocal, multiphoton, total internal reflection and light sheet geometry. We now discuss these one by one.

3.1.4.2 The confocal geometry

The confocal detection geometry is the most common optical scheme employed in FCS. A typical confocal imaging device takes rays emanating from a spot in the sample under investigation and focuses them onto a different spot in front of the detector. The focused spot is therefore an image of the initial spot. Since light paths (usually, through linear media) are reversible, any ray emanating from the second

Math box 3.3

Volume contrast in FCS

A faithful report of number fluctuations would assume that the detectivity of a molecule is '1' when the molecule is inside the probe volume and '0' when it is outside. Unfortunately, it is not possible to optically create such a volume with such steep detection boundaries. However, the closer one gets to this ideal limit, the smaller is the effective volume. If the peak-normalized detectivity profile is given by $W(\vec{r})$, then the probe volume can be defined by $\int_{-\infty}^{\infty} W(\vec{r})d\vec{r}$. However, this is not the effective volume V_{eff} of FCS, which can be defined as $V_{\text{eff}} = N/C$, where C is the average concentration of the species in solution and N is the inverse of $G(0)$. The effective volume is given by [8] and is derived in math box 5.2 (chapter 5):

$$V_{\text{eff}} = \int_{-\infty}^{\infty} W^2(\vec{r})d\vec{r} \Big/ \left(\int_{-\infty}^{\infty} W(\vec{r})d\vec{r} \right)^2 = \gamma \Big/ \int_{-\infty}^{\infty} W(\vec{r})d\vec{r}$$

where

$$\gamma = \int_{-\infty}^{\infty} W^2(\vec{r})d\vec{r} \Big/ \int_{-\infty}^{\infty} W(\vec{r})d\vec{r}$$

Obviously, $\gamma = 1$, if we have an ideal probe volume, but that is not obtainable. For conventional excitation, the best contrast is obtained with a diffraction limited focus. There are other ways to create a volume in the near field or with STED excitation, for example, that can give better contrast than that allowed by diffraction limited far field approaches.

spot would also be focused at the first spot—i.e. the two spots are 'confocal' (figure 3.2). Now, if we put a bright object at one of the confocal spots (e.g. p in figure 3.2), and place a pinhole at the other one (see p′ in figure 3.2), then a detector behind that pinhole will preferentially detect the light coming from that spot p. Light from a neighbouring spot, say r in front of p, or q laterally away from p, will be mostly excluded from the detector (figure 3.2). Thus, the effective volume of detection will be limited in three dimensions.

Figure 3.3 shows how it is usually implemented in a commercial confocal microscope. The excitation source is a laser. The laser beam is expanded by a pair of lenses so that the diameter of the beam can overfill the back aperture of the objective lens (appendix 3.1 explains the reason behind doing this). The beam is then reflected by a long-pass dichroic mirror onto the objective, which focuses it onto a particular point inside the sample. It is passed through a pair of scanning mirrors before it reaches the objective to ensure that the beam can be focused to any chosen spot inside the sample. The emitted fluorescence is collected by the objective and proceeds backwards along the same path until it reaches the dichroic. The longpass dichroic separates it from the excitation and a 'tube' lens focuses it onto a pinhole.

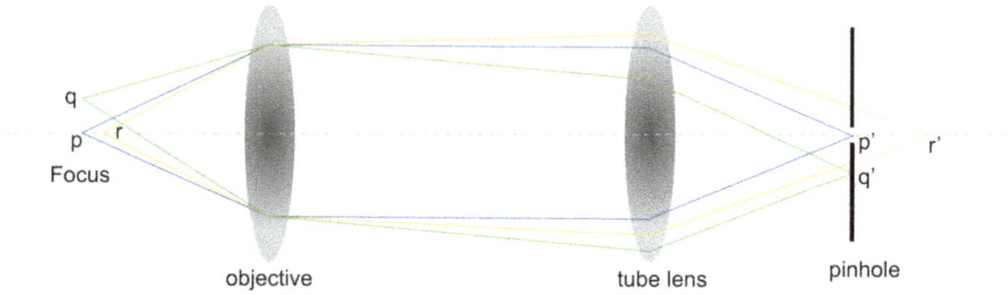

Figure 3.2. The confocal detection geometry. p, q and r are three fluorescent points in the object area, which are excited by a laser beam (not shown) focused by the objective lens. p is at the centre of the focus of excitation, q is laterally away from it, and r is in front of it. The objective and tube lens pair collects the fluorescence emitted by these points and makes images p′ and q′ on the image plane and r′ behind it. A screen is placed at the image plane with an aperture at p′, with a detector behind it (not shown). p′ is 'confocal' to p, and this arrangement preferentially collects fluorescence from p.

Different wavelengths of emission can then be separated, if required, by another dichroic. The light is then focused by a lens onto a detector (this focusing is necessary because the APD detectors typically have a small detection area). An emission filter in front of the detector selects the desired range of emission wavelengths and excludes any scattered excitation light, as well as any other background. The electrical signal from the detector then travels to the correlator which computes the autocorrelation.

What matters for confocal detection is the probability that a molecule placed at a certain position r in the object will get illuminated, and then photons from this molecule will get detected by the detector. The illumination depends on the position of this molecule with respect to the pattern of illumination defined by the diffraction limited excitation spot (figure 3.2).

The probability to detect photons from it depends on the collecting lens, and the location as well as the size of the confocal aperture or pinhole. As the molecule moves near the focal volume by diffusion/flow (e.g. from p to q to r in figure 3.2, we assume that the sample is a dilute solution of molecules), the probability of both excitation and detection changes. Thus the total fluorescence collected at the detector fluctuates, as it depends on the instantaneous location of all the molecules at or near the excitation volume. It is this fluctuation which constitutes the raw data for FCS.

The calculation of this fluctuation has to start with the probability of detection of a molecule at an arbitrary point near the focus. This is a simple product of the probability of excitation and the probability of detection. Both the geometry and the mode of excitation defines this probability. The effective description of such a volume involves diffraction theory. The mathematical form of this has been described by several authors, e.g. [9]. Appendix 3.1 provides a conceptual overview, and some numerical calculations (mostly for a planar 2D object). Briefly, the light is diffracted twice. The first diffraction is that of the excitation beam by the objective lens. Here the effective diffraction element is the circular back aperture of the objective. The result of this diffraction is shown in figure 3.A1.1 in appendix 3.1.

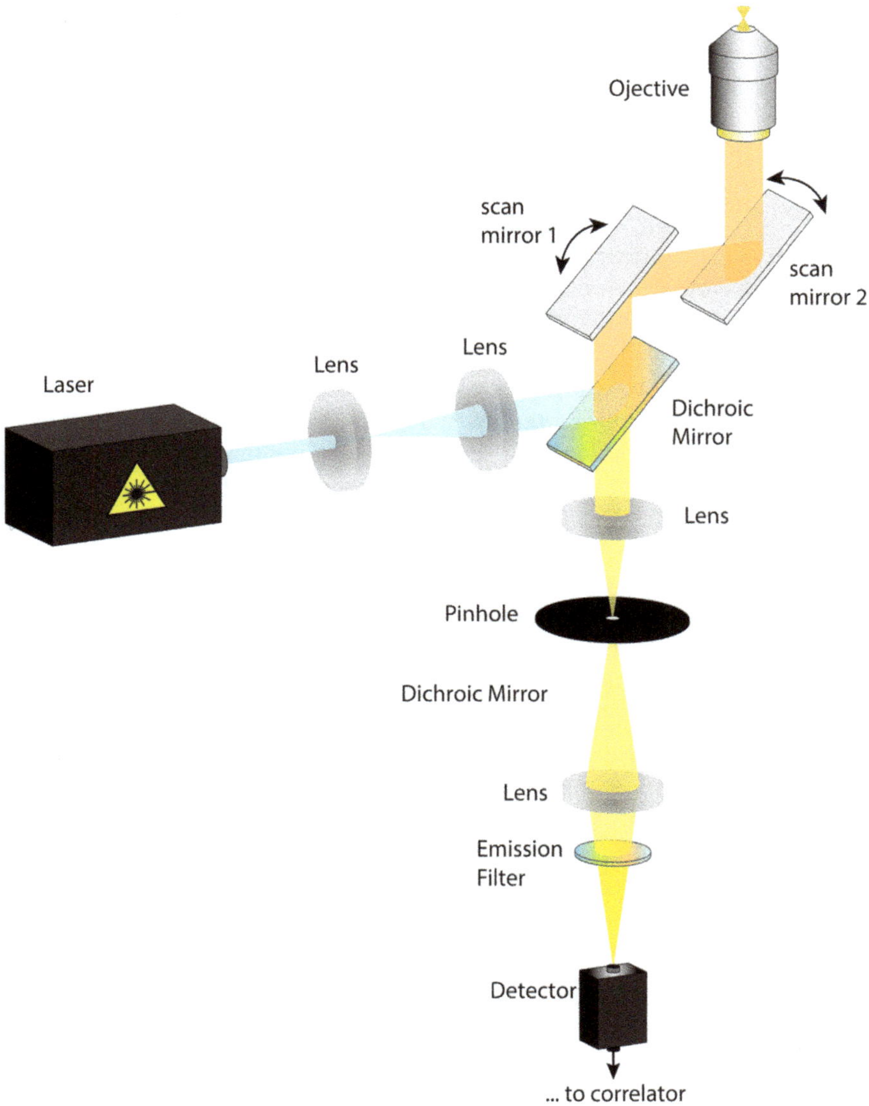

Figure 3.3. A schematic of an actual fluorescence correlation spectrometer, as usually implemented in a commercial confocal microscope. The laser beam is passed through a pair of scanning mirrors which can tilt it about the center of the back aperture of the objective lens, which can focus it at any desired point within the sample (not shown). The fluorescence travels back through the same path until the dichroic mirror, where it separates and gets focused to the detection pinhole, and further to the detector.

This diffracted and focused beam then excites fluorescence in the sample, and this is collected back by the same objective which collimates the fluorescence signal. This undergoes a second diffraction by the same back aperture, but it is at a wavelength which is usually longer than the wavelength of excitation (in the case of single-photon excitation, while it is blue shifted for multiphoton excitation). This light is then collected by another longer focal length lens, usually termed the tube lens, and focused onto the image plane (figure 3.2). The diffraction by the tube lens can be neglected, as it is easy to take this lens to be much larger than the back aperture of the objective lens, which at this point approximately defines the diameter of the collimated beam. It is important to realize that each point in the object space emits fluorescence independently, and thus the image consists of an overlap of the diffraction limited images of all the points with non-zero brightness at the object plane. Finally, this fluorescence is collected through the confocal pinhole placed in front of the detector, which collects fluorescence from a small region of this image plane. The pinhole is usually circular, and so the collected amount of light is an integral over this pinhole area. It of course depends on where in the image plane this aperture is located and how big its diameter is, as discussed in appendix 3.1.

The Gaussian volume approximation: while the confocal volume is well described by diffraction theory, this does not lend itself to analytical solutions for an FCS correlation function. Fortunately, approximate 'equivalent' volumes can produce simple analytical functions, which provide quite accurate effective parameters for the quantities we are interested in. The most popular fitting functions for confocal FCS use the Gaussian approximation to the diffraction limited focal volume. The validity of this model is apparent from the discussion in appendix 3.1. A Gaussian focal volume is described in the Gaussian Focus section in the accompanying interactive CDF file available at https://iopscience.iop.org/book/978-0-7503-2080-1.

How bad is this approximation? When does it fail? It turns out that it is critical to understand these two points in order to obtain a reliable estimation of the parameters from an FCS measurement. The answer to the first of these two questions is that the approximation is surprisingly good under high NA focusing and small aperture detection (i.e. when the detection aperture is less than or equal to the Airy disk size for the focused fluorescence. Please see appendix 3.1 for a more detailed explanation, and the definition of the Airy disk). The acceptability of a Gaussian approximation is theoretically shown, and can be experimentally verified as shown in appendix 3.1. The Gaussian volume is assumed in most of the FCS experiments, and the mathematical form of the fitting function is derived in chapter 5. However, if the detection aperture is large compared to the Airy disk, then this approximation can be wrong, and the standard equation would give rise to spurious components, whose origin is obvious from math box 3.1 and the figures therein [10]. This artefact, including others, has been discussed in chapter 7. One should always keep this common artefact in mind when choosing an aperture for the confocal detection.

3.1.4.3 Two-photon excitation

The contribution of the small non-Gaussian features (sometimes referred to as 'side lobes') can be minimized by using the 'multiphoton' excitation technique, which has the added advantage of not requiring a pinhole. In multiphoton excitation, the probability of excitation depends nonlinearly on the excitation intensity. This would mean that the term $I(\vec{r}, t)$ in the equation for fluorescence signal $F(t)$ is replaced by its nth power for an n-photon excitation process (see equation (3.5) given below for two-photon excitation). The relative contribution of any ‘side lobe’, which is typically a small fraction of the main peak, becomes negligible. We provide a short description of the multiphoton excitation process here, but we will not explicitly discuss it in the remainder of the book as the fit functions are essentially the same for one- and multi-photon excitation. This is a result of the fact that multi-photon excitation uses the same Gaussian approximations of the observation volume (albeit with a different standard deviation than the one-photon case, see the ‘drawbacks’ paragraph below). All the treatments developed for the single-photon excitation in the latter chapters of this book can be applied to the multiphoton case, as long as the standard deviation of the Gaussian is known.

Multiphoton excitation involves simultaneous excitation of a fluorophore with multiple (say n) photons, each of which have wavelengths n times that of the effective excitation wavelength. This process implies that the probability of excitation is proportional to the nth power of the excitation intensity. In two-photon excitation (the most common variant of multiphoton excitation), the excitation volume is the square of the focused intensity.

$$F_{2-\text{photon}}(t) = \kappa \int I^2(\vec{r})\delta(\vec{r})C(\vec{r}, t)d\vec{r} \tag{3.5}$$

The variables are as defined in equation (3.1), except for two major differences. First, the factor κ now contains the two-photon absorption cross section instead of the one-photon one, and second, I is a function of time for a pulsed laser, which is typically used in two-photon excitation. This time dependence is in the femto to nanosecond timescale, and is not explicitly incorporated in equation (3.5), which deals with fluorescence fluctuations in the μs timescale. Since the two-photon cross section is small, the instantaneous intensity should be high to obtain a measurable rate of excitation. This is why a femtosecond pulsed laser needs to be used for two-photon excitation. In the focused intensity distribution for two-photon excitation, the diffraction lobes which contributed in the single-photon confocal excitation case become proportionally smaller. It is easy to see this in a relative sense. Suppose the peak intensity is 100, and the side lobe has a peak intensity of 10 (of course, as can be seen from figure 3.A1.2, this would be clearly an exaggeration). When two-photon excitation is used for the same intensity distribution, then the contribution of the peak intensity is 100/10 000 = 1%, instead of the 10% contribution made by the side lobe in the single photon case. The profile at once becomes cleaner, and the Gaussian approximation becomes better. There is another advantage of this mode of excitation: the detection pinhole can be removed from the path altogether. In reality, a detection pinhole of the size of the Airy disk does increase the resolution to some extent [11]. However, this increase is at the cost of rejecting some of the photons, which is especially important if the sample is a scattering one, as described below.

Multiphoton excitation is the method of choice for probing scattering samples as the excitation beam experiences much lower scattering. Scattering typically has a 1/(wavelength)^4 dependence, and the longer wavelength of excitation for multiphoton means that scattering is typically more than an order of magnitude lower compared to one-photon excitation. Of course, this advantage is there only for the excitation wavelength, and not for the emission wavelength, which remains the same as single-photon excitation. In a real object, the photons do not take the path that is dictated by the optics. Due to scattering, their path gets somewhat randomized. In such a case, closing the pinhole will lose a lot more photons that actually arose from the excitation volume. Since the excitation volume is defined by the excitation light (in the absence of a pinhole), the emission scattering only reduces the brightness of the object, and does not affect the probe volume in multiphoton excitation. Another major advantage of multiphoton excitation is that the excitation is truly localized near the focus. So for a long-time experiment, the effect of photo-induced chemistry (photodamage or photobleaching) along the excitation beam path can be more limited compared to single-photon excitation.

Drawbacks of two-photon excitation: The multiphoton scheme is not without its drawbacks. The first drawback is that the excitation happens at a higher wavelength, so the resolution is worse by a factor of $\sqrt{n}$, for an n-photon excitation process. However, this would assume that the excitation wavelength is really exactly the double (for two-photon, or n-fold for n-photon) of the single-photon excitation wavelength. In reality, this is seldom the case. The n-photon excitation maximum wavelength is frequently less than n-times the single-photon excitation wavelength. For example, fluorescein has a one-photon excitation peak at 488 nm, while rhodamine B has a peak at 543 nm. However, their two-photon excitation peaks are at 775 nm (secondary peak at 920 nm) and 700 nm (secondary peak at 830 nm), respectively. In this case, the resolution is worse only by 12% for fluorescein and it is actually better than one-photon excitation by a factor of 10% for rhodamine B. On the other hand, GFP excites well at 920 nm, which is not too far from 976 nm, which would be the double of its one-photon excitation wavelength, and the resolution is therefore worse by about 33%. Another restriction of multiphoton excitation is that the power requirement is much higher than the one-photon counterpart as this has a much lower cross section. The biological samples tend to be much more transparent in the NIR wavelengths which would correspond to most two-photon excitation imaging, so usually it is less damaging even at these higher powers. However, if some component of the sample by chance has a small absorbance at these excitation wavelengths, then the sample can be significantly damaged. Another problem is that the lasers typically used for multiphoton excitation are mode-locked lasers, which produce pulses every 12 ns or so. This means that the sample is idle for the time interval between the pulses. To be more accurate, if the lifetime of the fluorophore is short compared to the inter-pulse time interval, then the sample is in the ground state for an inordinately long part of the duty cycle. This makes the highest obtainable photon emission rates lower than that in the single-photon case. A possible remedy for this is to split each pulse in the pulse train into two or more parts, and to delay the pulses with respect to each other, making the effective repetition rate higher.

3.1.4.4 Total internal reflection

The confocal detection has many advantages, but at least one glaring disadvantage. It can monitor only one point in space (one diffraction limited spot to be exact) at a time. Of course, one can use multiple beams which are separately monitored by different detectors, but creating additional probe volumes becomes progressively harder. Modern instrumentation has begun to make this easier. For example, a diffractive optical element provides a way to produce a large number of parallel beamlets for this purpose [12]. The detection can be performed by a SPAD array detector (see section 3.2.2). However, such a scheme not only requires additional costly hardware and more complex alignment, but can lead to cross talk between the channels if the probe volumes are not sufficiently far apart.

Can we monitor a uniformly illuminated large area simultaneously, e.g. with a camera type of detector, where each pixel can be used as a detector? A simple fluorescence camera ordinarily performs such wide-field detection, and each point of the sample is detected by a corresponding pixel of the camera. The requirement of a small volume provides the main challenge for such a wide area detection scheme, since a large area illumination at the sample also implies a large depth of field, i.e. a low z-resolution. So each pixel will end up monitoring a large volume, which would appear to negate the possibility of performing wide field FCS, unless the observed sample area somehow has a very low depth of field. This is automatically satisfied by a two-dimensional sample. But even for a three-dimensional sample there are certain excitation geometries, e.g. total internal reflection (TIRF) and light sheet illumination, which offer such a possibility.

In total internal reflection microscopy, the excitation beam is reflected from the inside surface of a glass (typically) substrate, and the sample is placed over it (figure 3.4). In this geometry, the angle of incidence of the excitation beam is larger than the critical angle (i.e. the angle at which Snell's law predicts that the refracted light will be grazing the surface). Then the reflection at the glass-solution boundary is complete, but the light field does not go to zero abruptly at the surface, which would imply non-physical field gradients. Instead, the excitation beam has an exponentially decaying field amplitude profile in the other 'forbidden' side of the interface. So no light ultimately escapes from the surface and no energy is lost. But if we measure the electric field of light at the interface as a function of distance from the interface, then we would obtain a finite distance-dependent value. The light intensity decays as an exponential with a characteristic length described in the following equation:

$$I(z) = I_0 e^{-\left(\frac{z}{d}\right)} \tag{3.6}$$

where $I(z)$ is the intensity of the evanescent field at a distance z above the interface, I_0 is the intensity at the interface ($z = 0$), and d is the characteristic depth given by:

$$d = \frac{\lambda_0}{4\pi\sqrt{\left(n_1^2 \sin^2 \phi_i - n_2^2\right)}} \tag{3.7}$$

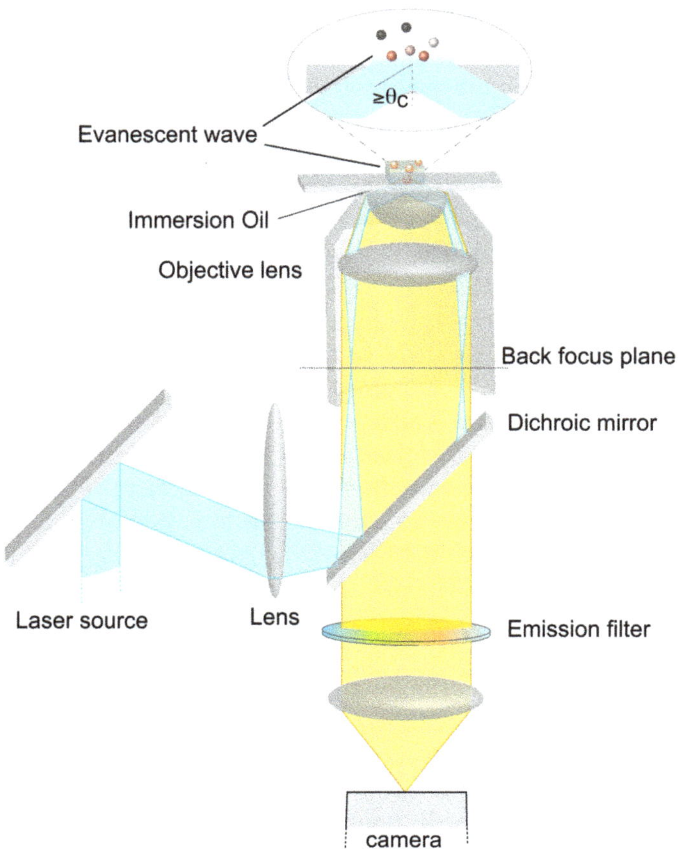

Figure 3.4. Schematic representation of an objective- based TIRF set-up. The excitation light beam (blue) is focused at the back aperture and is total internally reflected off the horizontal top plane of the cover glass (gray). It creates an evanescent field above it at the interface. Any fluorophore within the range of this field can get excited. The fluorescence from this area is imaged by a high numerical aperture objective on to a camera.

where λ_0 is the wavelength of light in vacuum, ϕ_i is the incidence angle, n_1 is the refractive index of the denser medium, and n_2 is the refractive index of the lighter medium. Please see figure 3.4 for a pictorial representation of the geometry of a TIRF set up. Obviously, d can be much less than λ_0. This is the key factor that gives a limited probe volume for FCS. If we have a wide area focus, and the fluorescence from this wide area can be monitored by an area detector (more on that later), then we would be able to perform FCS simultaneously at all pixels. Of course, the technique will be able to monitor only those molecules which come close to the surface (within about 100 nm), but the advantage would be that the molecules far from the surface would not provide appreciable background. So, for example, it would be possible to monitor the basal membrane of a cell which adheres to the surface of a coverslip through which the excitation beam arrives.

As already mentioned, there are two popular geometries for performing TIRF. The more common one is the epi-fluorescence geometry, where the excitation goes through the objective lens which also collects the fluorescence (figure 3.4). The objective needs to have a numerical aperture (*NA*) large enough so that the condition for total internal reflection from glass to water can be met by a thin excitation beam coming through the edge of the back aperture. For an oil immersion lens, the NA has to be considerably greater than 1.33. This ensures that the beam impinges on the glass water interface at an angle larger than the critical angle. The critical angle is (see figure 3.4)

$$\theta_c = \sin^{-1}(n_2 / n_1) \tag{3.8}$$

At the glass–water interface, $\theta_c = 61.7°$. Since $NA = n_1 \sin \alpha$, where α is the half angle subtended by the objective at the focus, this would imply that the excitation beam has to go through the outer 11% of the back aperture for a lens with $NA = 1.49$ (since 1.33/1.49 = 0.89). The main advantage of this is the ability to simultaneously monitor the surface over a large area, so the excitation beam is made to be a thin one focused at the back aperture of the objective lens. This ensures that the beam emerging from the objective is nearly collimated (it is difficult to get a perfectly collimated beam out from the objective, but that is not a necessary condition for TIRF as long as all angles are above the critical angle). The area covered on the cover slip can be of the order of 100 μm in diameter. A thin beam is necessary, since the inner part of the beam will necessarily have a lower angle of incidence, and beyond a certain beam thickness, the inner part would not obey the critical angle condition. Special objective lenses with very high *NA* (e.g. *NA* of 1.49, or even $NA = 1.65$, which requires special immersion oil and special cover glass) are available for performing TIRF, and these allow a thicker pencil of rays, and correspondingly, a larger area can be imaged. Unfortunately, multiple reflections from different optical surfaces, imperfections, and dirt on the optics typically give rise to many interference fringes in the TIRF field of view. Also, these aberrations imply that some excitation light emerges from the objective in the far field, though this can be minimized to a large extent, especially with large *NA* objective lenses and clean aberration free optics.

Another way of performing TIRF that avoids the multiple lens objective altogether involves exciting the sample through a prism (figure 3.5). In this case, the objective is used just for collecting the fluorescence. The objective does not need to have a very high *NA*, since the TIRF is not performed through it. However, the requirement of placing the sample on a prism substrate makes this a somewhat inconvenient, so this is a much less prevalent way of performing TIRF microscopy or TIRF-FCS.

3.1.4.5 Lens-less FCS probe volume created with a fibre tip

All the above mentioned techniques are regularly used for FCS, but each of them has one major drawback: they are part of a bulky instrument. A modern day microscope, which usually houses the objective, the dichroic and some other optical elements, have dimensions which are a good fraction of a meter in all three dimensions. Even if the microscope is somehow minimized, the objective itself is a

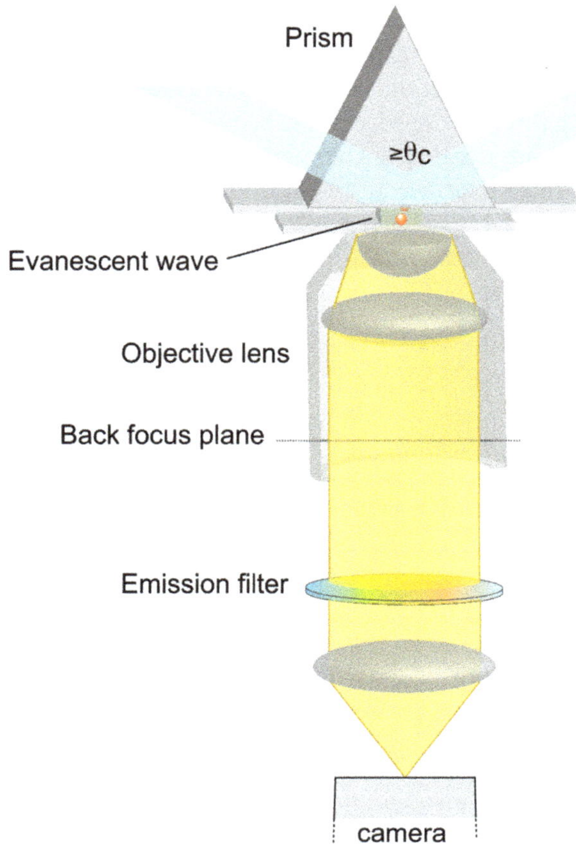

Figure 3.5. Schematic representation of a prism-based TIRF set up. The excitation light beam (blue) is total internally reflected off the horizontal bottom plane of the prism, and creates an evanescent field outside of the prism at the interface. Any fluorophore within the range of this field can get excited. The fluorescence from this area is imaged by a high *NA* objective on to a camera.

few centimetres long, and this seems to set a limit for how small an FCS spectrometer can be. Fortunately, if space is at a premium, and some part of the signal can be traded for it, then it is possible to make an FCS instrument which is sub-millimetre in size, at least at the front end of the instrument. Maiti and co-workers have shown that the tip of an optical fibre can be used to replace the microscope [13]. A quasi confocal volume can be created at the tip of a single mode fiber, by using the same fiber for excitation and collection of emission [14]. This strategy also obviates the need to align the instrument, as the rest of the optical parts can be pre-aligned, fixed and sealed in a box containing just prealigned fiber junctions and optics. In principle this allows FCS to be performed at a remote location which is difficult to reach with a standard microscope, or is too dangerous to reach or handle (e.g. a toxic slurry). The penalty one pays for this miniaturization is a low *NA* for excitation and collection (typical *NA* of a visible wavelength single mode fiber is 0.13), making the volume large and the sensitivity somewhat low [13].

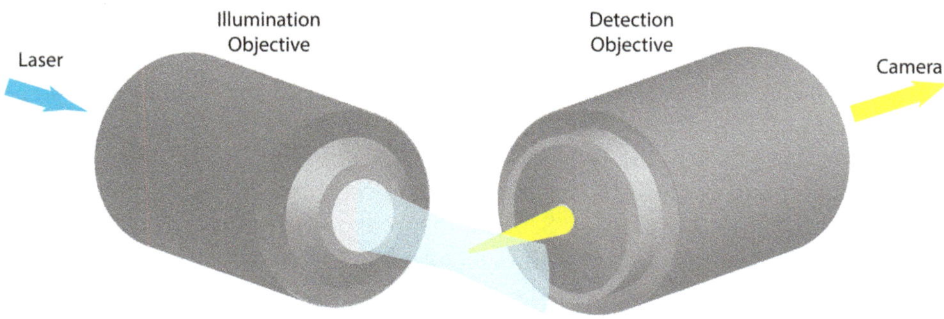

Figure 3.6. Schematic representation of a planar light sheet set up for imaging FCS. The low numerical aperture illuminating objective projects a propagating sheet of light which illuminates a plane of the object (not shown). The high numerical aperture detection objective images the central area of this sheet onto a camera (not shown).

3.1.4.6 Light sheet microscopy

A major drawback of the TIRF microscopy scheme is its inability to probe any location that is not in the immediate vicinity of the glass–water interface. This has been addressed by light sheet microscopy, where a sheet of light is projected into a three-dimensional sample. The sheet can be placed at any height inside the sample, and the objective lens, usually placed orthogonally, produces an image of the sample plane excited by the sheet (figure 3.6). This image is recorded by a scientific camera (EM-CCD or sCMOS), just as in the case of TIRF. A typical geometry would involve a cylindrical lens that focuses light in only one lateral dimension but not the other, producing a line, and a low numerical aperture microscope objective which further projects this on to the sample as a light sheet. There is a trade-off between how thin the light sheet can be at the centre versus how long it remains relatively thin. Please see chapter 5 for details.

3.2 Photon detection

3.2.1 Photon counting

For computing the correlation function, one needs to record the time varying fluorescence signal from the sample. The earliest FCS set-ups used by Magde, Elson and Webb [15] used analog photomultiplier tube detectors. However, computation of the correlation function required them to convert this to a digital signal using an analog-to-digital converter. From about the early nineties, as new types of 'photon counting' detectors became available, it has become standard practice to record photon counts as a function of time, which is sent to the correlator (usually a piece of hardware specially designed to compute an auto- or cross-correlation function from the signal). Of course, the signal in its primary form is analog. However, the photon counting detectors have become sensitive enough that a distinct electrical pulse is generated for each photon that the detector senses. This is usually a short current pulse, which is ultimately detected by the internal electronics and processed when it is above some background threshold. The electronics then produces a digital logic pulse (TTL or NIM, usually) in response to every pulse that is above this background. The timing of the outgoing pulse is closely related to the arrival time of the fluorescence photon at the detector, so the train of pulses given out by the detector provides a representation of the temporally varying fluorescence signal. This representation is of course not fully accurate. First, the QE of detection (which depends on the wavelength) is finite and so the outgoing pulse train has some information missing. As long as this is a random occurrence, it does not introduce a systematic artefact in the analysis. Second, the time of the outgoing pulse has some 'jitter', i.e. two pulses arriving exactly 1 μs apart may result in an outgoing pair of pulses that are apart by 1 ± 0.01 μs, say. Usually for an avalanche photodiode (APD) detector, this jitter is of the order of a few ns. However, high time resolution APDs can reduce the timing jitter to <100 ps. This level of accuracy is not required for measuring diffusion, as the diffusion time scales are typically >10 μs. However, they are important for some other processes which involve the fluorescence lifetime, as we will see later. APDs have small detector areas, typically between 20–50 μm. This requires that the fluorescence signal be focused, and the focused signal has to be aligned precisely to the APD photodetector. While this appears to be an inconvenience, in practice the small detector areas of the APDs are often themselves used as the detector pinholes in a confocal setup. There are other APD-like detectors (the so called Si photomultipliers) which have bigger detector areas, but they usually have lower QEs.

Though in almost all cases an APD is the best choice for performing FCS at a point in a 3D solution, it need not be the best choice for all such cases. The primary consideration should be the QE of the detector at the wavelength of emission. The APDs are Si based photo-detectors, and they have a high QE which can exceed 70% at their peak wavelength, which is typically around 550 to 600 nm (figure 3.7). Their efficiencies fall drastically at wavelengths in the blue, and they are mostly unusable in the ultraviolet. On the other hand, photomultiplier tubes, which are relatively

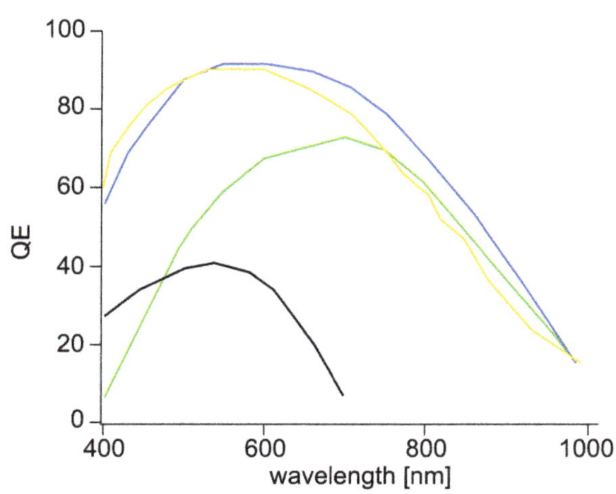

Figure 3.7. Typical QE of different detectors as a function of wavelength. Black: a Bi-alkali photomultiplier tube (PMT), Blue: back-illuminated EM-CCD, green: avalanche photodiode (APD), yellow: sCMOS. Note: this is indicative only, different variants of the same type of detector may have different QEs.

noisy and have lower peak QE compared to the APDs, actually have higher efficiency in the UV compared to the visible (e.g. see a comparison of a photomultiplier tube QE curve versus that of an APD presented in figure 3.7, which shows the QE for a so-called 'hybrid PMT', which has the best QE among the PMTs.). So for near UV fluorophores with emission around 300–350 nm, the detector of choice would typically be a PMT.

High-resolution photon arrival time recording: Typically, a detector would send a pulse to the data processor a certain time after the arrival of the photon. So the electrical pulse always has a relationship with the arrival time of the photon, though the accuracy of this information depends on the detector type, and can vary between a few picoseconds to many nanoseconds. A typical data processor bins these arrivals in time bins, i.e. it creates a histogram of the number of photons arriving within a certain time interval, without keeping an exact measure of the arrival time of each photon. Would it be useful to record the arrival time of individual photons? As long as the timescale of interest is much longer than the bin widths (e.g. diffusion can take 10s of μs, while the typical bin width of a modern processor will be <1 ns), there does not appear to be any particular need for this extra information. However, many experiments require information at a much higher time resolution. For example, experiments with anti-bunching require time resolution of the order of 100 ps. Even here, as long as the time bins are of the appropriate size (say a hundred ps), it is adequate to just bin the arrival times. However, photon arrival times must be recorded by the detectors with an accuracy equal to or better than the bin times, for it to be meaningful. Also, there are other related methodologies that require the recording of the arrival times of individual photons, especially when the excitation is pulsed. For example, in FLCS, the arrival times of individual photons are recorded with respect to the excitation pulse. This provides a time correlated single-photon counting estimate of the fluorescence life time in the same experiment as the FCS, and allows the segregation of species on the basis of both size and lifetime (see chapter 4).

3.2.2 Array detectors

As we have already mentioned, TIRF or light sheet imaging offer detection that is limited in the z-dimension, which can be suitable for FCS. In this case each pixel of the detector acts as a FCS detection pinhole. This would require an array detector with high sensitivity, such as a CCD camera. In general, a camera is suited for wide-field detection, where the object plane is focused on the camera. Although neighbouring pinholes may not be exactly independent, that is usually not a major issue. The real limitation of this technique is the frame rate of the camera. This is usually at least a few milliseconds for a scientific CCD camera. Using a CMOS camera can get it down to about a ms, but even that would mean that the sampling interval, or the basic time bin for the experiment is about 1 ms. This is at least a factor of 10–100 slower than what would be required for performing diffusion measurements of freely diffusing molecules in an aqueous medium, or even in a living cell, where the diffusion is a few times slower than that in water. So typically,

FCS with camera-based detectors look at molecules which are moving rather slowly, e.g. those on a membrane. This is the arena where imaging FCS has made a large contribution, as described in a different chapter of this book.

The array detector needs to have a few characteristics which would make it appropriate for FCS. As already mentioned, the frame rate should be as high as possible, and >100 frames per second is required for a meaningful FCS for most needs. The other characteristics of course are the sensitivity and the noise characteristics of these detectors. The QE of detection can be up to ~90% for the best CCD cameras. However, the frame rate comes at the cost of noise. Each time the frame is 'read', i.e. the data is transferred from the CCD chip, there is an error that is associated with the process of reading which gets added to the signal at each pixel. The point to note is that the noise added is per read, not per unit time. If the read rate is low, the number of photons per pixel would be correspondingly high, and this noise may not matter. However, this error may become significant in comparison with the recorded photons per pixel, if the frame rate is high. There is a variant of the CCD, called electron gain multiplying CCD (EM-CCD), which addresses this issue to some extent. It has an 'on-chip' gain multiplier, which increases the number of counts of photons before the photons are read out. Unfortunately, this 'on-chip gain' has its own error. However, at large read rates, it still provides substantial advantages over the conventional CCD. Another type of array detector is the scientific CMOS camera. Here the read rates can be much higher as each pixel is read out individually. However, the CMOS cameras typically have a lower fill factor and hence a lower QE for detection. Also, the noise of CMOS sensors tends to be higher. However, recently, there has been a new class of scientific CMOS cameras that have overcome some of these limitations to a great extent. These have been used for FCS, both for TIRF and for light sheet based applications.

The most exciting recent addition to the collection of array detectors for FCS is an APD array. The commercially available current APD array detectors have a 32 × 32 pixel array. This is of course rather small compared to the typical 1000 × 1000 pixels of the CCD or CMOS detectors, but it is still enough for imaging an area inside the cell (or of a sample of interest). An additional feature they typically offer is the ability to perform time correlated single-photon counting (TCSPC), since they contain circuitry which provides the arrival time of the photons at each pixel with reasonably high accuracy. The current limitations of these devices is the fill factor, which can be as low as a few percent, since the on-chip circuitry occupies part of the space on the chip. One hopes that soon these will mature to have better fill factors, and optical tool builders will utilize technologies which will make sure that the array of excitation spots will match the array of detectors, such that light is not wasted in between the detector elements. These then can truly become FCS/TCSPC 'cameras' (e.g. [12, 16]).

3.2.3 Photomultiplier tubes

Several other types of detectors have been used for FCS. As already mentioned, for UV detection, APDs are quite inefficient. However, photo multiplier tube (PMT) detectors have a QE that typically goes up to >20% in the 300–400 nm region (depending on the cathode type). So single-photon counting PMTs are the preferred devices for UV FCS. The QE of PMTs typically falls drastically at wavelengths below 300 nm, due to the glass used in front to the photocathode. One can get UV sensitive PMTs which do not have this problem, but there is rarely any need for a deep or mid-UV fluorophore. PMTs work by electrostatically accelerating the initial photon-ejected electron from a suitable photocathode surface. A stage-by-stage acceleration is achieved through a series of 'dynodes' which are held at progressively higher potentials (each separated by ~100–200 V). Each time an electron accelerated by the inter-dynode potential collides with the next with a high kinetic energy, that dynode emits multiple electrons, thereby amplifying the original signal. At the end of the dynode chain (typically ~10 dynodes), the PMT anode receives a reasonable-sized current pulse for each photon detected at the photocathode. This current pulse can then be detected in the data acquisition system and read as the arrival of a photon. PMTs typically provide an analogue signal, but there are PMT modules which have a discriminator and a digitizer built in, and essentially function the same way as APDs. One disadvantage of PMTs is their wide detector face. A separate pinhole must be placed in front of the PMTs and the unused area of the detector needs to be well-shielded from extraneous light. Also, PMTs are typically rather noisy as their dark counts can be an order of magnitude higher than that of APDs, and of course even at the peak wavelength, their QE is nowhere near the peak QE of the APDs (>70% in the green/orange, see figure 3.7). So their use in FCS has been rather limited. One major exception are the amino acids. If one could perform FCS measurements of proteins using just the fluorescent fluorophores such as Tryptophan, FCS could have become a label-free technique and could access native molecules. UV-FCS was first demonstrated on 2-aminopurine [17], followed by two-photon FCS of Trp [18]. These measurements are, however, rather difficult. The major reason is not the low quantum yield or low cross section, or even the requirement for UV optics (especially required for single-photon excitation), but the photo-stability of the natural chromophores. Since one has to ensure the detection of more than one photon on average from each individual particle during their stay in the probe volume, photo-bleaching is a major problem. Low QE or low absorption cross section can in principle be taken care of by using a more powerful excitation laser, but this becomes impractical if the fluorophore itself is prone to photo-bleaching. Even then, the lure of a 'label-free' experiment has drawn several researchers into the game of trying to perform FCS with Tryptophan as the fluorophore. The typical trick has been to target particles which have many tryptophans, such as a protein aggregate. This way, even if a few of the Trp photo-bleach, the protein will remain fluorescent as a whole, and FCS can be performed. Maiti *et al* have shown that the aggregates of the protein barstar can be measured with FCS using the intrinsic Trp fluorescence [19]. This has been improved upon later by Gratton and co-workers who have shown that as few as 200 tryptophans are enough for reasonable FCS data collection [20]. However, for the vast majority of proteins, some kind of fluorescent labelling remains the only choice for obtaining reasonable quality FCS data.

Appendix 3.1

Theoretical estimation of the probe volume based on diffraction theory

The probe volume in conventional diffraction limited observation is the result of a double diffraction: first, of the excitation beam by the back aperture of the objective, and next, of the collected fluorescence signal by the same aperture. As we have already discussed, typically we would prefer the smallest possible probe volume, with the sharpest possible edges, for the best possible FCS measurement. It turns out, that a property of light, named 'diffraction', comes in the way of arbitrarily reducing the probe volume or increasing the sharpness of the edges.

What is diffraction, and how does it limit the best probe volume obtainable in FCS? Diffraction is a wave property, and light of course is an electromagnetic wave. When a lens focuses a laser beam and creates a high intensity at the focus and a low intensity around it, it does so by adding the electromagnetic waves arriving along the different paths from the lens to the focal region. Along the path, the phases of the wave changes continuously from 0 to 2π for every segment that equals the wavelength. The different paths are of such lengths that the phases add up near the focus, giving a constructive interference resulting in a high intensity, and they cancel away from it, making the intensity low there. If a lens in the path of a beam has a smaller aperture than the beam diameter, it removes the rays that are incident beyond its aperture. This results in incomplete cancellations at some particular distances even away from the focus, giving rise to ring shaped patterns of brightness around the central focus. These are commonly referred to as diffraction rings. However, making beams small compared to the size of the lens does not help either. To focus the beam to a very small spot, one needs to have destructive interference just slightly away from the central spot. A larger variety of angles leads to a more efficient cancellation. This requires a large beam diameter and a small focal length. So to obtain the smallest spot, we usually use a large diameter, small focal length lens (an objective lens) and a beam that is bigger than the lens diameter.

The smallest spot size that can be obtained by a lens is determined by the so called numerical aperture (NA) of the lens, given by

$$NA = n \sin \alpha$$

And the smallest spot size (radius) is given by

$$0.61\lambda/NA$$

where α is the half angle that the lens aperture makes at the focus, n is the refractive index of the medium, and λ is the wavelength of light in vacuum.

This estimate can be rigorously derived from diffraction theory. Here, for a simple conceptual understanding, we will describe a two dimensional case (i.e. intensity at the focal plane) under paraxial approximation [21]. The distribution of intensity at the focal plane is given by the square of the first order Bessel's function:

$$I(\nu) \sim \left[2\frac{J_1(\nu)}{\nu}\right]^2$$

where ν is the radial distance in optical units. Optical unit (radial) is a rescaled unit proportional to the actual radial distance and incorporates both the NA and λ:

$$\nu = \left(\frac{2\pi}{\lambda}\right) * NA * r$$

Since the first order Bessel function has zeroes at particular values of ν ($\nu = 1.22\pi$, 1.635π, ...), the pattern has zero intensity at particular radial distances, giving rise to a strong central circular region (the so-called Airy disk, after George Airy), and progressively weaker rings in the focal plane. These diffraction rings (shown in figure 3.A1.1 where intensity $I(\vec{r})$ is normalized, $\lambda = 0.5$ μm, and the radial distribution $\boldsymbol{r}$ is in μm) are unavoidable if we want the smallest possible size for the central spot.

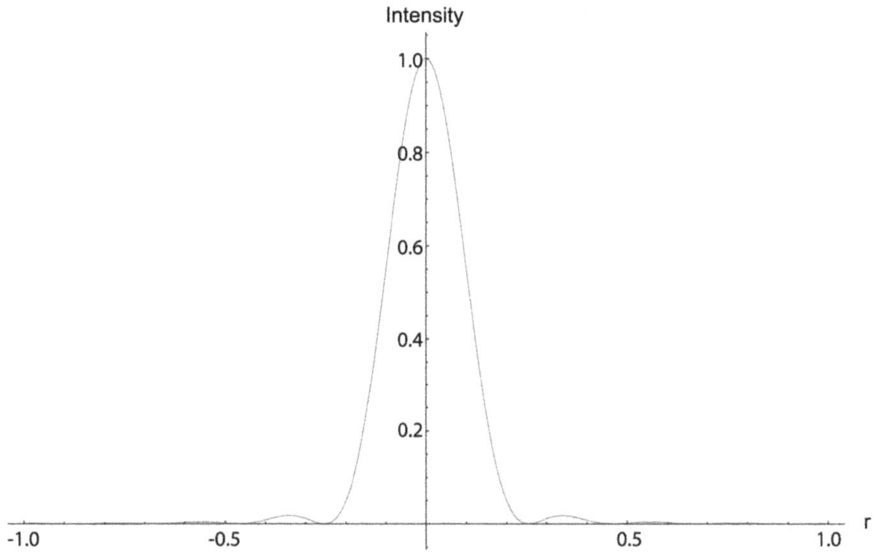

Figure 3.A1.1. Diffraction by a circular aperture. The intensity trace at the plane of focus of a large diameter excitation beam, as a function of radial distance from the centre of focus.

Now let us consider the confocal probe volume. Diffraction happens twice for confocal detection. In the first step, the laser beam is focused by a high *NA* objective lens. This creates the standard diffraction pattern created by a circular aperture, as already described for the focal plane (figure 3.A1.1). We call this intensity pattern the 'excitation profile' $I(\vec{r})$, which is just the normalized excitation intensity as a function of position. Any fluorophore molecule near the focus at a position $\boldsymbol{r}$ near the focus gets excited, with the rate of excitation proportional to the intensity at $\boldsymbol{r}$ (assuming a linear single-photon excitation process). This is shown in figure 3.2, where the molecule could be in position p (at the center of excitation) or at q. In the second step, a part of the isotropically emitted fluorescence signal from the molecule is collected by the objective lens. The collected beam propagates back through the objective and is again diffracted by the back aperture, which is the smallest effective aperture in the objective lens for the light beam. We note that the diameter of the front ('top') lens of the objective, though physically smaller, is not typically the limiting aperture. The back aperture, physically located inside the objective, has a diameter of approximately 1 cm for a typical high *NA* objective. The fluorescence signal is then focused by the tube lens on to the confocal pinhole placed in front of the detector. The focusing by the tube lens is not diffraction limited. The signal beam size is already limited by the objective back aperture (~1 cm), and so it is easy to make the tube lens much larger than the beam. This is why the *NA* of the tube lens does not enter into the consideration for determining the resolution. This then makes a diffraction limited image (with its own Airy disk and accompanying rings) on the pinhole (i.e. the image) plane at the focus of the tube lens at position p' or q', corresponding to the molecule at p or q, respectively (figure 3.2). q' is centred M*z away from the center of the pinhole, where the M is the magnification of the objective lens (M = ratio of the focal length of the tube lens to the focal length of the objective, which can be different from the value inscribed on the objective if a non-standard tube lens is used). The signal recorded by the detector placed behind the pinhole is simply the integral of this pattern over the pinhole area. Let's consider the probability of detecting a photon from q for a simple two-dimensional sample. This probability, for a given optical system, is determined by three geometric factors: the intensity profile $I(\vec{r})$ (see equation (3.4)) of the excitation at q (which depends on $\boldsymbol{r}$, also on $\lambda_{\text{excitation}}$ and NA), the diffracted image of q overlapping the pinhole (function of M, $\boldsymbol{r}$, $\lambda_{\text{emission}}$ and NA), and the diameter of the circular pinhole (which determines the portion of the diffracted image that can reach the detector). Of course, it also linearly depends on the average intensity of the laser beam, the probability of absorption and emission, the collection efficiency of the optical system and the efficiency of the detector, but these factors are independent of the position of the emitting molecule. The geometric factors together determine with what probability a photon from the molecule at q will be detected compared to that from a molecule at the center p. This probability function is called the 'detectivity profile' function or $W(\vec{r})$ and is related to the Point Spread Function used in confocal microscopy.

The detectivity function is easy to calculate for a two-dimensional object, and that provides a basic mathematical understanding of the process. This is shown in the series of figures below. The first one assumes that the pinhole size is equivalent to the diameter of the Airy disk (which is called 'one Airy unit'), i.e. the distance between the center and the first zero of the excitation profile. The first zero of the first order Bessel's function is at $\nu = 1.22\pi$, corresponding to 508 nm for $\lambda = 500$ nm, and $NA = 1.2$. This is of course magnified by the system by a factor of M, so the corresponding Airy disk size is about 20 μm for a 40X objective. figure 3.A1.2(a) shows that the side lobes are present in $W(\vec{r})$ (see the blue trace), but the full scale (zoomed out) version figure 3.A1.2(b) shows that its relative contribution is very small. This is important, since it implies that a Gaussian approximation of the probe volume, assumed in most of this book, is a reasonable one. However, if the size of the confocal aperture is bigger than this, the side lobes start becoming appreciable, and the resolution decreases. This is shown in the

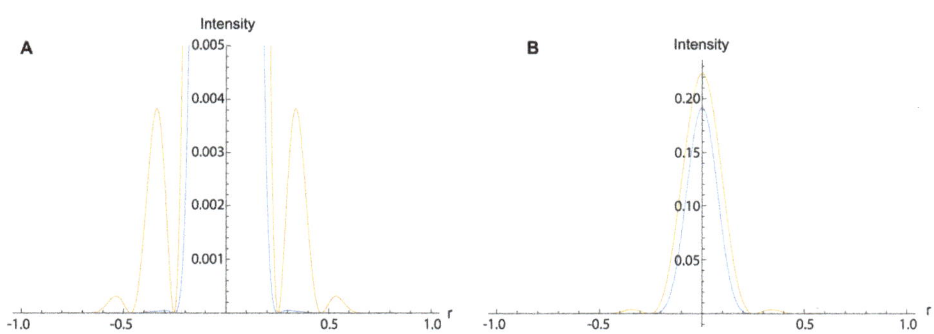

Figure 3.A1.2. Comparison of the confocal detectivity profile of a two-dimensional sample at focus at two different pinhole sizes: 1 Airy unit (blue) and 4 Airy units (orange). (a) shows a zoomed up version with pronounced sidelobes at the larger pinhole. (b) Full scale version showing that the total contribution of the sidelobes is still very small.

yellow trace in figure 3.A1.2, which corresponds to a pinhole opened to 4 Airy units. We note that the side lobe contributions are larger for a three-dimensional sample, but the concept remains similar.

What happens if we make the pinhole even smaller? The resolution does go up to some extent, but at the cost of a large loss of detected photons. This extra loss becomes too large a price to pay for the higher resolution, and usually the pinhole is set around 1 Airy unit for a typical confocal FCS experiment.

As we noted, the above calculation was for a two-dimensional sample only. The pinhole is effective in discriminating against light coming from the third dimension too, as is shown in figure 3.2 by the point *r*. A good description of the three dimensional confocal probe volume is given by Visser and Wiersma [9]. Here we provide numerically calculated values of the fluorescence signal at the focus of the tube lens (near the aperture plane) based on their treatment. We present this pattern together with a Gaussian approximation to it, which uses the standard deviation as the free parameter (figure 3.A1.3). We see that the Gaussian approximation still provides a reasonably good fit. This is the reason that FCS theoretical fits which assume a Gaussian volume actually work well (see chapter 5), if the aperture parameters are chosen well (for aperture related artefacts, see chapter 7).

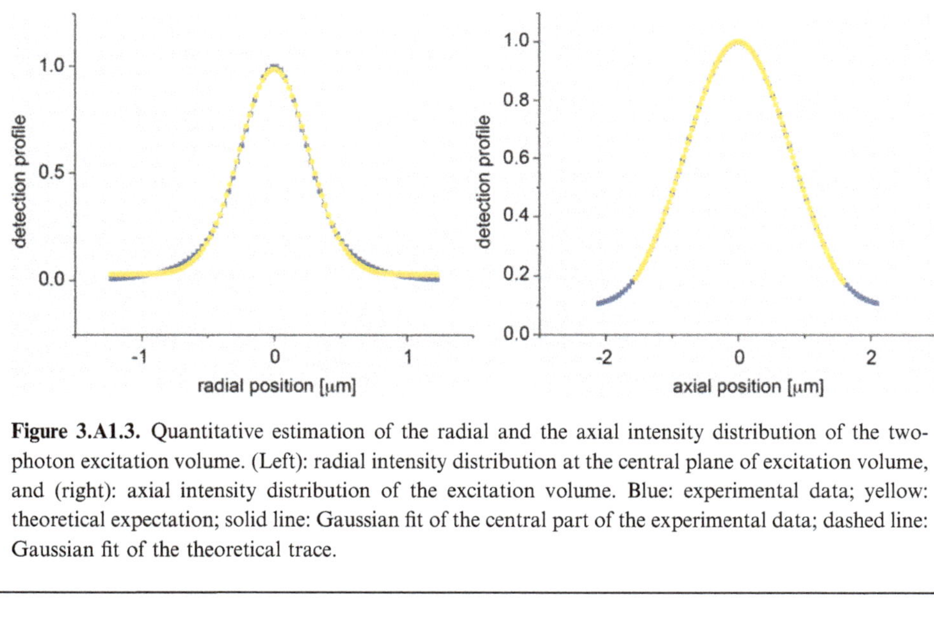

Figure 3.A1.3. Quantitative estimation of the radial and the axial intensity distribution of the two-photon excitation volume. (Left): radial intensity distribution at the central plane of excitation volume, and (right): axial intensity distribution of the excitation volume. Blue: experimental data; yellow: theoretical expectation; solid line: Gaussian fit of the central part of the experimental data; dashed line: Gaussian fit of the theoretical trace.

3.3 Exercises

1. Suppose you are building an FCS spectrometer yourself. You are planning to use an optical fiber as the confocal aperture. Your objective is 40×, 1.2 *NA*, and the excitation laser is a 543 nm helium–neon. The microscope for the objective has a 16 cm tube lens, but in your home-built set-up you are using a 25 cm one. What fiber diameter should you choose?

2. In a TIRF FCS experiment using 543 nm excitation and 1.49 *NA* objective lens, fluorescent beads of 1 μm diameter in aqueous solution are used for testing. The coverglass has a refractive index of 1.51, and the beam is incident at an angle of 70°. The viscosity of water is 1 mPa·s, at the laboratory temperature (20 °C). What would be the approximate diffusion time observed in the experiment?
   (Note that in TIRF geometry, the characteristic diffusion time in the direction away from the surface is given by ~3.3 d^2/D, where d = characteristic decay length of TIRF illumination, D = diffusion constant.)

3. Suppose you want to perform a TIRF based FCS experiment at 500 nm. You want the best resolution that your optics allows, in the lateral dimensions. You are using a 100 × 1.49 *NA* objective in a microscope. What is the best possible optical resolution (lateral) that you can achieve? Now, suppose your camera has a pixel size of 16 μm. What would be the resolution that you will actually get in your image? What can you do to improve it?

References

[1] Koppel D 1974 Statistical accuracy in fluorescence correlation spectroscopy *Phys. Rev.* A **10** 1938

[2] Qian H 1990 On the statistics of fluorescence correlation spectroscopy *Biophys. Chem.* **38** 49–57

[3] Wohland T, Rigler R and Vogel H 2001 The standard deviation in fluorescence correlation spectroscopy *Biophys. J.* **80** 2987–99

[4] Wenger J, Gérard D, Aouani H, Rigneault H, Lowder B, Blair S, Devaux E and Ebbesen T W 2009 Nanoaperture-enhanced signal-to-noise ratio in fluorescence correlation spectroscopy *Anal. Chem.* **81** 834–9

[5] Lakowicz J R 2006 *Principles of Fluorescence Spectroscopy* 3rd edn (Berlin: Springer)

[6] Valeur B and Berberan-Santos M N 2012 *Molecular Fluorescence* 2nd edn (Weinheim, Germany: Wiley)

[7] Parson W W 2015 *Modern Optical Spectroscopy* 2nd edn (Berlin: Springer)

[8] Mertz J, Xu C and Webb W W 1995 Single-molecule detection by two-photon-excited fluorescence *Opt. Lett.* **20** 2532

[9] Visser T and Wiersma S Electromagnetic description of image formation in confocal fluorescence microscopy *J. Opt. Soc. Am.* A **11** 599–608

[10] Hess S and Webb W 2002 Focal volume optics and experimental artifacts in confocal fluorescence correlation spectroscopy *Biophys. J.* **83** 2300–17

[11] Sandison D R and Webb W W 1994 Background rejection and signal-to-noise optimization in confocal and alternative fluorescence microscopes *Appl. Opt.* **33** 603–15

[12] Krmpot A J *et al* Functional fluorescence microscopy imaging: quantitative scanning-free confocal fluorescence microscopy for the characterization of fast dynamic processes in live cells *Anal. Chem.* **91** 11129–37

[13] Garai K, Muralidhar M and Maiti S 2006 Fiber-optic fluorescence correlation spectrometer *Appl. Opt.* **45** 7538–42

[14] Balaji J, Garai K, Chakrabarti S and Maiti S 2003 Axial resolution limit of a fiber-optic fluorescence probe *Appl. Opt.* **42** 3780–4

[15] Magde D, Elson E and Webb W 1972 Thermodynamic fluctuations in a reacting system—measurement by fluorescence correlation spectroscopy *Phys. Rev. Lett.* **29** 705–8

[16] Singh A P, Krieger J W, Buchholz J, Charbon E, Langowski J and Wohland T 2013 The performance of 2D array detectors for light sheet based fluorescence correlation spectroscopy *Opt. Express* **21** 8652

[17] Wennmalm S, Blom H, Wallerman L and Rigler R 2001 UV-Fluorescence correlation spectroscopy of 2-aminopurine *Biol. Chem.* **382** 393–7

[18] Lippitz M, Erker W, Decker H, van Holde K E and Basché T 2002 Two-photon excitation microscopy of tryptophan-containing proteins *Proc. Natl Acad. Sci. USA* **99** 2772–7

[19] Sahoo B, Balaji J, Nag S, Kaushalya S K and Maiti S 2008 Protein aggregation probed by two-photon fluorescence correlation spectroscopy of native tryptophan *J. Chem. Phys.* **129** 075103

[20] Ranjit S, Dvornikov A, Holland D A, Reinhart G D, Jameson D M and Gratton E 2014 Application of three-photon excitation FCS to the study of protein oligomerization *J. Phys. Chem.* B **118** 14627–31

[21] Born M and Wolf E 1999 *Principles of Optics* 7th edn (Cambridge: Cambridge University Press)

Thorsten Wohland, Sudipta Maiti and Radek Macháň

Chapter 4

Data structure, correlation and processing

FCS studies molecular processes by analysing and interpreting the correlation functions of fluctuations in fluorescence intensity traces collected from a defined observation volume in the sample. Chapter 2 has explained the theoretical relationship between the intensity fluctuations and their correlation functions and in chapter 3 we have seen how fluorescence intensity time traces are collected. This chapter will describe the practical process of obtaining correlation functions from experimentally measured fluorescence intensity time traces. This task can be performed either by a dedicated electronic device, a hardware correlator, or by a software correlator using a digitised intensity time trace as its input. We will introduce the hardware correlators only briefly and give more space to the software correlators which have become the preferred choice owing to the increased computational power of commonly available computers as well as the higher flexibility provided by such correlators. Several commonly used software correlator designs will be explained and their performance in processing data of different structures compared. In the last section, we will introduce the concept of statistical filtering and its implementation in software correlators. Applications of statistical filtering in eliminating artefacts of FCS and FCCS will be discussed further in chapters 6 and 7. After this chapter the reader should have a good understanding of strategies for generating correlation functions in FCS and their suitability for data obtained in different FCS modalities. This knowledge will be useful, among others, in understanding the origin of bias and standard deviations of correlation functions discussed in chapter 8.

4.1 Software correlators

In the case when the fluorescence intensity time trace is saved as the output of an FCS experiment, the ACF is calculated from these values by software. There exist several algorithms that can do the job and we will refer to them in general as software correlators. The performance and suitability of the individual software

doi:10.1088/978-0-7503-2080-1ch4

correlators depends strongly on the structure in which the raw data are saved. Let us here consider several commonly encountered cases and introduce the appropriate correlators for each of them.

4.1.1 Binned intensity trace—linear correlator

We will start from the most straightforward and intuitive scenario in which the total measurement time is divided into M bins of equal duration $\Delta\tau$ (the temporal resolution with which the data have been saved) and the raw data have the form of an array of M numbers $F(i)$ where i is the index of the time bin running from 1 to M. $F(i)$ is the detector reading in the i-th interval; in the case of data acquired by a photon counting detector, it is the number of photons detected within the i-th time interval. Note that even when using detectors that are not photon counting (such as cameras), it is important that the intensity trace values $F(i)$ (after background subtraction) can be considered proportional to the number of detected photons. Otherwise the ACF would not be correctly described by the theory derived in chapter 5. The ACF can then be calculated following the definition of the ACF given in chapter 2 as:

$$G(m\Delta\tau) = \frac{\frac{1}{M-m}\sum_{i=1}^{M-m} F(i)F(i+m)}{\left(\frac{1}{M}\sum_{i=1}^{M} F(i)\right)^2} \tag{4.1}$$

You may notice that the numerator is calculated only from $M-m$ samples while the denominator is calculated from M samples. The formula is suitable in the cases when m is negligible compared to M (long measurement times and short correlation lag times), but for longer lag times it can introduce errors in the ACF. For that reason, a modified formula with symmetrical normalization has been derived [1]:

$$\begin{aligned} G(m\Delta\tau) &= \frac{\frac{1}{M-m}\sum_{i=1}^{M-m} F(i)F(i+m)}{\left(\frac{1}{M-m}\sum_{i=1}^{M-m} F(i)\right)\left(\frac{1}{M-m}\sum_{i=1}^{M-m} F(i+m)\right)} \\ &= (M-m)\frac{\sum_{i=1}^{M-m} F(i)F(i+m)}{\left(\sum_{i=1}^{M-m} F(i)\right)\left(\sum_{i=1}^{M-m} F(i+m)\right)} \end{aligned} \tag{4.2}$$

The calculation following the formula above is straightforward. For each lag time $m\Delta\tau$ the correlator calculates three sums, which we will refer to as the Correlation (the numerator in the formula for the ACF), the Direct Monitor (the first sum in the denominator) and the Delayed Monitor (the second sum in the denominator). The starting values of the sums are 0; the correlator then calculates the products of the intensity values and in each step updates the three sums. Starting with $F(1)$ and $F(1+m)$ the correlator does the following assignment:

$$\text{Correlation} := \text{Correlation} + F(1)F(1+m) \tag{4.3a}$$

$$\text{Direct Monitor} := \text{Direct Monitor} + F(1) \tag{4.3b}$$

$$\text{Delayed Monitor} := \text{Delayed Monitor} + F(1 + m) \tag{4.3c}$$

An analogous calculation is then performed for $F(2)$ and $F(2 + m)$ and so on until all $M - m$ products have been summed.

Although straightforward, the calculation can be quite demanding because of the large number of products evaluated. Since it takes $M - m$ products to get a single point in the ACF $G(m\Delta\tau)$, the number of products needed to calculate the whole ACF to the maximum lag time $m_{\max}\Delta\tau$ ($1 \leqslant m \leqslant m_{\max}$) is given by the following formula (see exercises 1 and 2 for derivation):

$$\sum_{m=1}^{m_{\max}} (M - m) = Mm_{\max} - \frac{1}{2}m_{\max}(m_{\max} + 1) \tag{4.4}$$

Equation (4.4) shows that the computational complexity of a linear correlator scales with $M \cdot m_{\max}$. If we want to calculate the correlation function until the largest value of $m_{\max}$ permitted by the length of the intensity trace (which is $m_{\max} = M - 1$) the complexity is of the order M^2. However, in such a case the last ACF point $G(m_{\max}\Delta\tau)$ is calculated from a single product only; in practice, $m_{\max}$ is likely to be chosen much smaller to ensure a reasonable statistical accuracy of $G(m_{\max}\Delta\tau)$. If we choose to evaluate the correlation function only between the lag times $\Delta\tau$ and $m_{\max}\Delta\tau$, where $m_{\max}\Delta\tau$ is a certain fixed lag time that does not change with M, we get a much more favourable, linear, scaling of computational complexity with M.

This brings us to the question whether the number of operations involved in the calculation cannot be reduced. The simple answer is no, unless we reduce the number of correlation lag times for which the ACF is calculated. Looking back at the title of this section, we have called this correlator *linear*, by which we referred to the fact that the ACF is calculated for linearly spaced correlation lag times (with interval $\Delta\tau$). However, in FCS we usually plot correlation functions with logarithmic lag time scale, because it helps to visualize the correlation decay on various time scales spanning often several orders of magnitude (e.g. transitions between singlet and triplet states and translational diffusion). It is, therefore, a good idea to calculate the ACF for logarithmically instead of linearly spaced lag times. In the case of a logarithmic scale the interval between two consecutive lag time points for which the ACF values are calculated grows proportionally with the lag time and the number of ACF values needed to be calculated is, therefore, reduced (resulting in reduced computational complexity). Note that with logarithmically spaced lag times the ACF is calculated with a constant relative temporal resolution (interval between two consecutive lag time points relative to the lag time). Viewing it this way makes the logarithmic scaling a natural choice rather than a desperate measure to save computational time.

4.1.2 Binned intensity trace—multiple-tau correlator

As explained at the end of the previous section, logarithmic spacing of lag time points for which the ACF is calculated seems a better choice than linear spacing for reasons of computational efficiency. However, strictly logarithmic spacing would be rather difficult to follow in the ACF calculation from a binned intensity time trace. Therefore, quasi-logarithmic schemes known also as multiple-tau schemes (the name originating from a line of hardware correlators implementing such a scheme) have been introduced [2–4].

In a multiple-tau correlator the lag time points are divided into groups with equal spacing of the points within a group. The interval of the points in the first group, the group of the shortest lag times, is $\Delta\tau$ (the minimal interval possible), in the second group it is $2\Delta\tau$ and in each subsequent group the interval is twice longer than in the preceding group. In other words, the interval is $2^{k-1}\Delta\tau$, where k is the ordinal number of the group $k = 1, 2, \ldots, q$. Note that such a quasi-logarithmic scheme, which could be also called quasi-$\log_2$ correlator, is not the only one possible. In principle, other schemes could be derived from logarithms with different bases; for example, in a quasi-$\log_3$ correlator the intervals would increase as powers of 3, $3^{k-1}\Delta\tau$ [5]. However, because of its widespread use, we will stay with the quasi-$\log_2$ scheme in the following discussion. The groups contain the same number of points (let us denote the number as $p/2$) except for the first group which typically contains double the number of points (p). We will then refer to such correlator structure as (p, q), where q is the total number of groups. For example, a (16, 7) correlator calculates ACFs in 16 points with interval $\Delta\tau$, 8 points with interval $2\Delta\tau$ and so on until the last 8 points with interval $64\Delta\tau$. We can calculate the lag time points τ_m, where the index m describes the ordinal number of the point:

$$\tau_m = \begin{cases} m\Delta\tau & 0 < m \leqslant \frac{p}{2} \\ \tau_{m-1} + \Delta\tau 2^{\left\lfloor \frac{m-1-p/2}{p/2} \right\rfloor} & \frac{p}{2} < m \leqslant (q+1)\frac{p}{2} \end{cases} \tag{4.5}$$

where the bracket $\lfloor \ldots \rfloor$ represents the integer part of the enclosed division.

The temporal resolution with which the ACF is calculated at each lag time point follows the spacing of the lag time points; thus, for the points of the k-th group it is $2^{k-1}\Delta\tau$. Therefore, when progressing to each subsequent group, the binned intensity trace is binned further to make the bin width equal to the time resolution of the respective group. The first p points in the ACF are calculated from the original trace with temporal resolution $\Delta\tau$. To calculate the second group ($p < m \leqslant \frac{3}{2}p$) the time trace is binned by summing up pairs of consecutive intensity values resulting in bins of $2\Delta\tau$ width, $F^{(2)}(i) = F(2i - 1) + F(2i)$. To calculate the third group, the intensity trace is binned with time resolution $4\Delta\tau$ per bin, $F^{(3)}(i) = F^{(2)}(2i - 1) + F^{(2)}(2i)$, and analogously for all the other groups. The principle of a multiple-tau correlator is illustrated in figure 4.1. The figure also shows that besides reducing the number of

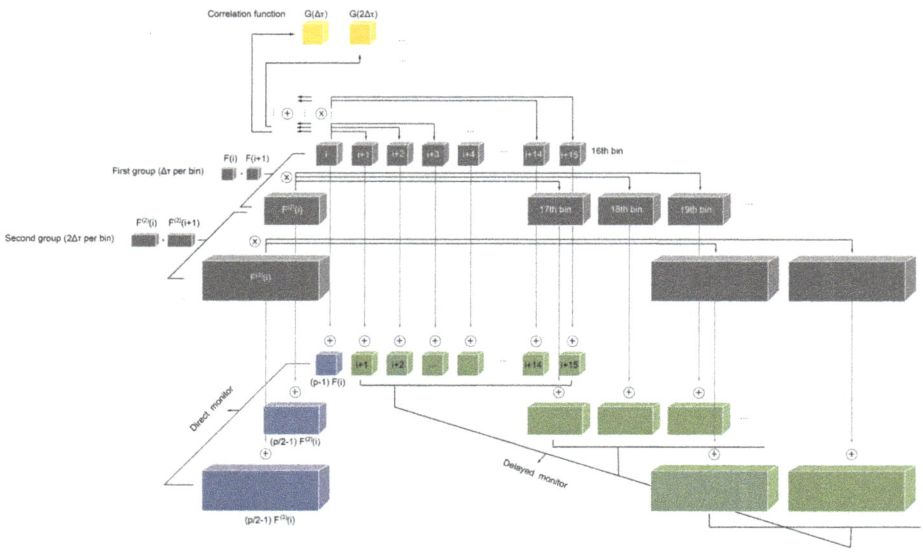

Figure 4.1. Illustration of the principle of a (16, q) multiple-tau correlator for real-time calculation of ACFs. The following tasks are performed upon registering every new point $F(i)$ in the intensity trace (which happens every $\Delta\tau$) : (i) all values stored in the registers are moved by one bin to the right; the value in the last bin is discarded and the last bin is filled with the value from the penultimate bin and so on until the value from the first bin is moved to the second bin and the newly registered value $F(i)$ is stored in the first bin. (ii) The value in the first bin is multiplied by the values in all the other bins (indicated as $\otimes$ in the figure) and the products are added (indicated as $\oplus$ in the figure) to the respective correlations; product of the values in the first and second bin is added to $G(\Delta\tau)$, product of the values in the first and the third bin to $G(2\Delta\tau)$ and so forth. (iii) The direct and delayed monitors are updated for each lag time; the value in the first bin ($F(i)$) is added to the direct monitor for each lag time and the value in the j-th bin is added to the delayed monitor for $(j-1)\Delta\tau$. For example, when we calculate the product of the values in the first and the fourth bin, we add the product $F(i)F(i+3)$ to the correlation $G(3\Delta\tau)$, $F(i)$ is added to the corresponding direct monitor and $F(i+3)$ to the corresponding delayed monitor. The procedure outline so far is exactly true only for the first p bins (the bins having width of $\Delta\tau$, the first 16 bins in the example shown) and is slightly modified for the bins belonging to the other groups. The values in the second group (bin width of $2\Delta\tau$) are shifted to the right only upon registering every second intensity value (with the period $2\Delta\tau$). The value in the 17th bin is moved to the 18th bin and the 17th bin is filled with the sum of the values in the 16th and 15th bin. Analogously to what was done for the first 16 bins, the correlations and direct and delayed monitors need to be updated for all the lag times of the second group. However, because of the double bin width, the values cannot be directly multiplied by the value in the first bin $F(i)$; instead, they are multiplied by the sum of the values in the first and second bin ($F^{(2)}(i) = F(i) + F(i+1)$). The sum of the first and second bin is also added to the direct monitors for each lag time of the second group. Analogously, the values in the bins belonging to the third group are shifted and multiplied with a period of $4\Delta\tau$, the values in the bins of the 4th group with a period of $8\Delta\tau$ and so forth until the last group.

data points for which the correlation is calculated, multiple-tau correlators reduce the number of intensity values needed to be accessible at each time during the calculation, reducing the memory requirements of the calculation as well.

The binning of the intensity traces ensures that coarsening of the time resolution with which the ACF is calculated does not imply skipping any parts of the intensity trace and that the whole trace is used in the correlation calculation. Besides, it decreases the noise in the correlation at long lag times because of the increased number of photons contributing to the signal in each bin.

However, there is also a downside to the coarsening of the intensity trace binning in the form of correlation dampening at longer lag times. This is best demonstrated at the ACF of a periodic signal such as the sine wave in figure 4.2. While the ACF calculated by a linear correlator is periodic with a constant amplitude as expected, the multi-tau correlator gives a wave decaying with increasing lag time, i.e. with increasing bin width. This is an artefact caused by triangular averaging of the correlation function. The correlation value $G(\tau_m)$ at each lag time point τ_m contains contributions from correlations with lag times ranging from $\tau_m - 2^{k-1}\Delta\tau$ to $\tau_m + 2^{k-1}\Delta\tau$. Therefore, with increasing bin width, the averaging is performed over a broader lag time range, resulting in increased smoothing of the ACF. While this is rather a serious artefact in the case of periodic signals, it is of minor importance in most real FCS applications in which non-periodic signals are correlated giving rise to monotonously decaying ACFs. The systematic error

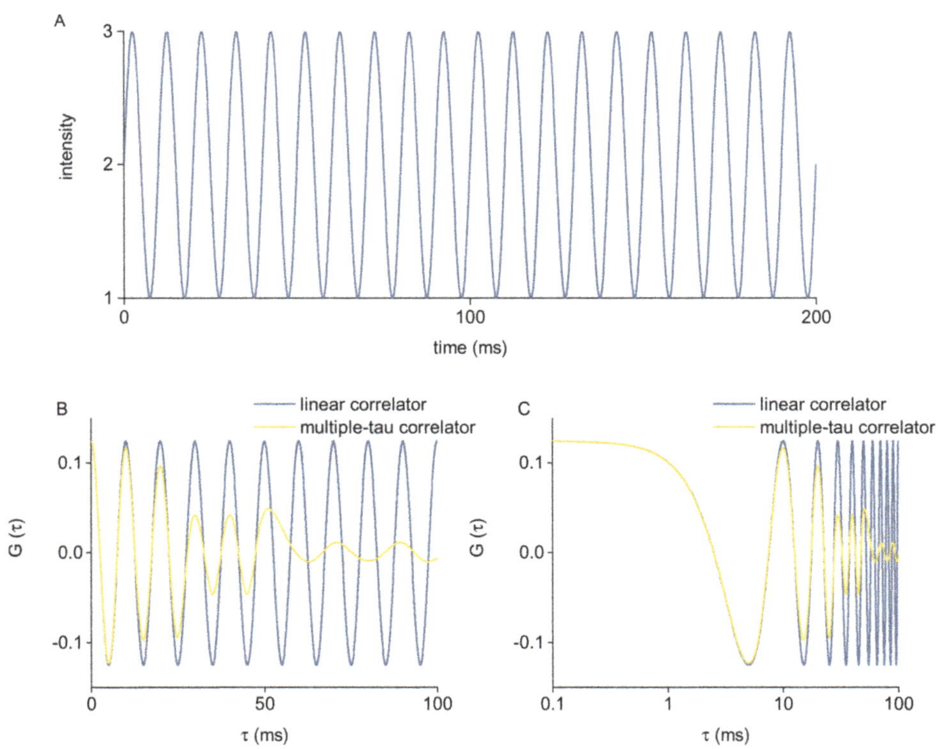

Figure 4.2. Illustration of ACFs of a periodic signal calculated by a linear and by a multiple-tau correlator. Panel A shows a portion of the signal which is a periodic function with the period of 10 ms. Panels B and C show the ACFs of the periodic signal calculated by a linear correlator and a (16, 8) multiple-tau correlator; the ACFs are plotted with linear (B) and logarithmic (C) lag-time axis.

introduced to the ACF by triangular averaging depends on the lag time-to-bin width ratio $\tau/\Delta\tau$. For exponentially decaying correlation functions it has been shown that the systematic error is smaller than 10^{-3} if the ratio is larger than 7, which is satisfied for all groups of lag time points except the first one (provided a reasonable choice of p and q) [5].

Let us look now briefly at the computational complexity of multiple-tau correlators. We will use the fact that a (p, q) multiple-tau correlator is a set of q linear correlators each of which processes an intensity trace with different binning (and, therefore, consisting of a different number of intensity values). From equation (4.4) it follows that $Mp - \frac{1}{2}p(p + 1)$ products are needed to calculate the ACF for the first p lag times (the first linear correlator operating on the original intensity trace). Since M is in this case much larger than p, we can approximate this number by Mp, which is an upper estimate of the actual value. All the other linear correlators calculate the ACF for $\frac{p}{2}$ lag times only; the correlator for the k-th group of lag times processes a time trace consisting of $\frac{M}{2^{k-1}}$ intensity values. Therefore, the number of products needed by that correlator can be approximated (by an upper estimate) as $\frac{p}{2}\frac{M}{2^{k-1}} = \frac{Mp}{2^k}$. The total number of products will be then obtained by adding the contributions of all q correlators. Let us again, for simplicity, use an upper estimate and calculate the sum for $q = \infty$ (unrealistic as it may sound). In that case we get the following upper estimate of the total number of products that need to be evaluated (see exercise 7 for derivation):

$$Mp + \sum_{k=2}^{\infty} \frac{Mp}{2^k} = \frac{3}{2}Mp \tag{4.6}$$

The complexity of a (p, q) multiple-tau correlator scales, therefore, as Mp. We can also observe that calculating the ACF for the first p lag times involves twice as many products as calculating the rest of the ACF (for arbitrarily high q).

4.1.3 Time-tagged intensity trace

While the binned intensity trace as discussed in the previous two sections is a very straightforward and intuitive way to record and store FCS data, it may be very wasteful in terms of data size in many frequently encountered situations. Why? Because for typical FCS datasets recorded with high temporal resolution by a photon counting detector, the majority of the bins would contain zeros; no photon has been registered at the given time interval. For example, a dataset recorded with 100 ns resolution at a photon count rate of 100 kHz would contain non-zero values only in 1% of the bins. If we record such data for 60 s and store them as a binned trace, we would end up with 6×10^8 numbers or around 300 MB of data (with a modest depth of 4 bit per number) out of which 99% are zeros. Therefore, a different scheme of storing such photon counting data has become common, called time-tagged-time-resolved (TTTR) data by some authors [6]. In this scheme the photon arrival time (in units of the temporal resolution $\Delta\tau$) is stored for each photon. Hence, for the example above we would need only 6×10^6 numbers. This for a 2-byte long number format translates to only 1.2 MB. As we will see later much, more information can be saved in a TTTR dataset by storing more than a single number for each detected photon. However, at present we will content ourselves with a single number per photon which gives the time of its detection measured from the beginning of the experiment in units of $\Delta\tau$. The raw data would in this case consist of a vector of photon arrival times $\{t_i\}$, where the index i goes from 1 to the total number of photons N. Note that an equivalent representation of the intensity trace is to store for each photon its arrival time $\{\tilde{t}_i\}$ measured in units of $\Delta\tau$ since the arrival time of the preceding photon. The size of the resulting dataset is again proportional to the number of photons N, which makes it an equally efficient way of representing a time trace consisting of sparse photon detection events. The vector of time intervals between photon detection events $\{\tilde{t}_i\}$ can be easily obtained from the absolute photon arrival times $\{t_i\}$ as: $\tilde{t}_i = t_{i+1} - t_i$, $0 \leqslant i < N$.

It is easy to see that to calculate the ACF from TTTR data by the above described multiple-tau correlator is rather inefficient. Disregarding the computational time needed to translate the TTTR dataset into a binned intensity trace, the correlator will have to calculate a large number of products resulting in zero, which although contributing nothing to the ACF take an equal time to evaluate as any other product. Binning the intensity trace with bin width larger than the actual temporal resolution of the TTTR data acquisition would help to reduce the number of empty bins and, thus, improve the speed of the ACF computation, albeit at the loss of temporal resolution of the ACF. It is, therefore, preferable to calculate the ACF directly from the TTTR data. Note that when using lag times following the multiple-tau scheme (section 4.1.2), the temporal bins are broader for larger lag times, which implies that for longer lag times the probability of a bin containing no photon arrival event diminishes to zero. Therefore, several authors have calculated ACF directly from TTTR data only for short correlation lag times, switching to binned trace multiple-tau correlator at longer lags [7, 8]. Other authors have proposed algorithms for calculating the entire ACFs directly from photon counting data recorded either as sequences of photon arrival times [9] or as sequences of intervals between subsequent photon detection events [10]. We will outline here one such algorithm [6].

To calculate $G(\tau)$ we take the original photon arrival time vector $\{t_i\}$ and generate a shifted one $\{t'_i\}$ by adding τ to all its components ($t'_i = t_i + \tau$). Then, analogously to the correlator described in section 4.1.1, we calculate three sums denoted as Correlation, Direct Monitor and Delayed Monitor. We set the three sums to zero in the beginning and then we search the original vector $\{t_i\}$ from the first element t_1 until we find t_j such that $t_j \geqslant t'_1$. If $t_j = t'_1$, we have found a pair of photons separated by a time lag τ and we increment the Correlation by 1. Then we search the shifted vector $\{t'_i\}$ from t'_2 until we find t'_k such that $t'_k \geqslant t_j$. If $t'_k = t_j$, we increment the Correlation by 1. Then we search the original vector starting with t_{j+1} until we find t_m such that $t_m \geqslant t'_k$ and so forth until we reach the end of one of the vectors. While searching the vectors we keep incrementing by 1 the Delayed Monitor for every element t_i of the original vector larger than τ and the Delayed Monitor for every element t'_i of the shifted vector smaller than or equal to the total measurement time T. The principle of the correlation calculation from a time-tagged intensity trace is illustrated in figure 4.3.

In order to further reduce the computational complexity and improve the ACF noise at long lag times we will again follow the multiple-tau scheme introduced in section 4.1.2. This means we will calculate the ACF for lag times defined by equation (4.5) and at the same time coarsen the temporal resolution of the intensity trace for longer lag times. The time coarsening step, analogous to summing up neighbouring bins when working with binned intensity traces (see section 4.1.2), is in this case accomplished by dividing the vector $\{t_i\}$ by factor 2 and rounding all its elements to the nearest lower integer. The time coarsening is performed always before progressing to the next group of $\frac{p}{2}$ lag time points τ_m. Thus, the ACF for the lag time points of the second group ($p < m \leqslant \frac{3}{2}p$) are calculated from vector $\{t_i^{(2)}\}$, which is obtained from the original vector $\{t_i\}$ in the following way:

$$t_i^{(2)} = \left\lfloor \frac{1}{2} t_i \right\rfloor \tag{4.7}$$

where the bracket $\lfloor \ldots \rfloor$ represents the integer part of the enclosed number. The rounding to the nearest lower integer can result in consecutive arrival times having

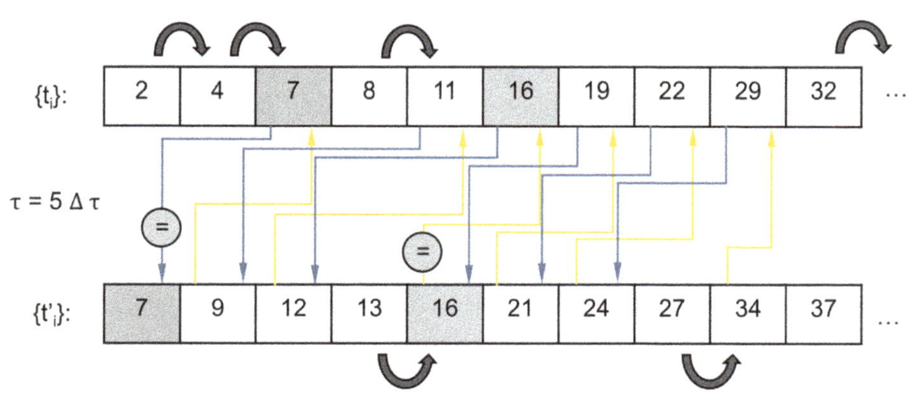

Figure 4.3. Illustration of the principle of ACF calculation from photon arrival times following the method of Wahl *et al* [6]. In the example shown, we are constructing ACF for lag time $\tau = 5\Delta\tau$, which means that the elements of the shifted arrival time vector $\{t'_i\}$ are obtained by adding 5 to the elements of the original vector $\{t_i\}$. We are looking for photon pairs with arrival times spaced by $5\Delta\tau$, which is equivalent to searching for pairs of identical numbers in $\{t_i\}$ and $\{t'_i\}$. We start by searching the original vector from the beginning looking for an element t_j equal to or larger than the first element of the shifted vector t'_1. In this case it is t_3 which is equal to t'_1. Having found a pair of equal numbers, we increment the Correlation sum by 1. Then we continue by searching the shifted vector, starting from t'_2, looking for an element larger than or equal to t_3. Since 9 is lager than 7, we switch back to the original vector and search it, starting from t_4, for an element larger than or equal to t'_2. 8 is smaller than 9, so we make one step to the right to t_5, where 11 satisfies the condition. We switch to the shifted vector again and search it from t'_3, looking for an element equal to or larger than t_5. The condition is satisfied for t'_3; therefore, we switch again to the original vector and search it from t_6. In this way we proceed till we reach the end of the original vector.

the same value ($t_k^{(2)} = t_{k+1}^{(2)}$). In other words, by reducing the temporal resolution of the time trace, the arrival of more photons can fall into the same unit time interval. To remove the duplicity of arrival times, we associate one more value with each arrival time t_i its weight w_i. In the beginning we set all weights equal to 1. The temporal resolution coarsening (equation (4.7)) does not affect the weights associated with each photon arrival time:

$$w_i^{(2)} = w_i \tag{4.8}$$

However, when a duplicity in arrival times appears ($t_k^{(2)} = t_{k+1}^{(2)}$), the weight of $t_k^{(2)}$ is increased:

$$w_k^{(2)} := w_k^{(2)} + w_{k+1}^{(2)} \tag{4.9}$$

and the other identical arrival time ($t_{k+1}^{(2)}$) is removed from vector $\{t_i^{(2)}\}$. We illustrate this procedure in figure 4.4. The ACF for lag time points of the third group are calculated from the vector $\{t_i^{(3)}\}$ obtained from $\{t_i^{(2)}\}$ in a manner analogous to the procedure outlined above and the analogy holds for all subsequent groups of lag times. Removing the duplicities in photon arrival times reduces the size of the vector $\{t_i^{(k)}\}$ for higher groups of lag times (higher k), reducing the computational time for the lag times of the higher groups, analogously to the situation of binned intensity traces. Without introducing the photon weights, it would not be possible to remove the duplicities without losing information on some photons. Therefore, the authors who have not introduced the weights into their algorithms typically use the photon

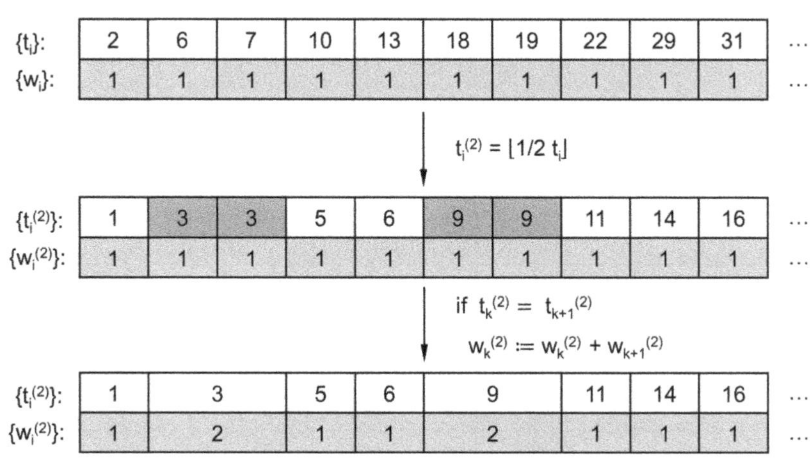

Figure 4.4. Illustration of time resolution coarsening which introduces multiple-tau quasi-logarithmic lag time spacing to ACF calculation from photon arrival times [6]. We take the original vector of photon arrival times $\{t_i\}$ with all photon weights $\{w_i\}$ set to 1. In the first step we divide each arrival time by two and round the values to the nearest lower integer (equation (4.7)). In the next step we identify all photons that have the same arrival time on the coarsened timescale, which means all pairs of identical elements in vector $\{t_i^{(2)}\}$. In the last step we replace each pair of identical elements ($t_k^{(2)} = t_{k+1}^{(2)}$) with a single one having weight equal to the sum of weights of the two elements ($w_k^{(2)} + w_{k+1}^{(2)}$).

arrival times to calculate the ACF only for the lag times of the first group (with the maximum time resolution $\Delta\tau$) and bin the intensity trace to calculate the ACF for the lag times of the other groups [7, 8].

The weights also need to be taken into account when calculating the ACF, $G(\tau_m)$; whenever we find a pair of arrival times separated by τ_m $(t'_k = t_j,)$, we increment the Correlation by $w_j w'_k$. The Delayed Monitor and the Direct Monitor are incremented by w_j and w'_k for every $t_j > \tau_m$ and every $t'_k \leqslant T$, respectively. Note that we have assumed that there were no duplicates in arrival times in the original vector $\{t_i\}$, that is that we could not detect more than a single photon per the time resolution $\Delta\tau$. While this is true if $\Delta\tau$ is the actual time resolution of the photon counting setup, in some situation this may not be the case and then we need weights larger than 1 already for the original vector. Furthermore, the weights allow us to incorporate additional information into the ACF calculation by an approach called statistical filtering (see section 4.4).

The complexity of the calculation depends on the total number of photons N instead of on the total measurement time as is the case for binned trace correlators (at constant temporal resolution $\Delta\tau$). Calculating the ACF directly from the TTTR data is, thus, particularly advantageous at low photon count rates, in which case the binned intensity trace would contain few non-zero values. It has been shown that at low to moderate photon count rates standard computers can correlate photon counting data in real time even with a linear correlator scheme [10]. However, even at high photon count rates calculating the ACF directly from the TTTR data is not slower than from the binned time trace; both approaches converge to the same complexity at the limiting case of all bins containing non-zero values. The binned trace multiple-tau correlator is, therefore, used mainly in situations where the raw data are directly saved in the form of a binned intensity trace. As a typical example we can consider the data in camera-based FCS modalities that are saved in the form $F_{xy}(i)$, where x and y are pixel coordinates and i is the frame index in the temporal image stack; $\Delta\tau$ is then determined by the camera time per frame.

4.1.4 Cross-correlation calculation and correlation function amplitude

Although the previous sections have explicitly mentioned ACF calculation, the same procedures with little or no modification can be used to calculate CCFs of two different intensity time traces $F_A(i)$ and $F_B(i)$. The only significant difference between ACF and CCF calculation lies in the way in which $G(0)$ is treated. You may have noted that all the algorithms outlined in the previous sections of this chapter have started from a non-zero lag time. The reason for that is the presence of shot noise in any real intensity time trace (see math box 3.1). The shot noise is not correlated in time; it, therefore, affects only the $G(0)$ value, resulting in an artificial increase in the ACF at zero lag time. Consequently, the $G(0)$ value is usually estimated by extrapolating the ACF $G(\tau)$, $\tau \neq 0$ to zero lag time. On the other hand, the shot noise in two independent time traces (recorded by different detectors) is not correlated even for $\tau = 0$, which means that in the case of CCFs it makes sense to start the calculation from zero lag times. Equation (4.2) then becomes:

$$G_{AB}(m\Delta\tau) = (M - m)\frac{\sum_{i=1}^{M-m} F_A(i)F_B(i + m)}{\left(\sum_{i=1}^{M-m} F_A(i)\right)\left(\sum_{i=1}^{M-m} F_B(i + m)\right)} \tag{4.10}$$

Note that, as we mentioned above, lag times in this case start from $\tau = 0$, which means that the index m in equation (4.10) also starts from 0.

Analogously, we can define $G_{BA}(m\Delta\tau)$ by exchanging the roles of $F_A(i)$ and $F_B(i)$ in the equation above. Differences between $G_{AB}(m\Delta\tau)$ and $G_{BA}(m\Delta\tau)$ can reflect an asymmetric relationship between the processes underlying the fluctuations in $F_A(i)$ and $F_B(i)$, respectively, as discussed in section 2.6. However, in the common cases where no such asymmetry exists and assuming both intensity traces to be stationary, $G_{AB}(m\Delta\tau)$ and $G_{BA}(m\Delta\tau)$ are identical at the limit of infinite M. Therefore, a single CCF is often calculated as an average of the two:

$$G_X(m\Delta\tau) = \frac{1}{2}(G_{AB}(m\Delta\tau) + G_{BA}(m\Delta\tau)) \tag{4.11}$$

4.1.5 Correlation function calculation via Fourier transform

All the algorithms we have discussed so far were directly based on the correlation function definition (equation (4.1)) in the way that they calculate the averages of products of signals in time points separated by a specific lag time τ. A different approach to ACF calculation is based on the Wiener–Khinchin theorem, which says that (under conditions that can be considered satisfied for the intensity time traces in FCS) the autocorrelation function of a signal is equal to the inverse Fourier transform of the signal power spectrum $S(\omega)$ [11]. The power spectrum of a signal can be obtained as $S(\omega) = \mathcal{F}(F(t))\mathcal{F}^*(F(t))$ and it describes the decomposition of the signal time trace into individual frequencies (in terms of power associated with each frequency). $\mathcal{F}$ and $\mathcal{F}^{-1}$ denote the direct and inverse Fourier transform, respectively, and the asterisk signifies complex conjugation. Using what we have just said, the normalized autocorrelation function can be calculated as:

$$G(\tau) = \frac{\mathcal{F}^{-1}\{\mathcal{F}(F(t))\mathcal{F}^*(F(t))\}}{\langle F(t)\rangle^2} \tag{4.12}$$

The numerical complexity of calculating ACFs by this approach depends on the complexity of the Fourier transform algorithm used. The complexity of fast Fourier transform (FFT) algorithms, which are implemented in most data processing software packages scales with $M\log_2(M)$ [12]. This means that for a binned intensity trace the Fourier transform approach is much faster than linear correlators (the complexity of which scales with M^2); however, for long time traces (large M) it would be outperformed by (p, q) multiple-tau correlators, the complexity of which scales with pM. This also implies that for TTTR data recorded at low photon count rates the direct calculation as described in section 4.1.3 is superior to the Fourier transform approach. Therefore, the Fourier transform approach is not commonly used in FCS to calculate temporal correlation functions; direct calculation from TTTR data or multiple-tau correlators being preferred. On the other hand, it is typically used in ICS (see section 5.7.4) to calculate two-dimensional spatial correlation functions [13–15]. No pseudo-logarithmic correlator scheme (such as the multiple-tau correlator) is easily applicable in these cases, while efficient implementations of two-dimensional FFT are available in most data processing packages. The normalized two-dimensional ACF can be then calculated as:

$$G(\xi, \psi) = \frac{\mathcal{F}^{-1}\{\mathcal{F}(F(x, y))\mathcal{F}^{*}(F(x, y))\}}{\langle F(x, y)\rangle^2} \tag{4.13}$$

where x and y are the spatial coordinates of a pixel in the image; ξ and ψ are the correlation distances (a spatial analogy to the lag time τ).

4.1.5.1 Image correlation spectroscopy of localisation microscopy images

Having mentioned that Fourier transform is used predominantly in ICS, we would like to briefly discuss the specific case of applying ICS on images obtained by localisation microscopy techniques [12, 16]. Localisation microscopy images differ fundamentally in their data structure from images obtained in other microscopy modalities. In most other imaging techniques, the images are stored as two-dimensional matrices of pixel values $F(x, y)$, which give the intensity detected in the given pixel within the image integration time. In localisation microscopy, the images are stored as lists of coordinates $\{x_i, y_i\}$ giving the positions of individual fluorophore molecules. A single fluorophore can be approximated as a point-like source of fluorescence; the image of the i-th fluorophore can be then described by a product of Dirac delta functions $\delta(x - x_i)\delta(y - y_i)$ which is equal to one when $x = x_i$ and $y = y_i$ and equal to zero elsewhere. Cross-correlation $G_{AB}(\xi, \psi)$ between images A and B will assume non-zero values only if the correlation distances ξ and ψ are equal to the differences in coordinates of a pair of fluorophore localisations (a localisation in image A and a localisation in image B). The correlation function is then given by [12]:

$$G_{AB}(\xi, \psi) = \frac{XY}{M_A M_B}\sum_{i=1}^{M_A}\sum_{j=1}^{M_B}\delta(\xi - x_j + x_i)\delta(\psi - y_j + y_i) \tag{4.14}$$

where X and Y are the dimensions of the images and M_A and M_B are the numbers of localisations in image A and B, respectively. The correlation function has non-zero values only at discreet points corresponding to vectors pointing from localisations in image A to localisations in image B. An ACF ($A = B$) will have a maximum at the origin corresponding to a zero vector contributed by each fluorophore. To account for finite localisation precision, the Dirac delta functions describing the localisations can be replaced with Gaussian functions centred at the fluorophore positions. The correlation function will then be also a sum of Gaussian functions (instead of Dirac delta functions) similarly to the correlation functions of pixel-based images (see section 5.7.4).

Correlation functions of localisation-based images can be calculated directly from the localisation coordinates following equation (4.14). The computational complexity of the calculation depends on the number of localisations and scales with $M_A M_B$, compared to the calculation by FFT in pixel-based representation which scales with $XY\log_2(XY)$. Depending, therefore, on the number of localisations and the size of the image, it is either advantageous to calculate the correlation directly from the localisation coordinates or to convert the image to a pixel-based representation and apply FFT. For images with sparse localisations, calculating the correlations directly from the localisation coordinates can be significantly faster. Consider, for example, a typical localisation microscopy experiment in which series consisting of thousands (or tens of thousands) of images are acquired, each containing only a sparse set of localisations. Using the localisation coordinates would be definitely preferable when calculating ACFs of individual images or CCFs between them; however, when calculating the ACF of the resulting super-resolved image (the sum of localisations in the whole series), FFT on a pixel-based representation of the image can be several orders of magnitude faster. The relationship of localisation-based image representation to pixel-based representation in ICS can be then seen as an analogy to the relationship of binned intensity traces to time-tagged traces (photon arrival times) in the case of temporal correlations.

4.2 Hardware correlators and their comparison with software correlators

Hardware correlators are electronic devices for correlating two input signals. If the same signal is fed to both inputs of the correlator the output is an ACF, if the two signals are different, the output is a CCF. Most hardware correlators used in FCS are multiple-tau correlators which work on a principle similar to the multiple-tau software correlator introduced in section 4.1.2 with the difference that instead of numerical calculation they employ dedicated electronic circuits to perform the task.

Hardware correlators were the preferred solution in the early days of FCS when the limitations in computing power of available computers made the possibility to delegate the relatively demanding task of correlation function calculation to a dedicate piece of hardware very attractive. Hardware correlators are able to perform the correlation in real time as the fluorescence intensity trace is recorded, which was at first beyond the possibilities of software calculation. Seeing the correlation functions in real time during the measurement is a great advantage, especially when setting up an experiment and optimizing the acquisition parameters, as anyone having practical experience with FCS can confirm. The reduction in the amount of stored data resulting from saving only a correlation function (calculated on a quasi-logarithmic scale) instead of the whole fluorescence intensity trace was another factor in favour of hardware correlators at those times.

The progress in computers in the past decades has rendered the above-mentioned advantages of hardware correlators practically obsolete. The increase in available computational power alongside efficient software correlator implementations allows real-time correlation calculation similarly to hardware correlators [3, 5, 6]. A complete ACF can be calculated as soon as a fluorescence intensity trace equal in duration to the longest correlation time lag has been recorded. The ACF (which at first is necessarily very noisy because of the very poor photon statistics especially at long time lags) is then updated in real time as the recording of the intensity trace proceeds.

In this respect the hardware and software correlators have become equal. Software correlators have, however, additional advantages. One of them is the flexibility in choosing arbitrarily the p and q values of a (p, q) multiple-tau correlator to suit the needs of the experiment (unlike in the case of hardware correlators where the p and q are determined by the design of the electronics). Another, more important, advantage lies in the possibility to recalculate the correlation function any time in the future as post processing of the stored intensity trace. There exist situations in which this is of great advantage. A typical example is the situation when a part of the intensity trace is corrupt, which, then, affects the whole correlation function. Several processes can lead to severe corruption of a part of the recorded intensity trace, for example, instrument instability or the transit of a highly scattering dust particle or large molecular aggregate through the detection volume, resulting in a burst of very high intensity (see section 7.2). These events result in large changes in detected intensity, much larger than the fluctuations, which are of interest in FCS, and, therefore, they are likely to dominate the ACF even though they affect only a very small fraction of the intensity trace. While the real-time correlation calculated from the whole intensity trace would need to be discarded, software correlators can rescue the data by recalculating the correlation offline only from the unaffected part of the intensity trace (see figure 4.5). Another situation which requires offline correlation calculation from the saved intensity trace is encountered, for example, in statistical filtering, which will be explained in section 4.4; in this case the whole intensity trace is used to find the statistical filter function which is then used during the correlation calculation. For this reason the more modern hardware correlators offer the option of saving the whole intensity time trace along with the correlation functions.

Software correlators have gradually rendered hardware correlators redundant for nearly all FCS modalities, imaging FCS being until recently the last domain where hardware correlators had held an advantage over their software counterparts. In imaging FCS modalities many intensity traces are recorded in parallel by two-dimensional spatially resolved array detectors. The highly parallelized intensity trace acquisition presented a serious challenge for real-time software correlators and, therefore, these FCS modalities have relied mostly on offline software correlation. The absence of instantaneous ACF display is decidedly an inconvenience for the experimenter, and for that reason specialized hardware correlators based on field programmable gate arrays have been developed, which can process highly parallelized intensity trace acquisition by some types of two-dimensional detector arrays [17]. However, a software correlator implementation using a graphics processing unit has recently made real-time software correlation calculation available in imaging FCS with any type of two-dimensional detector arrays (http://www.dbs.nus.edu.sg/lab/BFL/imfcs_image_j_plugin.html).

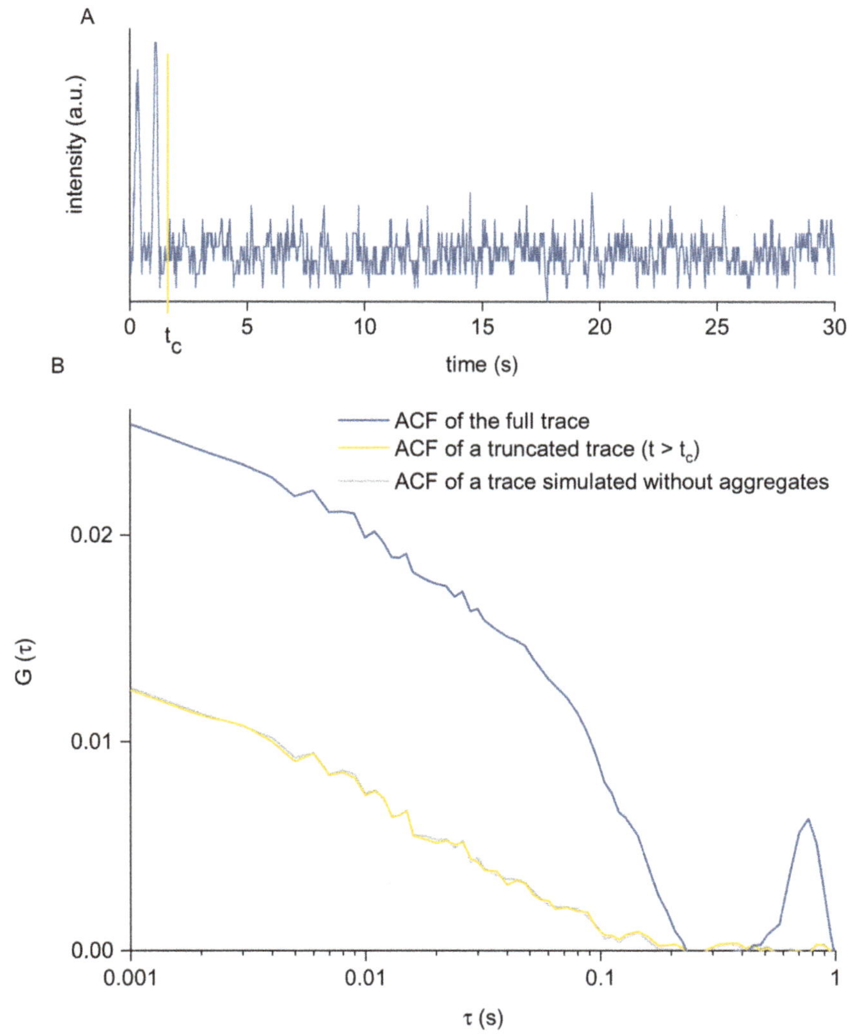

Figure 4.5. An example of a simulated fluorescence intensity time trace (A) containing bursts of high intensity, which could represent, for example, the transit of a large fluorescent aggregate through the observation volume, and the corresponding ACF (B). The brightness of the aggregate in the simulation was 20 times higher than the brightness of the individual fluorophores. As a result, the contribution of the aggregate dominates the ACF (particles contribute to the ACF with the square of their brightness). Calculating the ACF only from a part of the intensity trace free from the influence of the aggregate (for time larger than cut-off time t_C) returns an ACF describing the diffusion of non-aggregated particles. ACF calculated from an intensity trace simulated without aggregates is shown for comparison.

4.3 Temporal resolution of correlation functions

The time resolution of the correlation functions is limited either by the temporal resolution of the hardware correlator used or by the resolution with which the fluorescence intensity traces are recorded (in cases when the traces are processed by a software correlator). When the data are recorded in the form of a binned intensity trace, it is equal to the time per bin $\Delta\tau$; if the data are saved in the time-tagged format, it is the temporal resolution with which the detection time of each photon is saved.

Modern hardware correlators as well as modern photon counting cards (used to digitise the outputs of photon counting detectors) can have temporal resolution on the order of ps. However, detectors as well as the photon counting cards have certain dead times, time intervals after each photon detection event during which they are blind to any subsequent photons. The component with the longest dead time (be it the detector or the photon counter) is the bottleneck and its dead time is the dead time of the whole system. The dead time is usually more than an order of magnitude longer than the temporal resolution of high-resolution photon counters. Since there can be no correlation with lag times shorter than the system dead time (there are no two photon detection events separated by a shorter time interval), the ACF temporal resolution achievable with a single detector is effectively limited either by the detection system dead time or the photon counter temporal resolution, whichever is longer. To utilize fully the high temporal resolution of modern photon counting cards, we need to employ at least two detectors with independent photon counters to circumvent the dead time of each individual detection channel. The fluorescence signal is divided in this case by a 50/50 beamsplitter onto the two detectors and the ACF is obtained by cross-correlating the two detectors' outputs [18]. Typical FCS experiments are, however, concerned with processes that happen on μs or slower timescales, for which purposes ps temporal resolution is more than sufficient. Most FCS data are, therefore, typically recorded with temporal resolutions of the order of 10 ns or coarser; in such cases simpler and less costly photon counters are sufficient [10].

The time resolution of some FCS modalities is rather limited by the temporal resolution of the intensity trace recording. This is especially the case of camera-based FCS modalities, where the temporal resolution is given by the time per frame of the camera. In the case of imaging FCS this is typically on the order of ms, although with fast cameras and small regions of interest time resolution on the order of μs can be achieved [19].

The longest time lag for which the correlation is calculated depends (for a multiple-tau correlator) on the (p, q) correlator structure (see section 4.1.2). Since the structure of a software correlator can be adjusted freely, the only limitation is, then, the total time for which the intensity trace has been recorded. The total recording time should ideally be more than an order or magnitude longer than the maximum correlation time lag, to get some statistics even for the longest time lags [20].

4.4 Statistical filtering in correlation function calculation

In section 4.1.3, we introduced photon weights w_i into the calculation of the ACF directly from TTTR data. There, we set the initial weights of all individual photons to 1, by which we accounted for the equal contribution of all photons to the correlation function $G(\tau)$. Only when the coarsening of the temporal resolution (as a part of the semi-logarithmic correlation scheme) resulted in detection times of multiple photons coinciding on the coarsened timescale, were the weights increased correspondingly. Here, we will discuss situations where additional information about each detected photon is used to weigh its contribution to the calculated ACF via unequal photon weights w_i. However, before going into details of the calculations, let us start with a motivation for this procedure and draw an analogy with dual-colour FCCS (DC-FCCS, see chapter 6 for more details).

In DC-FCCS, photons detected by two separate detectors are cross-correlated. By means of spectrally-selective optics (such as dichroic mirrors, band-pass filters or dispersion elements) each of the two detectors detects photons from a different spectral range. In DC-FCCS, the samples contain fluorophores of two distinct species with such emission spectra that one of the two detectors receives predominantly photons emitted by one of the fluorophore species and the other detector photons emitted by the other species. It is often assumed that knowing by which detector a photon has been detected tells us unambiguously which of the fluorophore species had emitted the photon. Although this is never perfectly true (as we will discuss in more detail in section 6.2.3), for fluorophores with well-separated emission spectra and for well-matched spectral ranges of the detectors, it can be considered an acceptable approximation. We can then look at the measured CCF as the CCF between the fluctuations in the fluorescence of the two fluorophore species. If diffusion is the chief source of the fluctuations, then the CCF amplitude increases with increasing fraction of co-diffusing molecules of the two species.

However, the requirement of the two fluorophore species having well-separated emission spectra is quite restrictive and it would be useful to be able to utilize more subtle differences in fluorescence emission characteristics to decide by which fluorophore species a photon has been emitted. We will first consider the use of differences in fluorescence lifetimes, which will lead us to an FCS modality called fluorescence lifetime correlation spectroscopy (FLCS), which was historically the first technique to use statistical filtering [21, 22]. The main difference from DC-FCCS outlined above lies in the fact that differences in fluorescence lifetimes do not allow us to decide unambiguously (not even in the crudest approximation as we will discuss in the following section) by which fluorophore species a photon has been emitted. Nevertheless, we can estimate the probabilities with which each photon has been emitted by the individual fluorophore species, which makes it possible to compute statistical filter functions. Those, when applied on a large ensemble of photons during the correlation functions calculation, give good estimators of the ACFs of the individual fluorophore species and CCFs between them. After explaining the principles of FLCS, we will show that the idea of statistical filters can be applied directly also on differences in shapes of otherwise strongly overlapping spectra (in a technique called fluorescence spectral correlation spectroscopy—FSCS) or on other differences in fluorescence emission characteristics. In its most general form, the use of statistical filtering in FCS can be referred to as filtered FCS (fFCS) [23].

Before discussing the actual statistical filtering, we will now briefly introduce fluorescence lifetime, its measurement and integration of fluorescence lifetime information into an FCS intensity time trace.

4.4.1 Fluorescence lifetime and its integration into FCS datasets

Fluorescence lifetime is defined as the average time the fluorophore molecules spend in the excited state. In other words, it is the average time interval between the excitation of a fluorophore molecule and the subsequent emission of a photon. A technique called time-correlated single-photon counting (TCSPC) measures this quantity in a very straightforward way. Its main ingredients are a pulsed laser (with pulses typically shorter than 100 ps) as the excitation source and precise timing of photon detection events with respect to the preceding laser pulse. The pulsed excitation determines a very narrow time interval at which the fluorophores may have entered the excited state. Timing the photon detection events relative to the preceding excitation pulse then gives a direct measure of the time the fluorophores have spent in the excited state. Accumulating such measurements for a large ensemble of photons, then, allows us to construct a histogram of photon detection times $F(\mu)$ (see figure 4.7) which is the usual representation of TCSPC data. If all fluorophores spend on average the same time in the excited state (have the same fluorescence lifetime), the TCSPC histogram follows a mono-exponential decay (see math box 4.1), which appears as a straight line in the semi-logarithmic plot usually used to depict the TCSPC histograms. If the sample contains multiple fluorophore species with different fluorescence lifetimes, the TCSPC histogram is a linear combination of histograms of the individual species and, therefore, follows a multi-exponential decay. However, multi-exponential decays may be observed even for samples containing single fluorophore species as a result of fluorescence anisotropy, fluorophore solvation dynamics or excited state reactions [24].

Figure 4.7 shows an example of the TCSPC histogram of a multiexponential decay of a mixture of fluorophore species. Since fluorophores of all the species are excited simultaneously by the same excitation pulse, their fluorescence intensity decays start (that means have their maximum) at the same instant. Photons detected shortly after the excitation pulse could have been very likely emitted by fluorophores belonging to any of the species and, therefore, there is no chance to trace them unambiguously to their source.

The main challenge in TCSPC lies in the very short timescales involved. Most commonly used fluorophores have lifetimes on the order of ns (which is a few orders of magnitude shorter than the timescales of processes usually probed by FCS); the temporal resolution of TCSPC should be, therefore, at least an order of magnitude finer. In most setups it is much finer than that (on the order of ps) to resolve even subtle differences in fluorescence lifetimes. Further details of TCSPC instrumentation and experiments exceed the scope of this chapter and can be found, for example, in references [24, 25].

The complete raw TCSPC data consist of detection times of individual photons, which are usually saved in the form of bin numbers μ in the photon detection times histogram $F(\mu)$, where the bin width is equal to the temporal resolution of the TCSPC instrument. This information can be stored in the TTTR data format introduced in section 4.1.3; the 'time-resolved' in the data format name actually

Math box 4.1

The evolution of fluorescence intensity after a short excitation pulse
Exposing a sample containing fluorophores to a very short (the duration of which can be neglected in the first approximation) excitation pulse of an appropriate wavelength results in the exciting of a subset of the fluorophores in the illuminated volume. Let us denote the number of excited fluorophores as $n*$. Following the end of the excitation pulse, the fluorophores will spontaneously return to the ground state either by emitting a photon or by a non-radiative process. The photon emission as well as the non-radiative de-excitation pathways are stochastic processes and can be characterised by their respective rate constants k_f and k_{nr}. If we introduce a summary de-excitation rate constant as $k = k_f + k_{nr}$, the average time a fluorophore stays in the excited state before it undergoes de-excitation by one of the possible pathways is $\frac{1}{k}$.

The evolution of the number of fluorophores in the excited state $n*(t)$ following the excitation pulse (by which we define the beginning of the time axis) is described by the following differential equation:

$$\frac{dn*(t)}{dt} = -kn*(t)$$

which says that the number of fluorophores that undergo de-excitation at any given infinitesimally short time interval dt is equal to the product of the number of excited fluorophores at the given time $n*(t)$, the de-excitation rate constant k and the duration of the interval dt.

The solution of the differential equation for $n*(t)$ has the form of:

$$n*(t) = Ce^{-kt}$$

where the constant C is determined from the initial condition as the number of fluorophores that were excited by the excitation pulse $n*(0)$.

The number of excited fluorophores $n*(t)$ is translated to detected fluorescence intensity $F(t)$ as:

$$F(t) = \varphi k_f n*(t)$$

where the constant φ accounts for the intensity contribution of each photon and the detection efficiency. This leads finally to the exponential fluorescence decay observed after a short excitation pulse:

$$F(t) = F(0)e^{-kt}$$

where $1/k$ is the fluorescence lifetime. It can be easily shown that in the case of a monoexponential decay, $1/k$ is indeed the average time spent by the fluorophores in the excited state. Detecting a photon at time t means that a molecule has spent time t in the excited state. The intensity decay $F(t)$ normalized to unit area can be, therefore, viewed as the probability density function of a fluorophore staying for time t in the excited state. The average time $\bar{t}$ spent in the excited state can be then calculated as:

$$\bar{t} = \frac{\int_0^\infty tF(0)e^{-kt}dt}{\int_0^\infty F(0)e^{-kt}dt} = \frac{1}{k}$$

In reality, the excitation pulses have always a finite duration. Each time point within the duration of the excitation pulse can be regarded as the beginning of an exponential fluorescence intensity decay $F(t)$. The overall decay is then described as a convolution of the exponential decay with the temporal profile of the excitation pulse. The measured decay is further broadened by the uncertainties in timing of the photon detection events and it can be represented as a convolution of the theoretical exponential decay with so-called instrument response function (IRF), which envelopes all the influences of the measurement setup on the fluorescence intensity decay. In order to extract fluorescence lifetimes accurately from TCSPC histograms, the IRF of the system should be known, ideally determined experimentally.

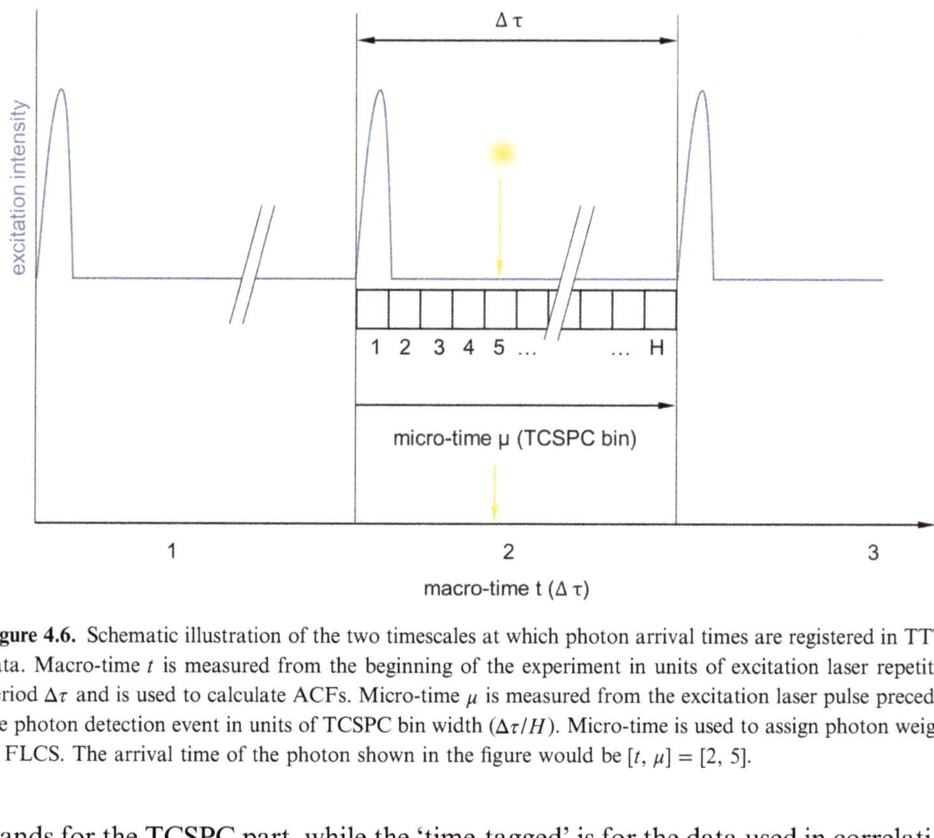

Figure 4.6. Schematic illustration of the two timescales at which photon arrival times are registered in TTTR data. Macro-time t is measured from the beginning of the experiment in units of excitation laser repetition period $\Delta\tau$ and is used to calculate ACFs. Micro-time μ is measured from the excitation laser pulse preceding the photon detection event in units of TCSPC bin width ($\Delta\tau/H$). Micro-time is used to assign photon weights in FLCS. The arrival time of the photon shown in the figure would be $[t, \mu] = [2, 5]$.

stands for the TCSPC part, while the 'time-tagged' is for the data used in correlation functions calculation. When we include the TCSPC data, there will be two numbers, t_i and μ_i, stored for each detected photon which represent the photon detection time on two different time scales. This is schematically depicted in figure 4.6. The TCSPC information, sometimes called the micro-time, measures the time interval between the photon detection and the preceding excitation pulse; it is measured in the units of the TCSPC time resolution and it is in the range from 0 to $\Delta\tau$, the interval between subsequent excitation pulses. The other time information, also called the macro-time, measures the interval between the photon detection and the beginning of the experiment in units of $\Delta\tau$. $\Delta\tau$, thus, has the same meaning as in the previous sections of this chapter, the temporal resolution of the FCS intensity trace. With typical laser pulses repetition rates on the order of 10 MHz, $\Delta\tau$ is on the order of 10 ns, which is more than sufficient for usual FCS measurements.

4.4.2 Generation of statistical filters in fluorescence lifetime correlation spectroscopy (FLCS)

Let us assume a sample containing S fluorophore populations (in the example in figure 4.7 $S = 3$, including the background) differing in their TCSPC histograms. We purposely say here 'TCSPC histograms' instead of fluorescence lifetimes, because any difference in the histograms can be utilized for statistical filtering; for example, differences caused by fluorescence anisotropy have been used in the case of fluorophores with identical lifetimes [23]. We also say fluorophore populations rather than species here, since the fluorescence lifetimes (or TCSPC histograms) of some fluorophore species are highly sensitive to the local environment, which may depend, for example, on the conformation of a macromolecule labelled by the fluorophore. The TCSPC histograms characterising individual populations normalized to unit area will be called here the populations' patterns $\pi^{(k)}(\mu)$, with the index $k = 1, 2, \ldots, S$ identifying the fluorophore population and $\mu = 1, 2, \ldots, H$ being the photon arrival time expressed as bin number in the TCSPC histogram (H being the number of bins in the histogram or, in other words, $\Delta\tau$ divided by the TCSPC time resolution). The patterns can be acquired experimentally by TCSPC measurements in samples containing individual fluorophore populations. This is the preferable method of finding the patterns as it is free of any assumptions (the patterns can be arbitrarily complex TCSPC histograms); however, it is not always possible (for example, if several of the fluorophore populations coexist in a dynamic equilibrium). Alternatively, the patterns can be obtained under some assumptions from the overall TCSPC histogram $F(\mu)$ of the mixture of all populations. If we, for example, assume that each fluorophore population has a single lifetime (and, consequently, a mono-exponential fluorescence decay), we can get the fluorescence lifetimes of all individual populations by fitting the overall fluorescence decay with a sum of mono-exponential functions, or more exactly with the convolution of a sum of mono-exponential functions with the instrument response function (IRF) of the measurement setup. The IRF describes what would be the response of the setup to an infinitesimally short photon pulse and the shape of the TCSPC histograms measured by the setup can be described as a convolution of the underlying exponential fluorescence decay with the IRF (see math box 4.1). The patterns of each individual population are then constructed by convolution of the corresponding mono-exponential decay with the IRF.

Our aim here is to obtain the ACFs $G^{(k)}(\tau)$ of the individual populations and CCFs between them $G_X^{(k)(m)}(\tau)$. For that purpose, we need to find statistical filter functions $w^{(k)}(\mu)$ that assign the probability with which each photon has been emitted by a fluorophore of the k-th population based on the photon's micro-time μ. When we, then, introduce such weights into the correlation algorithm outlined in

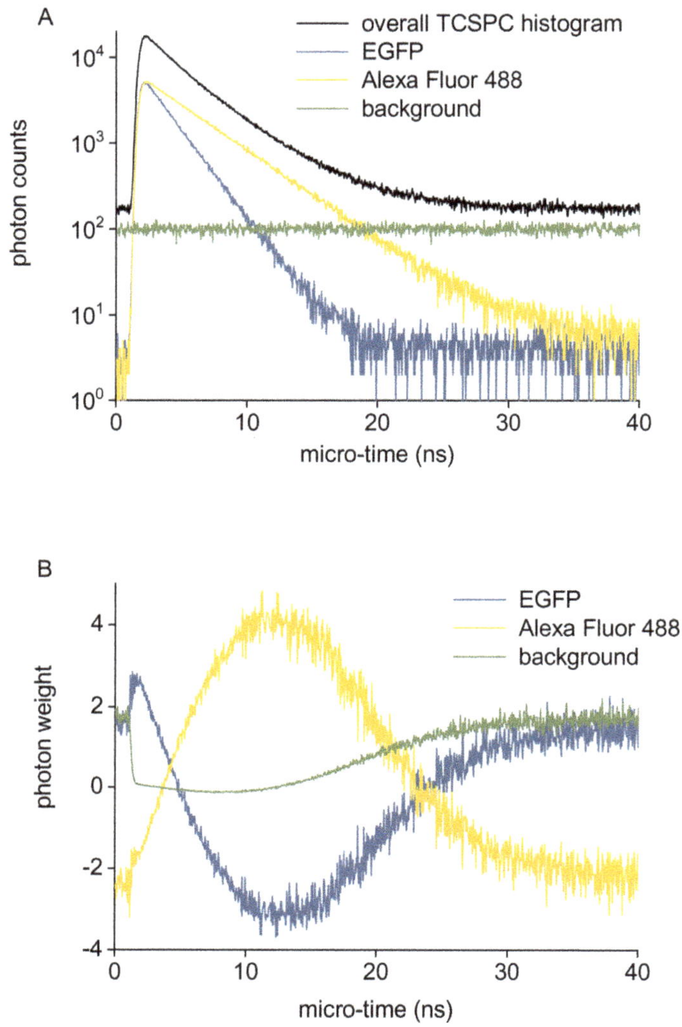

Figure 4.7. Examples of TCSPC decay patterns (a) and corresponding statistical filter functions (b) in FLCS. The fluorophore patterns were simulated as convolutions of monoexponential decays with a realistic IRF; the lifetimes of EGFP and Alexa Fluor 488 were taken as 2.1 ns and 4.1 ns, respectively [26]. The background was assumed to be evenly distributed across all TCSPC channels and Poisson noise was added to all patterns. The overall TCSPC histogram is a linear combination of the patterns contributing in the following ratio: 0.3/0.6/0.1 for EGFP/Alexa Fluor 488/background. The statistical filters shown in panel **B** were calculated for the patterns and the overall TCSPC histogram displayed in panel A.

section 4.1.3, we will get the ACF $G^{(k)}(\tau)$. Analogously, the CCF $G_X^{(k)(m)}(\tau)$ is obtained by weighting the original photon arrival vector $\{t_i\}$ by $w^{(k)}(\mu)$ and the shifted vector $\{t'_i\}$ by $w^{(m)}(\mu)$ and vice versa (see the symmetric definition of CCF in section 4.1.4). What we have just said can be written for the case of ACFs in the following way (using equation (4.2)):

$$\frac{G^k(m\Delta\tau)}{(M-m)} = \frac{\sum_{i=1}^{M-m} F^{(k)}(i)F^{(k)}(i+m)}{\left(\sum_{i=1}^{M-m} F^{(k)}(i)\right)\left(\sum_{i=1}^{M-m} F^{(k)}(i+m)\right)}$$
$$= \frac{\sum_{i=1}^{M-m}\sum_{\mu=1}^{H} w^{(k)}(\mu)F(i,\mu)\sum_{\mu=1}^{H} w^{(k)}(\mu)F(i+m,\mu)}{\left(\sum_{i=1}^{M-m}\sum_{\mu=1}^{H} w^{(k)}(\mu)F(i,\mu)\right)\left(\sum_{i=1}^{M-m}\sum_{\mu=1}^{H} w^{(k)}(\mu)F(i+m,\mu)\right)} \quad (4.15)$$

$F^{(k)}(i)$ is the intensity trace consisting only of photons emitted by fluorophores of the k-th population, which is not experimentally available. Therefore, we estimate it using the statistical filters $w^{(k)}(\mu)$ and the TSPC histograms recorded in each individual interval $F(i, \mu)$ as:

$$F^{(k)}(i) = \sum_{\mu=1}^{H} w^{(k)}(\mu)F(i, \mu) \quad (4.16)$$

For very short time intervals, in which no more than a single photon is detected, the histogram $F(i, \mu)$ contains only a single count in a single bin μ and the values $F^{(k)}(i)$ for such an interval then give the weights with which the single photon contributes to the correlation function for the k-th fluorophore population. At the same time, the overall TCSPC histogram at each time interval $F(i, \mu)$ is expected to be a linear combination of the patterns $\pi^{(k)}(\mu)$ weighted by the numbers of photons emitted at the given time interval by the respective fluorophore populations $F^{(k)}(i)$:

$$F(i, \mu) = \sum_{k=1}^{S} \pi^{(k)}(\mu)F^{(k)}(i) \quad (4.17)$$

Since equations (4.16) and (4.17) are expected to hold for any time interval, they must also hold for the whole experiment. We can, therefore, use the overall TSCPC histogram $F(\mu)$, constructed from all photon counts detected during the FLCS experiment to determine (according to equation (4.17)) the overall contributions $F^{(k)}$ of each population. The statistical filters $w^{(k)}(\mu)$ are then found by inserting $F^{(k)}$ and $F(\mu)$ into equation (4.16). This leads to the following formula for the statistical filters (see math box 4.2 for derivation):

$$\mathbf{W} = \left[\mathbf{\Pi} \cdot \mathbf{diag}\left(\overline{F}_{\mu}^{-1}\right) \cdot \mathbf{\Pi}^{\mathrm{T}}\right]^{-1} \cdot \mathbf{\Pi} \cdot \mathbf{diag}\left(\overline{F}_{\mu}^{-1}\right) \quad (4.18)$$

$\mathbf{W}$ and $\mathbf{\Pi}$ are matrices of statistical filters and patterns, respectively, $\left(\mathbf{W}_{\mu,k} = w^{(k)}(\mu), \mathbf{\Pi}_{\mu,k} = \pi^{(k)}(\mu)\right)$ and $\mathbf{diag}\left(\overline{F}_{\mu}^{-1}\right)$ is a diagonal matrix with the μ-th element on the diagonal equal to the reciprocal value of the overall TCSPC histogram bin $F(\mu)$. The superscript T denotes transposition and -1 matrix inversion. This calculation is easily implemented in any software package supporting matrix algebra.

Figure 4.7 shows an example of FLCS patterns and corresponding filter functions. The system contains two fluorophore populations differing in their

Math box 4.2

Derivation of statistical filter functions
Equation (4.17) can be written in matrix form as:

$$\vec{F}_\mu = \mathbf{\Pi}^\mathrm{T} \cdot \vec{F}^{(k)}$$

$\vec{F}_\mu$ and $\vec{F}^{(k)}$ are column vectors containing the values $F(\mu)$ and $F^{(k)}$, respectively. This is a system of H linear equation for S unknown variables $F^{(k)}$, where H is larger than S. In a typical FLCS experiment H, the number of TCSPC histogram bins is of the order of hundreds or thousands, while the number of fluorophore populations S is of the order of units. The system is, therefore, overdetermined and we cannot expect to find $\vec{F}^{(k)}$ that will satisfy all the H equations. Instead, we will search for a maximum likelihood estimator of $\vec{F}^{(k)}$ that minimises the weighted squared deviation χ^2:

$$\chi^2 = \sum_{\mu=1}^{H}\left(\frac{F(\mu) - \pi^{(k)}(\mu)F^{(k)}}{\sigma(F(\mu))}\right)^2 = (\mathbf{\Pi}^\mathrm{T} \cdot \vec{F}^{(k)} - \vec{F}_\mu)^T\mathbf{V}^{-1}(\mathbf{\Pi}^\mathrm{T} \cdot \vec{F}^{(k)} - \vec{F}_\mu)$$

$\sigma(F(\mu))$ is the standard deviation of the experimental values $F(\mu)$; weighting by the reciprocal values of its square ensures that deviations from $F(\mu)$ values that have been measured with higher uncertainty contribute less to χ^2. $\mathbf{V}$ is the covariance matrix of the data and in this case it is assumed to be diagonal, which reflects the fact that the variations in photon counts in individual TCSPC bins are not correlated with each other (see section 8.6). The diagonal elements of $\mathbf{V}$ are the values of variance (standard deviation squared) of $F(\mu)$. Since the photon count numbers in each bin follow Poisson statistics, their variance is equal to their mean value, which (for a reasonably high number of photon counts) can be approximated by the measured value $F(\mu)$. This leads to:

$$\mathbf{V}^{-1} = \mathbf{diag}\,(\vec{F}_\mu^{-1})$$

Vector $\vec{F}^{(k)}$ corresponding to the minimum of χ^2 needs to satisfy the following condition:

$$\frac{\partial \chi^2}{\partial \vec{F}^{(k)}} = 0$$

leading to:

$$2\mathbf{\Pi}{\cdot}\mathbf{V}^{-1} \cdot \mathbf{\Pi}^\mathrm{T} \cdot \vec{F}^{(k)} - 2\mathbf{\Pi}{\cdot}\mathbf{V}^{-1} \cdot \vec{F}_\mu = 0$$

which is solved by:

$$\vec{F}^{(k)} = (\mathbf{\Pi}{\cdot}\mathbf{V}^{-1} \cdot \mathbf{\Pi}^\mathrm{T})^{-1} \cdot \mathbf{\Pi}{\cdot}\mathbf{V}^{-1} \cdot \vec{F}_\mu$$

Next, we can use equation (4.16), which can be written in matrix form as:

$$\vec{F}^{(k)} = \mathbf{W} \cdot \vec{F}_\mu$$

to obtain equation (4.18) for the statistical filters $\mathbf{W}$.

lifetimes and the third pattern corresponds to the background which is evenly distributed over all bins in the TCSPC histogram. As expected, the ratios of the filter functions $w^{(k)}(\mu)$ for the individual components change with the TCSPC histogram bin μ to reflect the changes in ratios of the patterns $\pi^{(k)}(\mu)$. It is quite intuitive that the filter for the background has the lowest values in those bins where the fluorescence intensity is the highest (the beginning of the decay) and increases as the fluorescence intensity decays. Reasoning along the same lines explains why the filter of the shorter-lifetime fluorophore has higher values at the beginning of the decay than the filter of the longer-lifetime fluorophore. The latter filter then gains dominance at later stages of the decay, which reflects the fact that the photons contributing to those TCSPC bins have been more likely emitted by the longer-lifetime fluorophore. The absolute values of the filters increase as the fluorescence decay progresses to account for the decreasing number of photons contributing to those bins. However, other features of the filters lack a simple intuitive interpretation, especially the negative filter values found in many bins. It is obvious that statistical filtering can deliver reliable estimates of the numbers of photons $F^{(k)}$ contributed by the individual fluorophore populations only when applied on sufficiently large ensembles of photons. Applying it on small numbers of photons can easily result in negative $F^{(k)}$ values, which lack any physical meaning. FCS, being itself a statistical method, always works with large ensembles of photons, which makes FCS an ideal candidate for statistical filtering applications. Figure 4.8 illustrates the dependence of the accuracy of $F^{(k)}$ values retrieved by statistical filtering on the total number of processed photons. In this way, it has been shown that for the patterns displayed in figure 4.7 5000 photons are enough to retrieve the $F^{(k)}$ with better than 10% accuracy [27].

Another observation that can be made based on figure 4.8 is the higher accuracy with which the number of background photons is retrieved at low total photon numbers compared to the retrieved fluorescence photon numbers. This fact can be explained by the larger difference between the background pattern and any of the fluorescence patterns than between the two fluorescence patterns. This question has been recently addressed by simulations by Benda *et al* who found that the error in the ACFs obtained by FLCS scales with the relative overlap of the patterns [28]. The robustness of statistical filtering in identifying background photons is frequently used in FCS to remove the influence of background on ACFs [22, 29]. An example of experimental ACFs with and without background removal by statistical filtering is shown in figure 4.9. It can be seen that the background in the example in figure 4.9 is correlated on sub-ms timescales, indicating that it is dominated by the contribution of detector after-pulses (see section 7.5)

4.4.3 Generalisation of the FLCS principle —fluorescence spectral correlation spectroscopy (FSCS) and filtered FCS (fFCS)

As we have mentioned earlier, statistical filtering, which forms the basic principle of FLCS, can be applied also to other fluorescence emission characteristics than the TCSPC histograms used in FLCS. For example, FSCS uses fluorescence emission

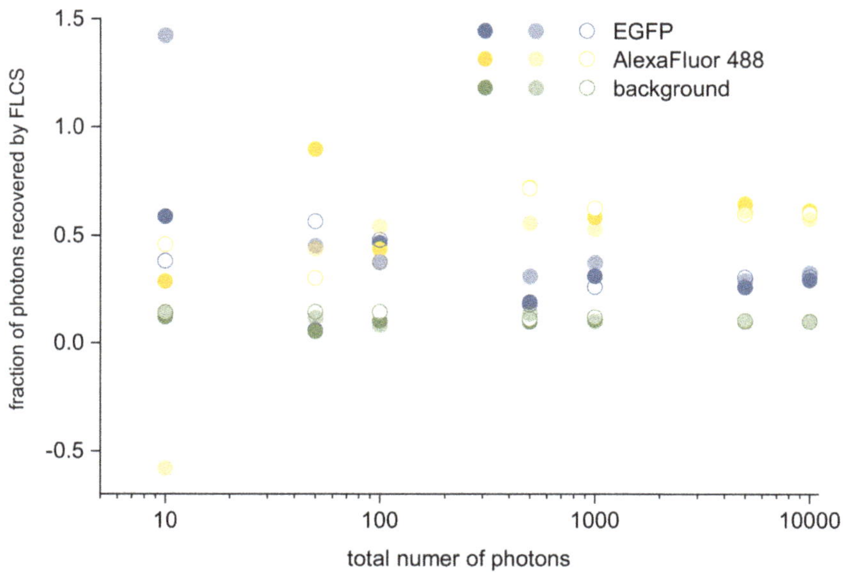

Figure 4.8. Illustration of the reliability of statistical filtering in retrieving the correct numbers of photons belonging to individual patterns as a function of the total photon number. For each total photon number 30% of the photons were simulated to originate from EGFP, 60% from Alexa Fluor 488 and 10% from background. Photon arrival times were simulated as random numbers following probability distributions described by the respective patterns shown in figure 4.7(a) and the contributions of each photon to the signals from individual fluorophores were weighted by the statistical filters shown in figure 4.7(b). Results of three independent simulations are shown for each total photon number.

spectra as patterns as is illustrated in figure 4.10. In this case the detected fluorescence is spectrally dispersed and collected by an array of detectors, each of which collects photons belonging to a particular wavelength interval. The information stored for each detected photon consists of its detection time t (the macro-time measured form the beginning of the experiment) and the spectral bin λ (the number of the detector in the array by which it was detected). The spectral bin λ then plays the same role which the micro-time bin μ played in FLCS. The filters $w^{(k)}(\lambda)$ are then calculated analogously from the spectral patterns $\pi^{(k)}(\lambda)$ and used in the same way in calculating the ACFs and CCFs. It has been shown that fluorophore populations with spectra extending over identical wavelength ranges can be distinguished by FSCS as long as the shapes of their spectra are distinct. As few as six spectral detection channels are sufficient for this purpose [30].

The same principle can be extended further to include several fluorescence emission parameters such as fluorescence polarization together with spectral information and TCSPC histograms. In that case, for each of the two possible polarization states ϑ there would be several channels corresponding to distinct spectral ranges λ, each of the detection channels allowing time-resolved fluorescence detection by TCSPC. This would then result in a four-dimensional pattern matrix $\pi^{(k)}(\mu, \lambda, \vartheta)$. The calculation of the statistical filter functions proceeds as before, only the dimensionality of the matrices involved in the calculation is increased.

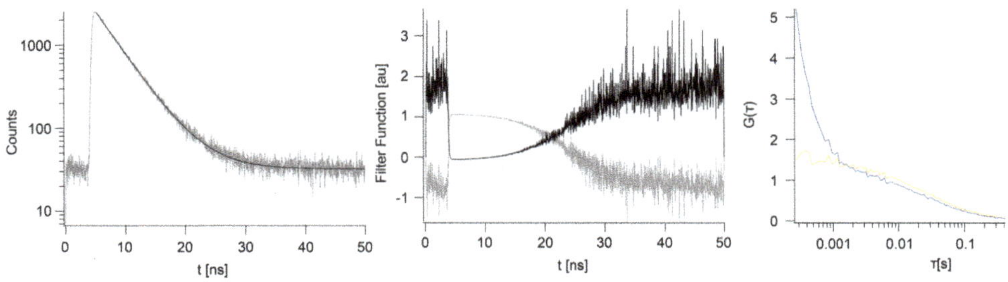

Figure 4.9. An example of the use of FLCS to obtain ACFs free from the influence of background. The overall TCSPC histogram is shown in the left-hand side panel. It is assumed to be a linear combination of patterns of the background and of the fluorescent tracer of interest. In such a case it is not necessary to know the patterns *a priori*; instead the assumption of background being evenly distributed across all TCSPC channels can be used. The contribution of background to the overall TCSPC histogram can be, thus, estimated by averaging the values in the TCSPC bins at the beginning (before the rising edge corresponding to the laser pulse) and at the end of the histogram (after the fluorescence has nearly completely decayed) and interpolating the average value to all bins. The contribution of the fluorescent tracer is then obtained by subtracting the constant background value from the overall histogram. The corresponding statistical filters are displayed in the central panel; background filter is shown in black and the fluorescent tracer filter in grey. The right-hand side panel shows the ACF calculated from all photon counts, the standard FCS ACF (blue curve) and the ACF calculated only from the counts identified by statistical filtering as originating from the fluorescent tracer, the FLCS ACF (yellow curve). The high correlation in the standard FCS ACF on sub-ms timescale indicates that the background is dominated by detector after-pulses, which would prevent quantifying any 'real' correlation on that timescale (caused, for example, by triplet state dynamics; see section 5.6).

Combining several fluorescence emission characteristics to obtain statistical filters allows distinguishing ACFs of fluorophore population which differ only very subtly in each of the individual parameters. The term fFCS has been adopted for this approach which realises the full potential of statistical filtering in FCS [23].

4.5 Synopsis

- The raw data in FCS can be stored in the form of fluorescence intensity time traces, which are afterwards correlated by specialised software (software correlator) or in the form of correlation function (ACF or CCF) obtained as the output of a hardware correlator.
- Efficient implementations of software correlators allow ACF calculation in real time during data acquisition. At the same time they can be used for offline correlation of stored intensity traces, which makes them more flexible than hardware correlators.
- One way of storing intensity time traces is in a binned form. The time trace duration is divided into short intervals and for each interval the number of detected photons is stored.
- Binned traces are a very inefficient way of storing data recorded with high temporal resolution at low photon detection rates (since most bins contain zeros). Storing detection time of each photon is more efficient in such cases. Software correlators have been developed for efficient ACF calculation directly from photon detection times.

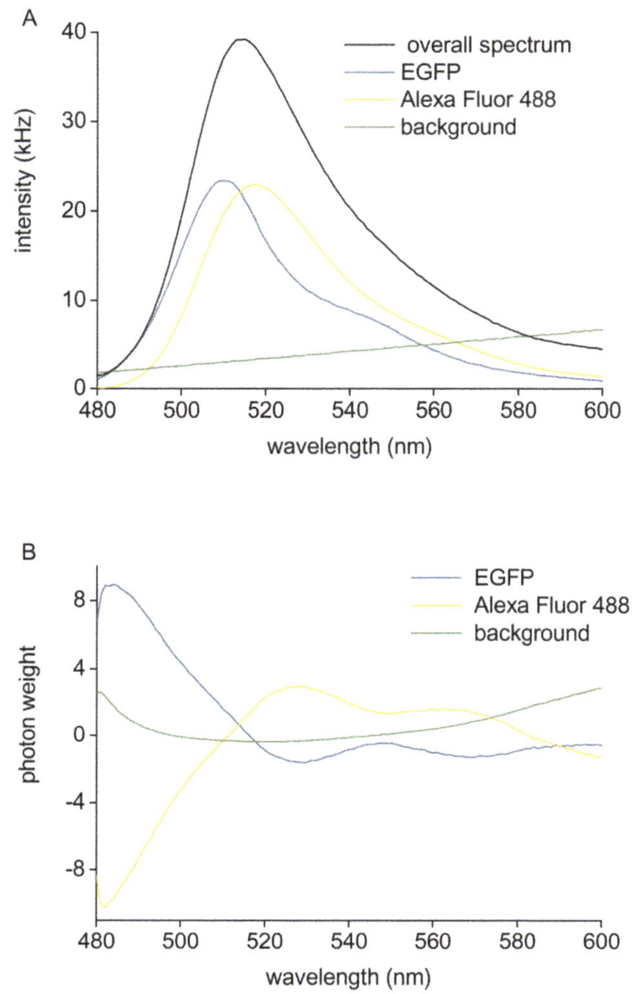

Figure 4.10. Examples of spectral patterns (A) and corresponding statistical filter functions (B) in FSCS. The fluorophore spectra were obtained from an online database [31]. The background simulates a situation in an FSCS experiment using a spectrograph with an EM-CCD camera. When EM-CCD cameras are operated with a high gain and fast read-out speed (both likely to be needed in FSCS), readout-induce charge gives rise to a background component that is not distributed evenly across all pixels; instead, it increases in the readout direction [30]. The overall spectrum is a linear combination of the patterns contributing in the following ratio: 0.3/0.6/0.1 for EGFP/Alexa Fluor 488/background. The statistical filters shown in panel B were calculated for the patterns and the overall spectrum displayed in panel A.

- To generate efficiently ACFs spanning several orders of magnitude of correlation time lags, the temporal resolution of the correlation calculation is coarsened quasi-logarithmically with increasing time lag. This correlator structure, called multiple-tau correlator, is applied in most hardware and software correlators and ensures approximately constant ratio of lag time to temporal resolution over the whole ACF.

- Additional information on each photon (for example, its wavelength, polarization or time after excitation at which it was emitted) can be used to weigh the photons contribution to the ACF. This approach, called statistical filtering, then, allows obtaining ACFs of individual fluorophore populations present in a mixture, even if the populations' fluorescence emission characteristics are very similar to each other and it is not possible to unambiguously trace individual photons to particular populations.

4.6 Exercises

1. Show that equation (4.4) for the number of products needed to calculate an ACF can be easily derived using the formula for triangular numbers. Triangular numbers can be written as sums of the form $\sum_{i=1}^{n} i$. For such sums it holds:

$$\sum_{i=1}^{n} i = \frac{1}{2}n(n+1).$$

2. Prove the formula for triangular numbers used in the previous exercise. *Hint*: use mathematical induction; show that the equation is valid for $n = 1$ and then that its validity for n implies its validity for $n + 1$. Alternatively use the fact that for an arithmetic series, the sum is equal to the number of terms times the average of the series.

3. Consider an FCS time trace recorded for 30 s with 1 μs temporal resolution (typical values in confocal FCS experiments). How many products need to be calculated to obtain the ACF for lag times up to 2 s? How does the number change if the measurement time is increased twice?

4. Show that equation (4.5) can be rewritten in a non-recursive manner as:

$$\tau_m = \begin{cases} m\Delta\tau & 0 < m \leqslant \frac{p}{2} \\ \Delta\tau\left(l + \frac{p}{2}\right)2^{k-1} & \frac{p}{2} < m \leqslant (q+1)\frac{p}{2} \end{cases}$$

where $k = 1, 2, \ldots, q$ is the number of the group of lag times and $l = 1, 2, \ldots, \frac{p}{2}$ is the number of the lag time point within its group and $m = k\frac{p}{2} + l$.

5. Show that the maximum correlation lag time of a (p, q) correlator with $p = 16$ is $2^{q+3}\Delta\tau$.

6. Use the result of the previous exercise to calculate the maximum correlation lag time (in units of $\Delta\tau$) achieved with a (16, 8) multiple-tau correlator. For how many lag time points τ_m is the value of $G(\tau_m)$ calculated? Compare the

number of points with that needed for a linear correlator to reach the same maximum lag time.

7. Prove equation (4.6). *Hint*: use mathematical induction to prove that:

$$\sum_{i=1}^{n} 2^{-i} = 1 - \frac{1}{2^n}$$

References

[1] Schätzel K, Drewel M and Stimac S 1988 Photon correlation measurements at large lag times: improving statistical accuracy *J. Mod. Opt.* **35** 711–18

[2] Schätzel K 1990 Noise on photon correlation data: 1. autocorrelation functions *Quantum Opt.* **2** 287–305

[3] Wohland T, Rigler R and Vogel H 2001 The standard deviation in fluorescence correlation spectroscopy *Biophys. J.* **80** 2987–99

[4] Sankaran J, Shi X, Ho L Y, Stelzer E H K and Wohland T 2010 ImFCS: a software for imaging FCS data analysis and visualization *Opt. Express* **18** 25468–81

[5] Magatti D and Ferri F 2001 Fast multi-tau real-time software correlator for dynamic light scattering *Appl. Opt.* **40** 4011–21

[6] Wahl M, Gregor I, Patting M and Enderlein J 2003 Fast calculation of fluorescence correlation data with asynchronous time-correlated single-photon counting *Opt. Express* **11** 3583

[7] Magatti D and Ferri F 2003 25 ns software correlator for photon and fluorescence correlation spectroscopy *Rev. Sci. Instrum.* **74** 1135–44

[8] Schaub E 2012 F2Cor: fast 2-stage correlation algorithm for FCS and DLS *Opt. Express* **20** 2184–95

[9] Laurence T A, Fore S and Huser T 2006 Fast, flexible algorithm for calculating photon correlations *Opt. Lett.* **31** 829–31

[10] Yang L-L, Lee H-Y, Wang M-K, Lin X-Y, Hsu K-H, Chang Y-R, Fann W and White J D 2009 Real-time data acquisition incorporating high-speed software correlator for single-molecule spectroscopy *J. Microsc.* **234** 302–10

[11] Mulgrew B, Grant P and Thompson J 2002 *Digital Signal Processing* (London: Macmillan)

[12] Schnitzbauer J, Wang Y, Zhao S, Bakalar M, Nuwal T, Chen B and Huang B 2018 Correlation analysis framework for localization-based superresolution microscopy *Proc. Natl Acad. Sci. USA* **115** 3219–24

[13] Petersen N O, Höddelius P L, Wiseman P W, Seger O and Magnusson K E 1993 Quantitation of membrane receptor distributions by image correlation spectroscopy: concept and application *Biophys. J.* **65** 1135–46

[14] Costantino S, Comeau J W, Kolin D L and Wiseman P W 2005 Accuracy and dynamic range of spatial image correlation and cross-correlation spectroscopy *Biophys. J.* **89** 1251–60

[15] Digman M A, Sengupta P, Wiseman P W, Brown C M, Horwitz A R and Gratton E 2005 Fluctuation correlation spectroscopy with a laser-scanning microscope: exploiting the hidden time structure *Biophys. J.* **88** L33–6

[16] Sengupta P, Jovanovic-Talisman T, Skoko D, Renz M, Veatch S L and Lippincott-Schwartz J 2011 Probing protein heterogeneity in the plasma membrane using PALM and pair correlation analysis *Nat. Meth.* **8** 969–75

[17] Buchholz J, Krieger J W, Mocsár G, Kreith B, Charbon E, Vámosi G, Kebschull U and Langowski J 2012 FPGA implementation of a 32x32 autocorrelator array for analysis of fast image series *Opt. Express* **20** 17767–82

[18] Steger K, Bollmann S, Noé F and Doose S 2013 Systematic evaluation of fluorescence correlation spectroscopy data analysis on the nanosecond time scale *Phys. Chem. Chem. Phys.* **15** 10435–11

[19] Singh A P, Krieger J W, Buchholz J, Charbon E, Langowski J and Wohland T 2013 The performance of 2D array detectors for light sheet based fluorescence correlation spectroscopy *Opt. Express* **21** 8652

[20] Sankaran J, Bag N, Kraut R S and Wohland T 2013 Accuracy and precision in camera-based fluorescence correlation spectroscopy measurements *Anal. Chem.* **85** 3948–54

[21] Böhmer M and Enderlein J 2003 Fluorescence spectroscopy of single molecules under ambient conditions: methodology and technology *ChemPhysChem* **4** 792–808

[22] Kapusta P, Macháň R, Benda A and Hof M 2012 Fluorescence lifetime correlation spectroscopy (FLCS): concepts, applications and outlook *IJMS* **13** 12890–910

[23] Felekyan S, Kalinin S, Sanabria H, Valeri A and Seidel C A M 2012 Filtered FCS: species auto- and cross-correlation functions highlight binding and dynamics in biomolecules *ChemPhysChem* **13** 1036–53

[24] Lakowicz J R 2006 *Principles of Fluorescence Spectroscopy* 3rd edn (Berlin: Springer)

[25] Becker W 2019 *The bh TCSPC Handbook* 8th edn (Berlin: Becker & Hickl)

[26] Chen J and Irudayaraj J 2010 Fluorescence lifetime cross correlation spectroscopy resolves EGFR and antagonist interaction in live cells *Anal. Chem.* **82** 6415–21

[27] Macháň R, Kapusta P and Hof M 2014 Statistical filtering in fluorescence microscopy and fluorescence correlation spectroscopy *Anal. Bioanal. Chem.* **406** 4797–13

[28] Štefl M, Herbst K, Rübsam M, Benda A and Knop M 2020 Single-color fluorescence lifetime cross-correlation spectroscopy *in vivo* bioRxiv 2020.01.23.917435

[29] Enderlein J and Gregor I 2005 Using fluorescence lifetime for discriminating detector afterpulsing in fluorescence-correlation spectroscopy *Rev. Sci. Instrum.* **76** 033102–105

[30] Benda A, Kapusta P, Hof M and Gaus K 2014 Fluorescence spectral correlation spectroscopy (FSCS) for probes with highly overlapping emission spectra *Opt. Express* **22** 2973

[31] Lambert T J 2019 FPbase: a community-editable fluorescent protein database *Nat. Meth.* **16** 277–78

IOP Publishing

An Introduction to Fluorescence Correlation Spectroscopy

Thorsten Wohland, Sudipta Maiti and Radek Macháň

Chapter 5

Theoretical FCS models

The heart of FCS analysis is the fitting of the experimental correlation functions to theoretical models that describe the process under investigation. This chapter will describe the derivation of theoretical FCS models and chapter 8 will explain how one can determine which model is most consistent with the data, taking into account the data and its variance, and how to fit. It derives models for several processes, including diffusion in 2D and 3D, flow and active transport, for one and multi-component processes, anomalous diffusion and photophysics. It discusses general characteristics of correlation functions and the influence of noise. In the later sections the temporal correlation functions are extended to spatial and spatiotemporal correlations with the discussion of scanning and camera-based approaches. After this chapter the reader should have a good understanding of how FCS models are derived and how FCS can be extended to the spatial domain.

5.1 The autocorrelation function for diffusion

The theoretical ACF model is determined by the process under investigation and the optics of the measurement apparatus. The derivation of the ACF for diffusion as measured by a confocal system is given in math boxes 5.1 and 5.2. Here, we will only give a general description of how the derivation is performed. First, one has to determine four characteristics of the experimental setup. (i) The size of the sample, $S(\vec{r})$, which determines where the molecules of interest can be found. This can be larger or smaller than the volume which is illuminated and from which fluorescence is detected (figure 5.1) and thus influences the effective observation volume. (ii) The intensity distribution of the excitation light $I(\vec{r})$. The signal a fluorophore emits will be proportional to $I(\vec{r})$. (iii) The collection efficiency of the optical system $\mathrm{CEF}(\vec{r})$. This describes how much of an emitted signal is detected by the system. (iv) And finally the fluctuation causing process, here the changes in concentration of fluorescent molecules, $C(\vec{r}, t)$, and how it influences the correlation. Using these, one can write the fluorescence detected from the sample as

doi:10.1088/978-0-7503-2080-1ch5

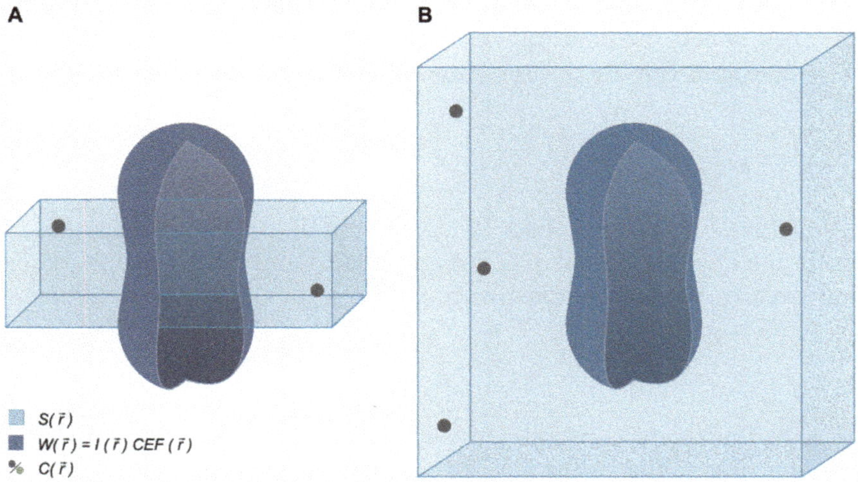

Figure 5.1. Schematic of the relation between the sample size $S(\vec{r})$, the extent of the observation volume $W(\vec{r})$, and the fluorescent particles and their concentration $C(\vec{r}, t)$. The derivation of the correlation function in math box 5.1 follows this naming scheme. Animation available at https://iopscience.iop.org/book/978-0-7503-2080-1.

$$F(t) = \kappa \int I(\vec{r})\mathrm{CEF}(\vec{r})S(\vec{r})C(\vec{r}, t)d\vec{r} \tag{5.1}$$

Here, the parameter κ incorporates the effects of fluorophore brightness and overall detection efficiency of the system. This can be inserted into the equation for the correlation function:

$$G(\tau) = \frac{\kappa^2 \int\int I(\vec{r})I(\vec{r}')\mathrm{CEF}(\vec{r})\mathrm{CEF}(\vec{r}')S(\vec{r})S(\vec{r}')\langle C(\vec{r}, t)C(\vec{r}', t+\tau)\rangle d\vec{r}d\vec{r}'}{\kappa^2\langle \int I(\vec{r})\mathrm{CEF}(\vec{r})S(\vec{r})C(\vec{r}, t)d\vec{r}\rangle^2} \tag{5.2}$$

It is this equation that needs to be integrated to lead to a model of the ACF for a particular process. Various simplifications have been introduced to be able to solve this equation. The simplifications include the assumption that the sample is large compared to the observation volume, i.e. $S(\vec{r})$ can be set to 1; and the product of illumination and collection efficiency $I(\vec{r})\mathrm{CEF}(\vec{r})$ has been approximated with a function that can be easily integrated. The change of any of these assumptions will lead to variations in the model. For example, models have been derived for situations in which the sample extent is small compared to the observation volume, e.g. when measuring transport in axons [1]. Or various functions for $I(\vec{r})\mathrm{CEF}(\vec{r})$ have been used depending on whether one works in a confocal or other optical instrument [2, 3] (see also chapter 3). Lastly, a model of the fluctuation creating process, $\langle C(\vec{r}, t)C(\vec{r}', t+\tau)\rangle$, has to be obtained. For diffusion, for instance, this can be solved by solving the differential equation for diffusion (see math box 5.3).

In math box 5.1, we show the solution for diffusion in a confocal setup (see chapter 3) in which the confocal observation volume is much smaller than the overall sample size. For 3D diffusion we obtain the following results:

$$G(\tau) = \frac{1}{\langle C \rangle \pi^{3/2} w_0^2 z_0} \left(1 + \frac{4D\tau}{w_0^2}\right)^{-1/2} \left(1 + \frac{4D\tau}{w_0^2}\right)^{-1/2} \left(1 + \frac{4D\tau}{z_0^2}\right)^{-1/2} + 1 \tag{5.3}$$

In this equation, we have introduced the $1/e^2$ radii of the 3D Gaussian observation volume for the radial and axial directions, w_0 and z_0, and the diffusion coefficient D, which is given by the Stokes–Einstein equation

$$D = \frac{kT}{6\pi\eta r_h} \tag{5.4}$$

Here, k is the Boltzmann constant, T the temperature, η the solution viscosity, and r_h the hydrodynamic radius of the labelled molecule under investigation.

Each dimension of diffusion has one term in this equation. As the system is rotationally symmetric, the first two terms for the x- and y-dimensions are the same. For practical purposes the function is mostly written as

$$G(\tau) = \frac{1}{N}\left(1 + \frac{\tau}{\tau_D}\right)^{-1} \left(1 + \frac{\tau}{K^2 \tau_D}\right)^{-1/2} + G_\infty \tag{5.5}$$

With N being the average number of particles in the observation volume. In the last equation we used the following definitions

$$\tau_D = \frac{w_0^2}{4D} \tag{5.6}$$

and

$$K = \frac{z_0}{w_0} \tag{5.7}$$

τ_D is known as the diffusion time of a particle. K is commonly known as the structure factor and describes the ratio between the length of the focal volume (along the optical axis) to the width of the focal volume (in the focal plane). G_∞ is the convergence of the ACF for long times. Theoretically this is 1 (or 0 depending on whether the signal or the fluctuation correlation function is used, see chapter 2, equation (2.30)) and this value would be reached for infinitely long measurements. However, practically the measurement time is limited and the convergence value can deviate from 1. Therefore, this value is used as a fit parameter. In solution measurements, the deviation from 1 is usually less than 1%. In more complex samples it can be somewhat higher. Strong deviations from the theoretical convergence value can indicate problems in the measurements (chapter 7).

Math box 5.1

The autocorrelation function for diffusion in a confocal setup

In fluorescence correlation spectroscopy (FCS), we record a fluorescence intensity time trace $F(t)$ from a defined observation volume. The fluorescence signal will be determined by the following parameters and functions, where $\vec{r}$ represents the spatial coordinates and t is the time:

(a) the intensity profile of the exciting light source $I(\vec{r}, t)$. We assume here that the light source is constant in time and can be described by $I(\vec{r})$;

(b) the collection efficiency function of the optical detection system $\mathrm{CEF}(\vec{r})$; typically this is determined by the point spread function (PSF) of the microscope and the transmission of either a pinhole in confocal FCS or, as will be discussed in math box 5.7, by the pixel size in camera based FCS;

(c) the product of the fluorophore absorption cross-section, its quantum yield, and the overall detection efficiency (this accounts for the detector quantum efficiency, and any losses within the system) κ;

(d) the extent of the sample $S(\vec{r})$;

(e) the distribution of the fluorophores $C(\vec{r}, t)$.

With these definition the fluorescence from a particular observation volume is given by

$$F(t) = \kappa \int I(\vec{r})\mathrm{CEF}(\vec{r})S(\vec{r})C(\vec{r}, t)d\vec{r}$$

The intensity fluctuations, which are due to concentration fluctuations, can then be defined as

$$\delta F(t) = \kappa \int I(\vec{r})\mathrm{CEF}(\vec{r})S(\vec{r})\delta C(\vec{r}, t)d\vec{r}$$

with

$$F(t) = \delta F(t) + \langle F(t)\rangle$$

Where $\langle F\rangle$ is the time averaged fluorescence and $\delta F(t)$ are the fluorescence fluctuations around $\langle F\rangle$. In particular, we can state that

$$\langle \delta F(t)\rangle = 0$$

Assuming again that we investigate stationary processes and thus that

$$\langle F(t)\rangle = \langle F(t+\tau)\rangle$$

The autocorrelation function is given by $G(\tau) = \frac{\langle F(t)F(t+\tau)\rangle}{\langle F(t)\rangle^2}$

Inserting this into the equation for the ACF results in

$$G(\tau) = \frac{\langle F(t)F(t+\tau)\rangle}{\langle F(t)\rangle^2} = \frac{\langle(\delta F(t) + \langle F(t)\rangle)(\delta F(t+\tau) + \langle F(t+\tau)\rangle)\rangle}{\langle \delta F(t) + \langle F(t)\rangle\rangle^2}$$

The numerator can be expanded and solved, remembering that the average over the fluctuations are 0, to yield

$$\begin{aligned}\langle(\delta F(t) + \langle F(t)\rangle)(\delta F(t+\tau) + \langle F(t+\tau)\rangle)\rangle &= \langle \delta F(t)\delta F(t+\tau)\rangle + \langle F(t)\rangle\langle F(t+\tau)\rangle \\ &= \langle \delta F(t)\delta F(t+\tau)\rangle + \langle F\rangle^2\end{aligned}$$

For the denominator it follows that

$$\langle \delta F(t) + \langle F \rangle \rangle = \langle F \rangle$$

This yields

$$G(\tau) = \frac{\langle \delta F(t)\delta F(t+\tau)\rangle + \langle F\rangle^2}{\langle F\rangle^2} = \frac{\langle \delta F(t)\delta F(t+\tau)\rangle}{\langle F\rangle^2} + 1$$

Note that sometimes the 1 as a constant factor is subtracted and the ACF will decay to 0 at long times. Both representations contain the same information. Here we will stick with the definition of the signal correlation function and the ACF at long times will decay to 1. We can use now the definition of the fluorescence fluctuations in the definition of the ACF:

$$G(\tau) = \frac{\kappa^2 \int\int I(\vec{r})I(\vec{r}')\mathrm{CEF}(\vec{r})\mathrm{CEF}(\vec{r}')S(\vec{r})S(\vec{r}')\,\langle \delta C(\vec{r}, t)\delta C(\vec{r}', t+\tau)\rangle d\vec{r}d\vec{r}'}{\kappa^2 \langle \int I(\vec{r})\mathrm{CEF}(\vec{r})S(\vec{r})C(\vec{r}, t)d\vec{r}\rangle^2} + 1$$

Assuming that the sample is much larger than the observation volume and thus $S(\vec{r}) = 1$, and using the fact that the time average over the integral in the denominator affects only the concentration resulting in the average concentration $\langle C\rangle = \langle C(\vec{r}, t)\rangle$, then we can write the ACF as

$$G(\tau) = \frac{\int\int I(\vec{r})I(\vec{r}')\mathrm{CEF}(\vec{r})\mathrm{CEF}(\vec{r}')S(\vec{r})S(\vec{r}')\,\langle \delta C(\vec{r}, t)\delta C(\vec{r}', t+\tau)\rangle d\vec{r}d\vec{r}'}{\langle C\rangle^2 \left(\int I(\vec{r})\mathrm{CEF}(\vec{r})S(\vec{r})d\vec{r}\right)^2} + 1$$

We now introduce some more simplifications. In confocal FCS we can approximate the laser focus from which fluorescence is observed and the spatial filtering of the pinhole as a three-dimensional Gaussian function. With the laser power denoted as P, this function is given by

$$W(\vec{r}) = I(\vec{r})\mathrm{CEF}(\vec{r}) = \frac{2P}{\pi w_0} e^{\frac{-2x^2}{w_0^2}} e^{\frac{-2y^2}{w_0^2}} e^{\frac{-2z^2}{z_0^2}}$$

In the case of pure diffusion, the concentration fluctuation correlation is given by the so-called diffusion propagator ([4] and math box 5.3).

$$\langle \delta C(\vec{r}, t)\delta C(\vec{r}', t+\tau)\rangle = \langle C\rangle \frac{e^{-\frac{(\vec{r}-\vec{r}')^2}{4D\tau}}}{8(\pi D\tau)^{3/2}} = \langle C\rangle \frac{e^{-\frac{(x-x')^2+(y-y')^2+(z-z')^2}{4D\tau}}}{8(\pi D\tau)^{3/2}}$$

Inserting these equations into the ACF yields

$$G(\tau) = \frac{\int\int\int\int\int\int e^{\frac{-2(x^2+x'^2)}{w_0^2}} e^{\frac{-2(y^2+y'^2)}{w_0^2}} e^{\frac{-2(z^2+z'^2)}{z_0^2}} \frac{e^{-\frac{(x-x')^2+(y-y')^2+(z-z')^2}{4D\tau}}}{8(\pi D\tau)^{3/2}} dxdx'dydy'dzdz'}{\langle C\rangle \left(\int\int\int e^{\frac{-2x^2}{w_0^2}} e^{\frac{-2y^2}{w_0^2}} e^{\frac{-2z^2}{z_0^2}} dxdydz\right)^2} + 1$$

The integration results in the following solution

$$G(\tau) = \frac{1}{\langle C\rangle \pi^{3/2} w_0^2 z_0}\left(1 + \frac{4D\tau}{w_0^2}\right)^{-1/2}\left(1 + \frac{4D\tau}{w_0^2}\right)^{-1/2}\left(1 + \frac{4D\tau}{z_0^2}\right)^{-1/2} + 1$$

Each dimension of diffusion has one term in this equation. As the system is rotationally symmetric, the first two terms for the x- and y-dimensions are the same. For practical purposes the function is mostly written as

$$G(\tau) = \frac{1}{N}\left(1 + \frac{\tau}{\tau_D}\right)^{-1}\left(1 + \frac{\tau}{K^2\tau_D}\right)^{-1/2} + G_\infty$$

In the last equation, we used the following definitions

$$\tau_D = \frac{w_0^2}{4D}$$

and

$$K = \frac{z_0}{w_0}$$

τ_D is known as the diffusion time of a particle, and K is commonly known as the structure factor and describes the ration between the length of the focal volume to the width of the focal volume (typical values for K lie between 3 and 8).

G_∞ is the convergence of the ACF for long times. Theoretically this is 1 and this value would be reached for infinitely long measurements. However, practically the measurement time is limited and the convergence value can deviate from 1. Therefore, this value is used as a fit parameter. In solution measurements the deviation from 1 is usually less than 1%. In biological samples the deviation can be as high as 10%.

In addition,

$$N = \langle C \rangle \pi^{3/2} w_0^2 z_0 = \langle C \rangle V_{\text{eff}}$$

We will show the calculation of V_{eff} in math box 5.2 below. Often N is used as a fitting parameter and has then to be corrected with a correction factor γ (for definitions and explanations see math box 5.2) to obtain number of particles or the effective volume V_{eff} to obtain absolute concentrations.

The equation can be readily simplified for diffusion in 2D in the focal plane of the microscope

$$G(\tau) = \frac{1}{N}\left(1 + \frac{\tau}{\tau_D}\right)^{-1} + G_\infty \tag{5.8}$$

or for the 2D case in which the system is not rotationally symmetric

$$G(\tau) = \frac{1}{N}\left(1 + \frac{\tau}{\tau_D}\right)^{-1/2}\left(1 + \frac{\tau}{K^2\tau_D}\right)^{-1/2} + G_\infty \tag{5.9}$$

In this case, it is easy to verify that the structure factor is just the ratio of the different radii of the observation volume in the x- and y- directions (short and long axis of the observation volume).

$$K = \frac{w_1}{w_0} \tag{5.10}$$

Math box 5.2

The effective observation volume for a confocal setup

The observation volume in FCS is given by the laser beam profile, $I(\vec{r})$, and the collection efficiency function, CEF($\vec{r}$), of the microscope. The CEF is determined by the collection optics and the confocal pinhole. We assume in the following that the sample is larger than the observation volume and S($\vec{r}$)=1. The product $W(\vec{r}) = I(\vec{r})\mathrm{CEF}(\vec{r})$ then determines the observation volume of the instrument. As all constant factors in W($\vec{r}$) cancel in the normalized correlation function, it is the shape of W($\vec{r}$) that determines the ACF. We thus use the normalize function W($\vec{r}$)/W(0) in the calculations. As shown above, the amplitude of the ACF is given by

$$G(0) = \frac{1}{\langle C \rangle} \frac{\int (W(\vec{r})/W(0))^2 d\vec{r}}{\left(\int W(\vec{r})/W(0) d\vec{r}\right)^2}$$

With the definition of

$$w_n = \int (W(\vec{r})/W(0))^n d\vec{r}$$

The ACF can be written as

$$G(0) = \frac{1}{\langle C \rangle} \frac{w_2}{w_1^2}$$

The number of particles N is equal to the concentration times an effective volume. Therefore, the effective volume and the average number of particles in FCS can be define as

$$N = \frac{w_1^2}{w_2} \langle C \rangle = V_{\mathrm{eff}} \langle C \rangle$$

It follows that

$$V_{\mathrm{eff}} = \frac{w_1^2}{w_2}$$

and thus

$$G(0) = \frac{1}{\langle C \rangle} \frac{w_2}{w_1^2} = \frac{1}{V_{\mathrm{eff}} \langle C \rangle} = \frac{1}{N}$$

Note: Sometimes, N is set equal to the product of the concentration times the focal volume w_1 instead of the effective volume w_1^2/w_2

$$N \equiv w_1 \langle C \rangle$$

In that case the definitions are different and a correction factor γ has to be introduced

$$\gamma = \frac{w_2}{w_1}$$

leading to

$$G(0) = \frac{1}{\langle C \rangle} \frac{w_2}{w_1^2} = \frac{\gamma}{N}$$

Both the correction factor γ and the effective observation volume, V_{eff}, can be readily calculated. The laser focus is a Gaussian function in the radial direction and a Lorentzian function in the axial direction. The pinhole has a strong effect in the axial direction and much less effect in the radial direction. The common assumption therefore is that the FCS observation volume can be described by a Gaussian function in all three dimensions which makes the calculations significantly easier. First, for the 2D case, the laser beam can be described by a 2D Gaussian function.

$$W(\vec{r}) = W(0)e^{\frac{-2\vec{r}^2}{w_0^2}}$$

$$w_1 = \int_{-\infty}^{\infty} \frac{W(\vec{r})}{W(0)} d\vec{r} = \int_{-\infty}^{\infty} e^{\frac{-2\vec{r}^2}{w_0^2}} d\vec{r} = \int_{-\infty}^{\infty}\int_{-\infty}^{\infty} e^{\frac{-2(x^2+y^2)}{w_0^2}} dxdy = \frac{\pi w_0^2}{2}$$

$$w_2 = \int_{-\infty}^{\infty} \left(\frac{W(\vec{r})}{W(0)}\right)^2 d\vec{r} = \int_{-\infty}^{\infty} \left(e^{\frac{-2\vec{r}^2}{w_0^2}}\right)^2 d\vec{r} = \int_{-\infty}^{\infty}\int_{-\infty}^{\infty} e^{\frac{-4(x^2+y^2)}{w_0^2}} dxdy = \frac{\pi w_0^2}{4}$$

The correction factor γ and the effective observation volume V_{eff} can be calcualted from these integrals. Note that either one uses the effective volume or one uses the correction factor in FCS [2].

$$\gamma = \frac{w_2}{w_1} = \frac{1}{2}$$

$$V_{\text{eff}} = \frac{w_1^2}{w_2} = \pi w_0^2$$

For the 3D case, the same calculations can be conducted.

$$W(\vec{r}) = W(0)e^{\frac{-2(x^2+y^2)}{w_0^2}} e^{\frac{-2z^2}{z_0^2}}$$

$$w_1 = \int_{-\infty}^{\infty} \frac{W(\vec{r})}{W(0)} d\vec{r} = \int_{-\infty}^{\infty}\int_{-\infty}^{\infty}\int_{-\infty}^{\infty} e^{\frac{-2(x^2+y^2)}{w_0^2}} e^{\frac{-2z^2}{z_0^2}} dxdydz = \frac{\pi^{3/2} w_0^2 z_0}{2\sqrt{2}}$$

$$w_2 = \int_{-\infty}^{\infty} \frac{W(\vec{r})}{W(0)} d\vec{r} = \int_{-\infty}^{\infty}\int_{-\infty}^{\infty}\int_{-\infty}^{\infty} e^{\frac{-4(x^2+y^2)}{w_0^2}} e^{\frac{-4z^2}{z_0^2}} dxdydz = \frac{\pi^{3/2} w_0^2 z_0}{8}$$

$$\gamma = \frac{w_2}{w_1} = \frac{1}{2\sqrt{2}} = \frac{1}{\sqrt{8}}$$

$$V_{\text{eff}} = \frac{w_1^2}{w_2} = \pi^{3/2} w_0^2 z_0$$

Math box 5.3

The diffusion propagator for translational diffusion
If a particle is at position $\vec{r}$ at time t, what is the probability that it is at position $\vec{r}'$ at time t'? This is the meaning of the expression

$$\langle C(\vec{r},\, t)C(\vec{r}',\, t + \tau)\rangle$$

Let's look at some simple examples first. If $\tau = 0$, i.e. we ask what is the probability to find a particle at the same time at position $\vec{r}$ and $\vec{r}'$, then that must be

$$\langle C(\vec{r},\, t)C(\vec{r}',\, t)\rangle = \langle C\rangle \delta(\vec{r} - \vec{r}')$$

The Dirac delta function $\delta(x)$ being everywhere 0 except at the position where $x = 0$. This means that at the same time a particle can only be found at the same position, which is not surprising. But what happens for values of $\tau \neq 0$? In this case we have to calculate the probability to find the particle at position $\vec{r}'$, if it started at position $\vec{r}$ a time τ earlier. This is achieved by solving the diffusion equation

$$\frac{\partial C(\vec{r},\, t)}{\partial t} = D\frac{\partial^2 C(\vec{r},\, t)}{\partial x^2}$$

A solution to the is equation, as can be verified by inserting it and calculating the derivatives, is

$$C(\vec{r},\, t) = \frac{\langle C\rangle}{(4\pi Dt)^{1/2}} e^{-\frac{x^2}{4Dt}}$$

A simple way to show this result has been given by, e.g., Berg [5].

If we make the following assumptions for a 1D random walk:

1) all particles start at position 0.
2) all particles step to the right or left with a step size δ per unit time.
3) the probability to go right or left is ½.
4) the particles move independently.

Then the nth position of a particle is given by

$$x_n = x_{n-1} \pm \delta$$

The average position after N steps is then

$$\langle x_n\rangle = \frac{1}{N}\sum_{i=1}^{N} x_i = \frac{1}{N}\sum_{i=1}^{N} x_{i-1} \pm \delta = \frac{1}{N}\sum_{i=1}^{N} x_{i-1}$$

The right side follows as the particle moves to the right and left with equal probability so that $\langle \delta \rangle = 0$. But this means that

$$\frac{1}{N}\sum_{i=1}^{N} x_i = \frac{1}{N}\sum_{i=1}^{N} x_{i-1}$$

So the average particle position doesn't change. As the average position at the start was 0, the average position remains 0.

$$\langle x_n\rangle = 0$$

What is the average distance moved, i.e. what is the mean squared displacement?

$$\langle x_n^2 \rangle = \langle x_{n-1}^2 + \delta^2 \pm 2x_{n-1}\delta \rangle$$
$$\langle x_n^2 \rangle = \langle x_{n-1}^2 \rangle + \delta^2$$

As all particles start at 0, it follows that

$$\langle x_0^2 \rangle = 0$$

All following positions can be calculated from the above recursive equation

$$\langle x_1^2 \rangle = \delta^2$$
$$\langle x_2^2 \rangle = 2\delta^2$$
$$\langle x_n^2 \rangle = n\delta^2$$

The number of steps n a particle takes is just given by the total time t divided by the time step Δt.

$$\langle x_n^2 \rangle = \frac{t}{\Delta t}\delta^2 = \frac{\delta^2}{\Delta t}t$$

We introduce here the diffusion coefficient as

$$D = \frac{\delta^2}{2\Delta t}$$

And thus have

$$\langle x_n^2 \rangle = 2Dt$$

Thus it follows that the variance is

$$\sigma^2 = \langle x_n^2 \rangle - \langle x_n \rangle^2 = 2Dt$$

The probability distribution that describes this process is the binomial distribution. The binomial distribution provides the probability that the particle ends up a certain number of steps m to the right after taking a total number of steps n, given the probability of stepping to the right p, and the probability of stepping to the left $q = 1 - p$.

$$P(m; n, p) = \frac{n!}{m!(n-m)!}p^m(1-p)^{n-m}$$

In diffusion we expect the number of steps n and the product np to be very large within the measurement time. Let's look at an example. The velocity v of a particle is given by

$$v = \sqrt{\langle v^2 \rangle} = \sqrt{\frac{kT}{m}}$$

where k is the Boltzmann constant, T is temperature and m the mass of the particle. For a protein of 60 kDa at room temperature this results in a velocity of about 6 m s^{-1}. With a diffusion coefficient $D = 10^{-12}$ m^2 s^{-1} one can calculate the step size δ and number of steps n taken per second.

$$\delta = \frac{2D}{v} \approx 10^{-13} m$$

$$n = \frac{v}{\delta} \approx 10^{13}$$

With n on the order of 10^{13} and $p = 0.5$, both n and np are large. In this case the binomial distribution is well approximated by a Gaussian function.

$$P(m)dm = \frac{1}{(2\pi\sigma^2)^{1/2}} e^{-\frac{(m-\langle m \rangle)^2}{2\sigma^2}} dm$$

We have already shown above that the mean position for diffusion is 0 and the mean squared position is $2Dt$. Therefore, the variance is equal to $2Dt$. Putting this into the equation of the Gaussian function we arrive at:

$$P(x)dx = \frac{1}{(4\pi Dt)^{1/2}} e^{-\frac{x^2}{4Dt}} dx$$

Note: For $\lim\limits_{t\to 0} \frac{1}{(4\pi Dt)^{1/2}} e^{-\frac{x^2}{4Dt}} = \delta(x)$, i.e. $P(x)$ converges to the Dirac delta function as expected.

So, the diffusion propagator for 1D is

$$\langle C(x, t)C(x', t + \tau)\rangle = \frac{\langle C \rangle}{(4\pi Dt)^{1/2}} e^{-\frac{(x-x')^2}{4Dt}}$$

For 2D, with $\vec{r} = \begin{pmatrix} x \\ y \end{pmatrix}$, it is

$$\langle C(\vec{r}, t)C(\vec{r}', t + \tau)\rangle = P(x)P(y) = \frac{\langle C \rangle}{4\pi Dt} e^{-\frac{(x-x')^2+(y-y')^2}{4Dt}}$$

And for 3D, with $\vec{r} = \begin{pmatrix} x \\ y \\ z \end{pmatrix}$, it is

$$\langle C(\vec{r}, t)C(\vec{r}', t + \tau)\rangle = P(x)P(y)P(z) = \frac{\langle C \rangle}{(4\pi Dt)^{3/2}} e^{-\frac{(x-x')^2+(y-y')^2+(z-z')^2}{4Dt}}$$

Note: It is enough to show that the correlation of the signal, $\langle C(\vec{r}, t)C(\vec{r}', t + \tau)\rangle$, and the correlation of the fluctuations, $\langle \delta C(\vec{r}, t)\delta C(\vec{r}', t + \tau)\rangle$, have the same functional dependence save a constant (see equation (2.30)).

Another, mathematically more rigorous, way to derive the diffusion propagator comes from the solution of the diffusion equation [6]:

$$\frac{\partial C(\vec{r}, t)}{\partial t} = D\nabla^2 C(\vec{r}, t)$$

To solve this equation, we take the Fourier transform to simplify the equation. For the left-hand-side (*lhs*) it is easy to show that

$$lhs: \int \frac{\partial C(\vec{r}, t)}{\partial t} e^{-i\vec{v}\vec{r}} d\vec{r} = \frac{\partial \int C(\vec{r}, t) e^{-i\vec{v}\vec{r}} d\vec{r}}{\partial t} = \frac{\partial \tilde{C}(\vec{v}, t)}{\partial t}$$

where we have used the fact that the Fourier transform for $C(\vec{r}, t)$ is defined as

$$\tilde{C}(\vec{v}, t) = \int C(\vec{r}, t) e^{-i\vec{v}\vec{r}} d\vec{r}$$

For the Fourier transform of right-hand side of the diffusion equation, we have to use the fact that for Fourier transforms

$$f(\nu) = \int f(x)e^{-ix\nu}dx \quad \text{and} \quad \lim f(x) = 0$$

it can be shown that the Fourier transform of the derivative has a simple relation

$$\int \frac{\partial^n f(x)}{\partial x^n} e^{-ix\nu}dx = (i\nu)^n f(\nu)$$

Using this relation it follows that

$$\begin{aligned} rhs &: \int D\nabla^2 C(\vec{r}, t)e^{-i\vec{\nu}\vec{r}}d\vec{r} \\ &= D\int \nabla^2 C(\vec{r}, t)e^{-i\vec{\nu}\vec{r}}d\vec{r} \\ &= D(\mathrm{i}\nu)^2 \int C(\vec{r}, t)e^{-i\vec{\nu}\vec{r}}d\vec{r} \\ &= -D\nu^2 \tilde{C}(\vec{\nu}, t) \end{aligned}$$

where we use the assumption that $\lim_{\vec{r}\to\infty} C(\vec{r}, t) = 0$. Putting the left- and right-hand side together again we obtain:

$$\frac{\partial \tilde{C}(\vec{\nu}, t)}{\partial t} = -D\nu^2 \tilde{C}(\vec{\nu}, t)$$

This can now be easily solved and you can verify the result by putting it back into the differential equation:

$$\tilde{C}(\vec{\nu}, t) = \tilde{C}(\vec{\nu}, 0)e^{-D\nu^2 t}$$

With the solution of the diffusion equation we can now proceed to determine the diffusion propagator. In the following we denote a Fourier transform in respect to a variable x as $\mathcal{F}_x\{\ \}$, and the inverse Fourier transform as $\mathcal{F}_x^{-1}\{\ \}$. And as we assume stationarity, we can write the concentration correlation as

$$\langle C(\vec{r}, t)C(\vec{r}', t + \tau)\rangle = \langle C(\vec{r}, 0)C(\vec{r}', \tau)\rangle$$

We can insert the solution of the diffusion equation into the concentration correlation

$$\langle C(\vec{r}, 0)C(\vec{r}', \tau)\rangle = \langle C(\vec{r}, 0)\mathcal{F}_{\vec{r}'}^{-1}\{\tilde{C}(\vec{\nu}, t)\}\rangle$$

Fourier Transforms and averaging can be exchanged, leading to

$$\begin{aligned} \langle C(\vec{r}, 0)C(\vec{r}', \tau)\rangle &= \mathcal{F}_{\vec{r}'}^{-1}\{\langle C(\vec{r}, 0)\tilde{C}(\vec{\nu}, t)\rangle\} \\ &= \mathcal{F}_{\vec{r}'}^{-1}\{\langle C(\vec{r}, 0)\tilde{C}(\vec{\nu}, 0)e^{-D\nu^2 t}\rangle\} \\ &= \mathcal{F}_{\vec{r}'}^{-1}\{\langle C(\vec{r}, 0)\tilde{C}(\vec{\nu}, 0)\rangle e^{-D\nu^2 t}\} \end{aligned}$$

The transition to the last line is possible as the averaging is over space and the exponential function is a constant factor. In the next step we transform the function $\tilde{C}(\vec{\nu}, 0)$ back to the spatial domain.

$$\langle C(\vec{r}, 0)C(\vec{r}', \tau)\rangle = \mathcal{F}_{\vec{r}'}^{-1}\{\mathcal{F}_{\vec{\nu}}\{\langle C(\vec{r}, 0)\tilde{C}(\vec{r}', 0)\rangle\}e^{-D\nu^2 t}\}$$

We assume that the particles are independent from each other and do not interact. In this case the concentrations are only correlated at the self-same point and we have

$$\langle C(\vec{r}, 0)\tilde{C}(\vec{r}', 0)\rangle = \langle C\rangle\delta(\vec{r} - \vec{r}')$$

Note that we dropped the functional dependence of $\langle C \rangle$ as it is the spatial average at one particular time. For this function we can calculate the Fourier transform

$$\mathcal{F}_{\vec{v}}\{\langle C \rangle \delta(\vec{r} - \vec{r}')\} = \langle C \rangle e^{-iv\vec{r}}$$

So, we finally arrive at

$$\langle C(\vec{r}, 0)C(\vec{r}', \tau)\rangle = \mathcal{F}_{\vec{r}'}^{-1}\{\langle C \rangle e^{-iv\vec{r}} e^{-Dv^2 t}\}$$

This can be solved by integration and we obtain finally

$$\langle C(\vec{r}, 0)C(\vec{r}', \tau)\rangle = \frac{\langle C \rangle}{(4\pi Dt)^{3/2}} e^{-\frac{(\vec{r}-\vec{r}')^2}{4Dt}}$$

$$= \frac{\langle C \rangle}{(4\pi Dt)^{3/2}} e^{-\frac{(x-x')^2+(y-y')^2+(z-z')^2}{4Dt}}$$

5.2 General characteristics of the ACF for diffusion

Let's first look at the 2D case. Assuming diffusion is the only process causing fluctuations, the ACF is given by equation (5.8)

$$G(\tau) = \frac{1}{N}\left(1 + \frac{\tau}{\tau_D}\right)^{-1} + G_\infty \tag{5.8}$$

A couple of simple trends can be discerned from the equation. Note that for long times,

$$\lim_{\tau \to \infty} G(\tau) = G_\infty \tag{5.11}$$

We have seen before that we expect the ACF to converge to 1 for long times. And while this is in general true, how well the ACF converges to this value depends on how long one measures. For infinitely long times G_∞ should be equal to 1. However, as we measure a finite time, G_∞ will only be close to 1. Therefore, we use G_∞ as a fit parameter. For a simple organic dye in solution, with a laser intensity below the saturation level of the dye, and a measurement time of 30 s, the error on τ_D can be as low as 1% but in general should be <3%; G_∞ as just mentioned is typically within 1% of the value of 1.

For positive τ, the ACF is a monotonic descending function with its maximum at $\tau = 0$

$$G(0) = \frac{1}{N} + G_\infty \tag{5.12}$$

We therefore define the amplitude of the correlation function as

$$G_0 = G(0) - G_\infty = \frac{1}{N} \tag{5.13}$$

Experimentally, G_0 cannot be directly measured due to shot noise but has to be approximated, e.g. from the value of $G(\tau)$ for the smallest τ available, or obtained from a fit as we will discuss in chapter 7 on FCS artefacts.

In the case that $\tau = \tau_D$, then

$$G(\tau_D) = \frac{1}{N}\left(1 + \frac{\tau_D}{\tau_D}\right)^{-1} + G_\infty = \frac{1}{2N} + G_\infty \tag{5.14}$$

So τ_D is the time at which the correlation dropped to half its amplitude. This is a simple way to estimate the value of τ_D in a measurement. In the case of 3D diffusion, the ACF is given by

$$G(\tau) = \frac{1}{N}\left(1 + \frac{\tau}{\tau_D}\right)^{-1}\left(1 + \frac{\tau}{K^2\tau_D}\right)^{-1/2} + G_\infty \tag{5.15}$$

Again, for positive τ, the ACF is a monotonic descending function with its maximum at $\tau = 0$, with amplitude G_0 (equation (5.13)).

In the case that $\tau = \tau_D$ we have

$$G(\tau_D) = \frac{1}{N}\left(1 + \frac{\tau_D}{\tau_D}\right)^{-1}\left(1 + \frac{\tau_D}{K^2\tau_D}\right)^{-1/2} + G_\infty = \frac{1}{2N\sqrt{1 + K^{-2}}} + G_\infty \tag{5.16}$$

With $K \sim 5$ being a typical value in confocal FCS we obtain

$$G(\tau_D) = \frac{1}{2N\sqrt{1 + K^{-2}}} + G_\infty \approx \frac{1}{2.04\ N} + G_\infty \tag{5.17}$$

So, the amplitude drop at τ_D is still very close to half the amplitude of the ACF (figure 5.2). By comparing equations (5.15) and (5.16), we can see that for increasing K, the 3D model converges to the 2D model:

$$\lim_{K\to\infty} \frac{1}{2N\sqrt{1 + K^{-2}}} + G_\infty = \frac{1}{2N} + G_\infty \tag{5.18}$$

A problem in FCS however, is that often the exact value of the size of the observation volume is not known. This is partly because often the PSF of the microscope has not been measured, or the alignment or the settings, e.g. the pinhole size in a confocal system, were changed, thus also changing the effective observation volume. Therefore, we cannot directly calculate the diffusion coefficient from an FCS experiment. However, we note that the product of diffusion coefficient D and diffusion time τ_D is constant (equation (5.6)), assuming the alignment of the system does not change:

$$D\tau_D = \frac{w_0^2}{4} = \text{const} \tag{5.19}$$

Therefore, we can relate measurements of different molecules to each other. It follows from equation (5.19) that

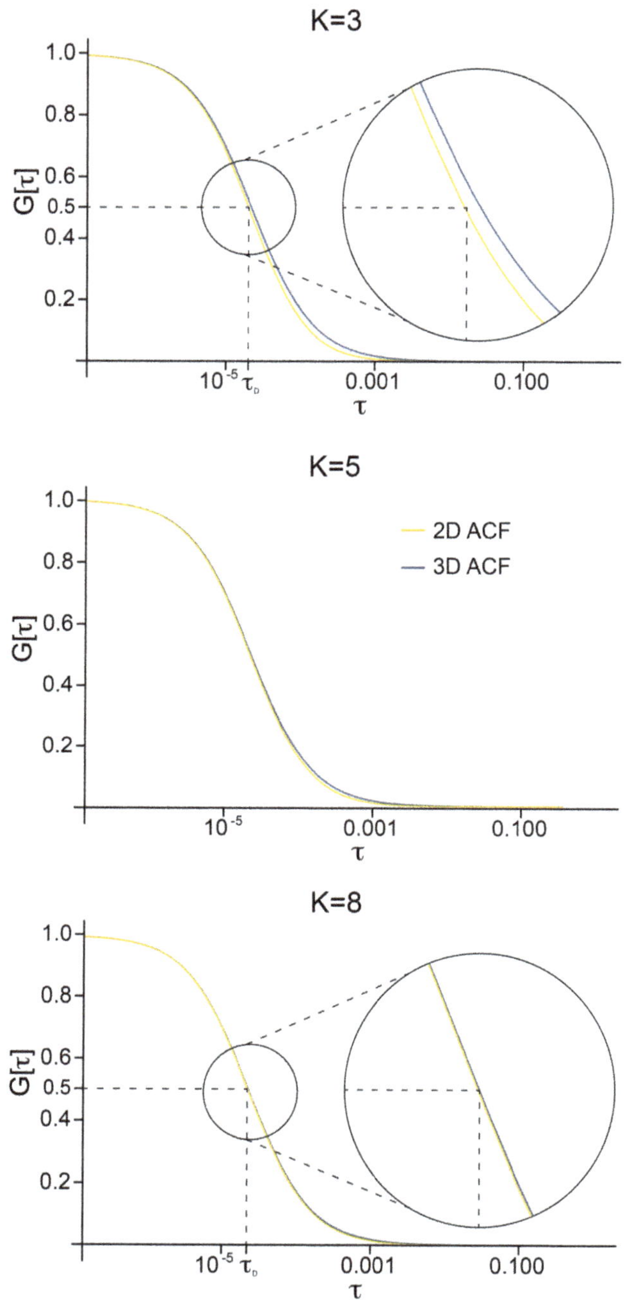

Figure 5.2. Comparison of one-component models in 2D (yellow) and 3D (blue) for particles with the same diffusion coefficient. For the 3D model the structure factor, K, is varied between 3 and 8, a typical range found in well-aligned systems. The differences between 2D and 3D models is quite small even for $K = 3$, and is almost negligible for $K = 8$.

$$D_1\tau_{D1} = D_2\tau_{D2} \tag{5.20a}$$

and thus

$$D_2 = \frac{\tau_{D1}}{\tau_{D2}} D_1 \tag{5.20b}$$

If we have a molecule of known diffusion coefficient D_1, we can calibrate the FCS system by measuring its diffusion time τ_{D1}. By measuring the diffusion time of an unknown molecule τ_{D2}, we can calculate its diffusion coefficient D_2 via equation (5.20b). In confocal FCS experiments, a good habit is to measure a calibration sample before and after the experiments to be sure the alignment did not change and to be able to extract absolute diffusion coefficients.

Furthermore, from the Stokes–Einstein relation (equation (5.4)), we know that the diffusion coefficient is inversely proportional to the hydrodynamic radius, r_h. As the hydrodynamic radius of a (spherical) molecule is proportional to the cubic root of the molecular mass, we can measure molecular weights by FCS:

$$D \sim \frac{1}{r_h} \approx \frac{1}{\sqrt[3]{M}} \tag{5.21a}$$

And here, too, by measuring a molecule of known mass we can then determine the mass of any unknown molecule by comparison, if we can assume they have similar densities:

$$\frac{D_1}{D_2} = \frac{\tau_{D2}}{\tau_{D1}} = \sqrt[3]{\frac{M_2}{M_1}} \tag{5.21b}$$

However, because of the cube root dependence on mass, any error in τ_D and thus r_h will result in a three times larger error in mass. So a 10% error in the diffusion time

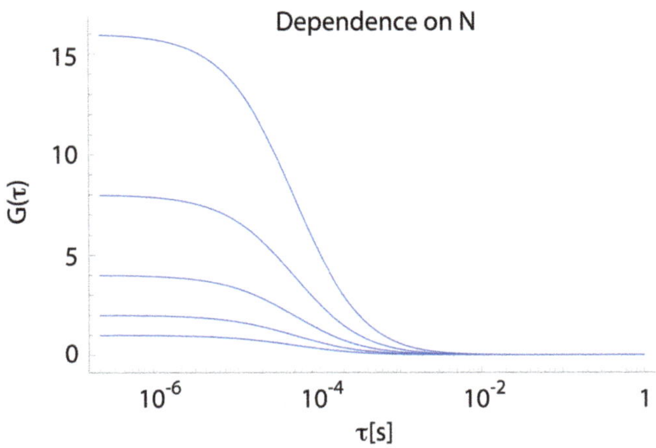

Figure 5.3. Dependence of the ACF on the number of particles. Shown is a 3D diffusion model with a single component with diffusion time τ_D = 50 μs. The bottom most ACF corresponds to N = 1 particle in the observation volume. Then the number of particles is halved for each following ACF resulting in N = 1, ½, ¼, 1/8, and 1/16 particles in the observation volume (figure 5.4).

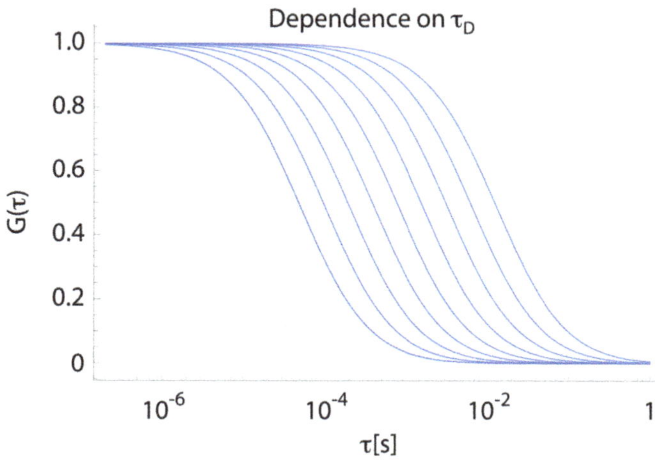

Figure 5.4. Dependence of the ACF on the diffusion time τ_D. Shown is a 3D diffusion model with a single component with $N = 1$. The leftmost ACF corresponds to $\tau_D = 50$ μs. The diffusion time is then increased by a factor 2 for each further ACF, the last one having a $\tau_D = 12.8$ ms.

results in a 30% error in the estimate of the mass. Furthermore, the approximation of molecules and particles as spherical objects has to be justified and might not always be appropriate. For non-spherical particles correction factors have to be calculated [7]. The dependence of the ACF on N and τ_D is shown in figures 5.3 and 5.4, respectively.

5.3 Including multiple particles

The inclusion of more than one diffusive species is straightforward for non-interacting particles. In equation (5.2) for the ACF we have two expressions which are dependent on the concentration of the particles. In the denominator this was $\kappa^2\langle C(\vec{r}, t)\rangle^2$ and in the numerator $\kappa^2\langle C(\vec{r}, t)C(\vec{r}\,', t + \tau)\rangle$. This can be easily extended if we introduce two different concentrations of particles, $C_1(\vec{r}, t) = fC(\vec{r}, t)$ and $C_2(\vec{r}, t) = (1 - f)C(\vec{r}, t)$, with their own brightness factors, κ_1 and κ_2, and f and $(1 - f)$ describing the mole fraction of species 1 and 2. The expression in the denominator is then extended to

$$\langle\kappa_1 C_1(\vec{r}, t) + \kappa_2 C_2(\vec{r}, t)\rangle^2 = \langle\kappa_1 fC(\vec{r}, t) + \kappa_2(1 - f)C(\vec{r}, t)\rangle^2 \tag{5.22}$$

The extension to multiple particles for the expression in the numerator is

$$\begin{aligned}&\langle(\kappa_1 C_1(\vec{r}, t) + \kappa_2 C_2(\vec{r}, t))(\kappa_1 C_1(\vec{r}\,', t + \tau) + \kappa_2 C_2(\vec{r}\,', t + \tau))\rangle\\ &= \kappa_1^2\langle C_1(\vec{r}, t)C_1(\vec{r}\,', t + \tau)\rangle + \kappa_2^2\langle C_2(\vec{r}, t)C_2(\vec{r}\,', t + \tau)\rangle\end{aligned} \tag{5.23}$$

The transition to the right side is possible as all cross-terms vanish as the particles are independent. Using the diffusion propagator as derived in math box 5.3 we obtain

$$\frac{\kappa_1^2\langle C_1\rangle}{(4\pi D_1 t)^{1/2}}e^{-\frac{(x-x')^2}{4D_1 t}} + \frac{\kappa_2^2\langle C_2\rangle}{(4\pi D_2 t)^{1/2}}e^{-\frac{(x-x')^2}{4D_2 t}} \tag{5.24a}$$

By replacing $\langle C_1 \rangle$ with $f\langle C \rangle$ and $\langle C_2 \rangle$ with $(1 - f)\langle C \rangle$ we obtain

$$\kappa_1^2 f \frac{\langle C \rangle}{(4\pi D_1 t)^{1/2}} e^{-\frac{(x-x')^2}{4D_1 t}} + \kappa_2^2 (1 - f) \frac{\langle C \rangle}{(4\pi D_2 t)^{1/2}} e^{-\frac{(x-x')^2}{4D_2 t}} \tag{5.24b}$$

Therefore, the expression for the ACF is just the sum of two expressions for single particles with different brightnesses, mole fractions and diffusion coefficients

$$G(\tau) = \frac{1}{N(\kappa_1 f + \kappa_2(1 - f))^2} \left[\kappa_1^2 f \left(1 + \frac{\tau}{\tau_{D1}}\right)^{-1} \left(1 + \frac{\tau}{K^2 \tau_{D1}}\right)^{-1/2} + \kappa_2^2 (1 - f) \left(1 + \frac{\tau}{\tau_{D2}}\right)^{-1} \left(1 + \frac{\tau}{K^2 \tau_{D2}}\right)^{-1/2} \right] + G_\infty \tag{5.25}$$

The amplitude of the ACF is now no longer simply the inverse of the number of particles as can be seen by setting $\tau = 0$

$$G(0) = \frac{\kappa_1^2 f + \kappa_2^2 (1 - f)}{N(\kappa_1 f + \kappa_2(1 - f))^2} + G_\infty \tag{5.26}$$

The contribution to the amplitude of the different species is proportional to the square of their brightness and thus if one wants to measure such a system to determine absolute concentrations, knowledge of the brightness of the different components is required. The extension to n different molecules is straightforward and yields

$$G(\tau) = \frac{\sum_{i=1}^{n} \kappa_i^2 f_i \left(1 + \frac{\tau}{\tau_{Di}}\right)^{-1} \left(1 + \frac{\tau}{K^2 \tau_{Di}}\right)^{-1/2}}{N\left(\sum_{i=1}^{n} \kappa_i f_i\right)^2} + G_\infty \tag{5.27}$$

And for the amplitude

$$G(0) = \frac{\sum_{i=1}^{n} \kappa_i^2 f_i}{N\left(\sum_{i=1}^{n} \kappa_i f_i\right)^2} + G_\infty \tag{5.28a}$$

Often the absolute brightness of the different particles is not known but one has a good estimate of the relative brightness. In this case one can introduce the relative brightness factor $q_i = \kappa_i / \kappa_1$. This does not fundamentally change the equation, except that the value of $q_1 = 1$.

$$G(0) = \frac{\sum_{i=1}^{n} q_i^2 f_i}{N\left(\sum_{i=1}^{n} q_i f_i\right)^2} + G_\infty \tag{5.28b}$$

Figure 5.5 shows how the two-component ACF changes with difference in diffusion time and mole fraction. Figure 5.6 depicts the same in the presence of noise to

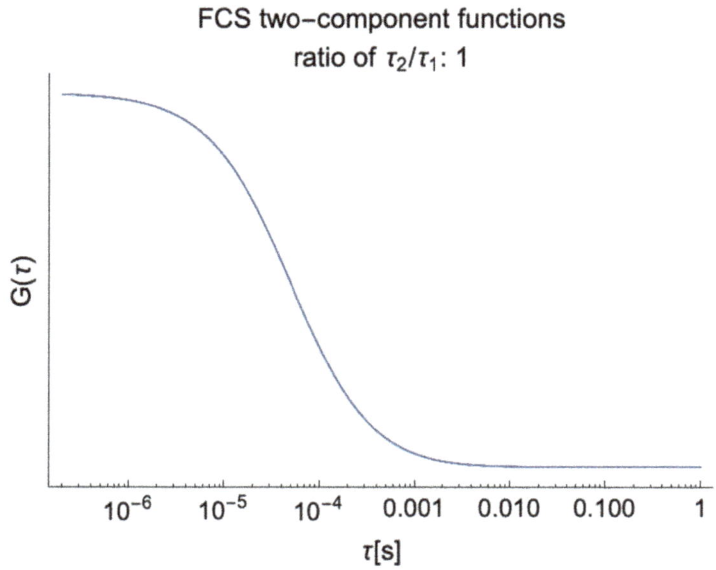

Figure 5.5. 3D ACF with two components with different diffusion times τ_1 and τ_2 but equal brightness. The value of τ_1 = 50 μs. The value of τ_2 varies between 50 μs ($\tau_2/\tau_1 = 1$) and 4.8 ms ($\tau_2/\tau_1 = 96$). Each panel contains all the ACFs for mole fractions of the second component, f_2, from 0–1 in steps of 0.1. In the first panel only 1 graph is seen as all particles have the same diffusion time ($\tau_2/\tau_1 = 1$). All subsequent panels contain 10 graphs with f_2 varying from 0 to 1 (figure 5.6). Animation available at https://iopscience.iop.org/book/978-0-7503-2080-1.

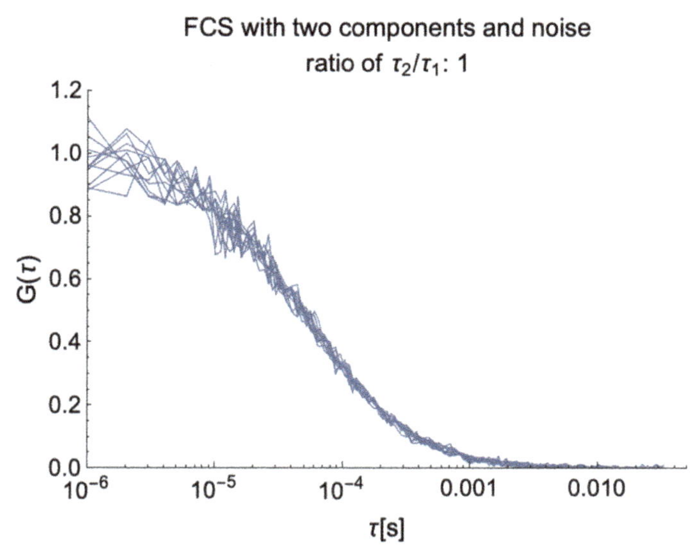

Figure 5.6. Influence of noise on two component 3D ACF functions. Shown are the same ACF graphs as in figure 5.5 but noise has been added. Animation available at https://iopscience.iop.org/book/978-0-7503-2080-1.

demonstrate how difficult it can be to differentiate two components, especially when the mole fraction of one component is small, and the diffusion times are close.

5.4 Anomalous diffusion

In math box 5.3, we have shown that the mean squared displacement (MSD) for a 1D random walk is a linear function of time:

$$\langle x_n^2 \rangle = \text{MSD} = 2Dt \tag{5.29}$$

This result can be extended to more dimensions, assuming the different directions are independent and separable:

$$\text{MSD} = 2mDt \tag{5.30}$$

where m is the number of dimensions. However, in complex systems, the MSD can deviate from the linear dependence on time (figure 5.7). In that case, the diffusion

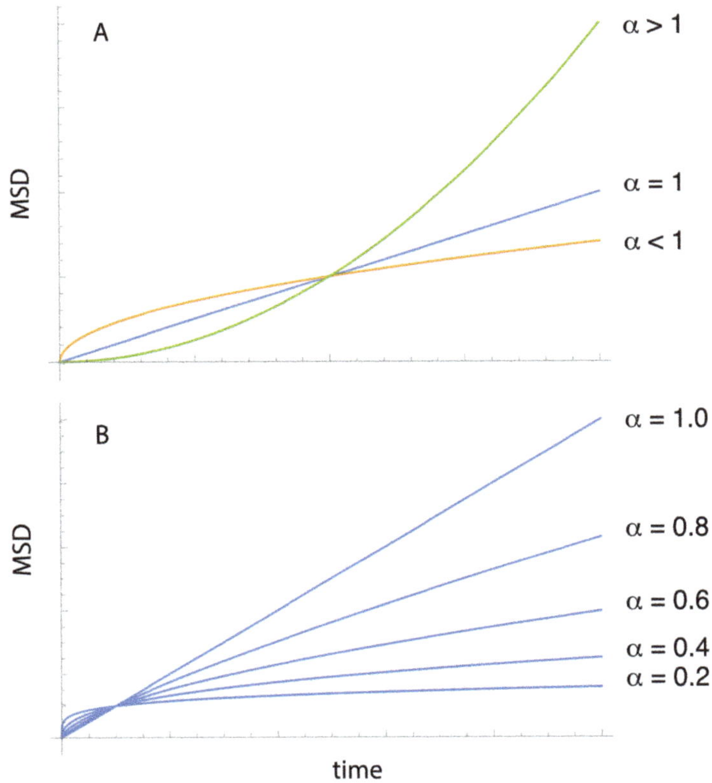

Figure 5.7. Mean square displacement (MSD) plots in dependence of α. (A) A comparison of free diffusion (α = 1), super-diffusion (α > 1) as, e.g. found in active transport, and anomalous sub-diffusion (α < 1) as can be found, e.g., in crowded media. Free diffusion shows a straight line in the MSD plot. For values of α > 1, the plot is convex, showing an MSD larger than expected from free diffusion. For values of α < 1, the plot is concave, showing an MSD smaller than expected from free diffusion. (B) A plot for varying α for the case of anomalous sub-diffusion.

coefficient is replaced by a transport coefficient, Γ, that has variable units, determined by the dependence on time, as denoted by the coefficient α [8]:

$$\mathrm{MSD} = \Gamma t^{\alpha} \tag{5.31}$$

However, reporting Γ can lead to confusion as the transport coefficient has variable units of m^2/s^{α}. It was therefore suggested that instead of reporting the transport coefficient, we report either a time-averaged diffusion coefficient

$$D_{\mathrm{app}}(t) = \frac{1}{2m}\Gamma t^{\alpha-1} \tag{5.32}$$

with the MSD given by

$$\mathrm{MSD} = 2mD_{\mathrm{app}}(t)t \tag{5.33}$$

or the instantaneous diffusion coefficient, defined as the diffusion coefficient at the particular measurement time, and calculated over the derivative of the mean-squared displacement with time (figure 5.8) [9].

$$D_t(t) = \frac{1}{2m}\frac{\partial \mathrm{MSD}}{\partial \mathrm{t}} = \frac{\alpha}{2m}\Gamma t^{\alpha-1} \tag{5.34}$$

Using the instantaneous diffusion coefficient, we obtain for the 3D anomalous diffusion propagator

$$\langle C(\vec{r},0)C(\vec{r}',\tau)\rangle = \frac{\langle C\rangle}{(2/3\pi\Gamma t^{\alpha})^{3/2}}e^{-\frac{(\vec{r}-\vec{r}')^2}{2/3\Gamma t^{\alpha}}} \tag{5.35}$$

And following our procedure from math box 5.1, we can evaluate the integral for a confocal volume to obtain the final autocorrelation function:

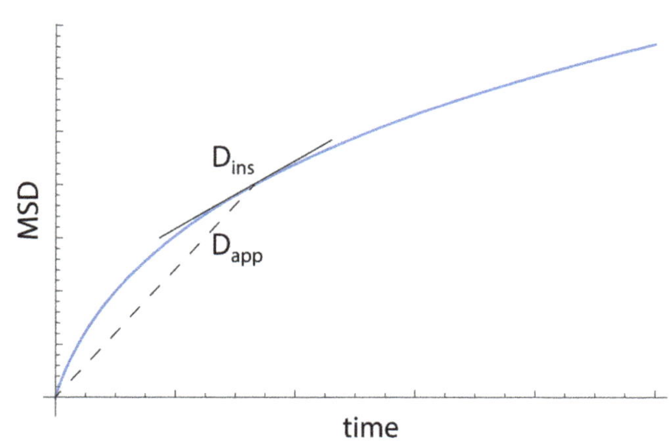

Figure 5.8. Apparent and instantaneous diffusion coefficients. The apparent diffusion coefficient is the diffusion coefficient that would have been required for free diffusion to produce a particular MSD at a particular time. This is shown by the dashed line that represents the freed diffusion MSD for D_{app}. The instantaneous diffusion coefficient is given by the slope of the MSD plot.

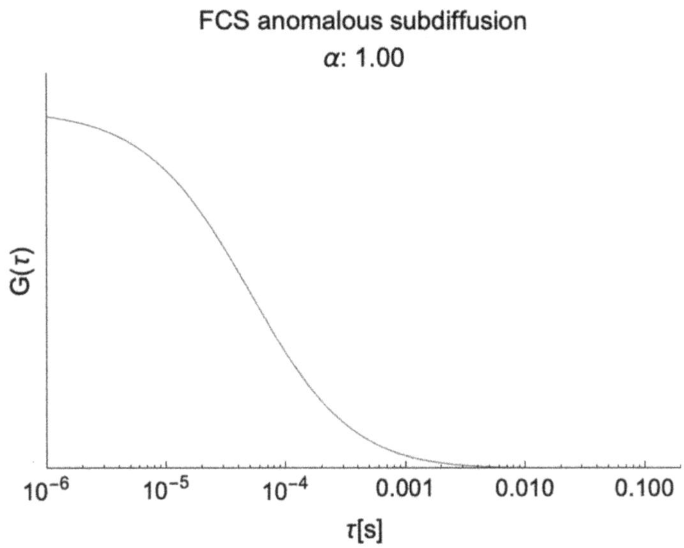

Figure 5.9. 3D anomalous sub-diffusion ACF. The diffusion times $\tau = 50$ μs. The value of α varies between 0 and 1 in steps of 0.1. Animation available at https://iopscience.iop.org/book/978-0-7503-2080-1.

$$G(\tau) = \frac{1}{\langle C\rangle \pi^{3/2} K w_0^3}\left(1 + \left(\frac{\tau}{\tau_D}\right)^{\alpha}\right)^{-1}\left(1 + \frac{1}{K^2}\left(\frac{\tau}{\tau_D}\right)^{\alpha}\right)^{-1/2} + 1 \tag{5.36}$$

and

$$\tau_D = \frac{\alpha w_0^2}{4D_t} = \frac{w_0^2}{2/3\Gamma t^{\alpha-1}} \tag{5.37}$$

The interconnection between MSD and ACF has been used, e.g., to investigate the kinetics of the random motion within polymer coils and anomalous diffusion [10–12]. In this book we will not follow up further on anomalous diffusion and instead refer the interested reader to the vast existing literature for random walks and diffusion, of which we can only provide some starting points [13–18]. It should be mentioned however, that it is often difficult to differentiate a fit with an anomalous sub-diffusion model and a two or more component model of free diffusion (figure 5.9). And thus extra information should be obtained to confirm the existence of anomalous sub-diffusion [19, 20].

5.5 Flow

If in addition to diffusion, also active transport or a flow component is present, then the ACF has to be modified. For flow within the *xy*-plane the ACF can be derived in the same way as shown before but by replacing the propagator for diffusion by a propagator that includes both diffusion and transport (math box 5.4). For an ACF in 2D and flow within the plane this results in

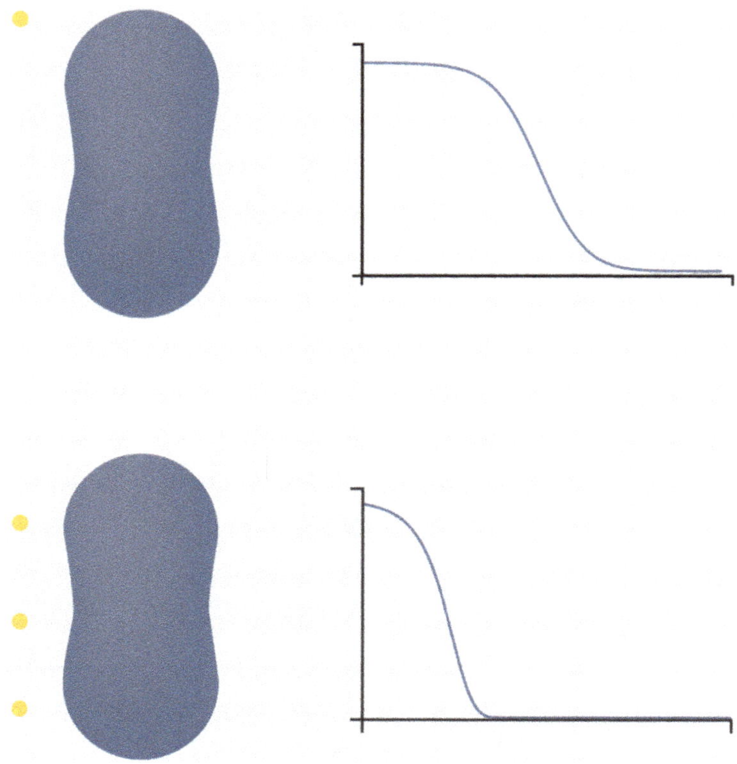

Figure 5.10. Comparison ACF for diffusion and flow/scanning. In the case of diffusion (upper panel), the various particles will have different trajectories through the observation volume. Some stay only a short time, some are detected for a longer time. Therefore, the correlation function will have a slope covering these different possibilities. In the case of flow (lower panel), the particles are driven through the observation volume and will spend a shorter and more uniform time in the observation volume. Therefore, the ACF will become steeper and will be narrower. Animation available at https://iopscience.iop.org/book/978-0-7503-2080-1.

$$G(\tau) = \frac{1}{\langle C\rangle \pi w_0^2}\left(1 + \frac{4D\tau}{w_0^2}\right)^{-1} e^{-\frac{(v_x^2+v_y^2)\tau^2}{4D\tau+w_0^2}} + G_\infty \tag{5.38}$$

The ACF including flow will decay faster than the ACF describing diffusion only (figure 5.10). This is evident from equation (5.38), which contains an extra decaying exponential function, but is also easily understood. The decay of the ACF in the case of diffusion takes account of all possible paths that particles take through the observation volume. These paths can be short or long and thus the ACF has a shape that has a plateau at short times, indicating the minimum time all particles stay in the observation volume. Then the ACF decays as particles with a short path in the observation volume will leave. Finally, even the particles that stay longest will leave the observation volume and the ACF converges to G_∞.

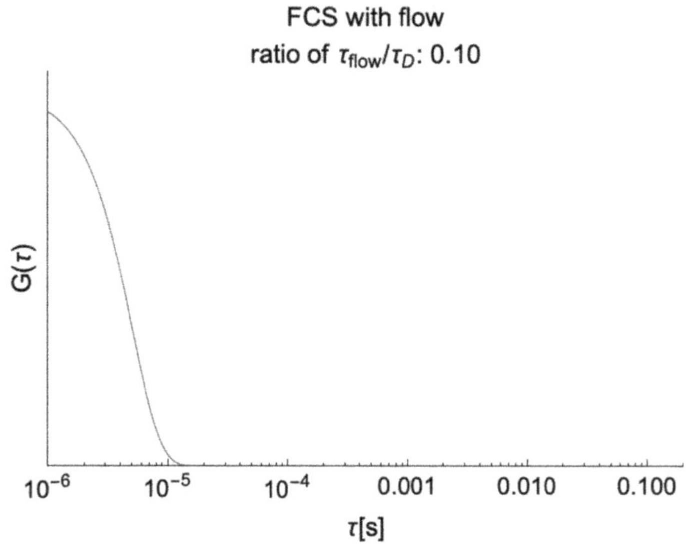

Figure 5.11. 2D ACF that exhibits both, diffusion and flow. The diffusion times $\tau_D = 50$ µs. The flow time of the particle though the observation area varies between $\tau_{flow} = 10$ µs to 1 ms. Depending on which of the two times is faster the ACF will be either more flow- or more diffusion-like. Note that the flow will increase the steepness of the ACF decay. Animation available at https://iopscience.iop.org/book/978-0-7503-2080-1.

Math box 5.4

The correlation function for flow and diffusion in 2D

The derivation of flow in FCS follows the same strategy as for diffusion. First we determine the propagator. The differential equation that describes the system includes now a flow part

$$\frac{\partial C(\vec{r}, t)}{\partial t} = D\frac{\partial^2 C(\vec{r}, t)}{\partial x^2} - \vec{v}C(\vec{r}, t)$$

However, instead of rederiving the full propagator as done in math box 5.3, we provide a simple argument of how the propagator needs to be modified. Flow is moving the whole samples by a distance $\vec{v}\tau$, with $\vec{v}$ being the velocity. So the propagator for the 2D case is just translated by that amount and we obtain

$$\langle C(\vec{r}, t)C(\vec{r}', t + \tau)\rangle = P(x)P(y) = \frac{\langle C\rangle}{4\pi Dt}e^{-\frac{(x-x'-v_x\tau)^2+(y-y'-v_y\tau)^2}{4Dt}}$$

We then insert the propagator into the definition of the ACF and perform the integration

$$G(\tau) = \frac{\iiiint e^{\frac{-2(x^2+x'^2)}{w_0^2}} e^{\frac{-2(y^2+y'^2)}{w_0^2}} \frac{e^{-\frac{(x-x'-v_x\tau)^2+(y-y'-v_y\tau)^2}{4Dt}}}{4\pi Dt} dxdx'dydy'}{\langle C\rangle\left(\iint e^{\frac{-2x^2}{w_0^2}} e^{\frac{-2y^2}{w_0^2}} dxdy\right)^2} + G_\infty$$

resulting in

$$G(\tau) = \frac{1}{\langle C\rangle \pi w_0^2}\left(1 + \frac{4D\tau}{w_0^2}\right)^{-1} e^{-\frac{(v_x^2+v_y^2)\tau^2}{4D\tau + w_0^2}} + G_\infty$$

Here we used again G_∞ as the convergence value for long times.

When flow is present particles are driven through the observation volume. This will lead to a shortening of the plateau as particles will stay a shorter time in the observation volume and a steepening of the slope as the times the particles stay in the observation volume become more uniform as all particles are moved through the observation volume with the same speed. ACFs for various flow speed are shown in figure 5.11.

5.6 Including multiple processes

In the case of fluorescence there are a number of other processes that can lead to fluorescence fluctuations in addition to diffusion in and out of the observation volume, and thus will give rise to other features in the ACF. In the case of FCS the main processes are antibunching, rotational diffusion, and photophysical fluctuations. Photophysical fluctuations lead to fluorophore blinking and can have various reasons, including triplet state transitions, isomerization, or protonation-deprotonation dynamics [21–23]. In the following we will discuss each of these fluctuations.

Antibunching is a quantum mechanical effect, which is best pictured by realizing that a fluorophore can only emit one photon at a time. The time between photons that are emitted is related to the lifetime of the fluorophore, which is typically on the order of 1–10 ns. Therefore, as correlation times decrease below the lifetime it is less and less likely that a second photon is detected. Therefore, at very short correlation times τ the photons are more likely to stem from two different molecules and are not correlated [24] leading to a contribution of the form

$$g_a(\tau) = 1 - e^{-\tau/\tau_A} \tag{5.39}$$

The function $g_a(\tau)$ converges to 0 for $\tau \to 0$, as one fluorophore cannot emit two photons simultaneously. However, if a particle carries more than one fluorophore then the detection of two simultaneous photons from a single particle becomes possible, and the probability of such an event increases the more fluorophores are bound to the particle. For N_b particles the function will take the form

$$g_a(\tau) = 1 - \frac{1}{N_b} e^{-\tau/\tau_A} \tag{5.40}$$

However, the accurate determination of N_b is not trivial and requires a high signal-to-noise ratio. For more information on antibunching in FCS and fluctuation spectroscopy see the literature [25–28].

At somewhat longer times, rotational diffusion can lead to blinking events. A fluorophore has an absorption and an emission dipole. Both can be exploited to measure rotational motion of the fluorophore. When using linear polarized light for excitation, light can be absorbed by the fluorophore when the polarization of light has a component along the absorption dipole. The more the polarization is aligned with the dipole the likelier will be the absorption. The strength of absorption, and consequently the emission, will change with the orientation of the fluorophore. Conversely, if the emission light is detected through polarizers, then the orientation of the emission dipole in relation to the polarizer will determine how much light will be detected. So using linear polarized light for excitation and/or detecting the light with polarization sensitivity will lead to fluctuations in the amount of detected light with molecular rotation.

Typical rotational diffusion times are on the order of tens of ns and thus a characteristic blinking of the fluorophore will be induced by its rotation on that time scale. The diffusion propagator and the derived autocorrelation function have been calculated by Berne and Pecora and solutions for rotational diffusion in FCS have been published by Ehrenberg and Kask [24, 29–31]. In the case that the rotational diffusion time is larger than the fluorescence lifetime but shorter than the time the molecule spends in the observation volume ($\tau_f < \tau_R < \tau_D$) it was shown that rotational diffusion can be modelled by an extra factor in the ACF ([2, 30] and references therein)

$$1 + Re^{-\tau/\tau_R} \tag{5.41}$$

With R being a normalization factor and

$$\tau_R = \frac{1}{6D_R} \tag{5.42}$$

with τ_R and D_R the rotational correlation time and rotational diffusion coefficient, respectively. The derivation for the rotational diffusion component is more complex because it requires in principle to take into account polarized excitation and/or detection and molecular rotations in a focused Gaussian laser beam in which the polarization has multiple components even for linear polarized lasers. However, in the fast time ranges on the order of nanoseconds, typically the signal-to-noise ratio is limited and a single component as described above is sufficient to model the rotational diffusion in FCS. Assuming spherical particles, the rotational diffusion coefficient is given by

$$D_R = \frac{kT}{8\pi\eta r_h^3} \tag{5.43}$$

with the parameters as defined for the translational diffusion coefficient (equation (5.4)). From the definition of D and D_R it is evident that the rotational diffusion coefficient is more sensitive to changes in particle size and thus molecular weight than the translational diffusion coefficient. Putting equation (5.43) into equation (5.42) we find that for spherical particles

$$\tau_R = \frac{8\pi\eta r_h^3}{6kT} \sim M_w \tag{5.44}$$

and thus

$$\frac{\tau_{R1}}{\tau_{R2}} = \frac{M_{w1}}{M_{w2}} \tag{5.45}$$

However, despite the fact that rotational diffusion is more sensitive to mass changes it is not often used, as the measurement of rotational correlations is more difficult than the measurement of translational diffusion. We can estimate a typical rotational correlation time for a protein, e.g. GFP, at room temperature ($kT \approx 4 \times 10^{-21} J$) in aqueous solution ($\eta \approx 1 mPa$) by assuming $r_h \approx 2.5\ nm$:

$$\tau_R = \frac{8\pi\eta r_h^3}{6kT} \approx 16\ ns$$

If this fast rotation is to be resolved, one needs to measure about 10 times faster than the rotational process, so one would need a time resolution of about 1–2 ns. There are two problems associated with that. First, fluorescence emission is a Poisson process. The variance of a Poisson process is proportional to the mean ($\sigma^2 = \mu$), which means the standard deviation is proportional to the square root of the mean ($\sigma = \sqrt{\mu}$). The signal-to-noise ratio can be estimated as the ratio of the mean to the standard deviation, i.e. by the value of $\mu/\sigma = \sqrt{\mu}$. As for shorter lag times, i.e. better time resolution, the mean decreases it follows that the signal-to-noise ratio also decreases. Second, as the signal-to-noise ratio increases only with $\sqrt{\mu}$, and the mean intensity μ is proportional to time, it means that we need to measure four times as long to increase the signal-to-noise ratio by a factor 2. Therefore, at better time resolution, one needs to measure much longer to obtain a sufficiently good signal-to-noise ratio to be able to determine the rotational correlation time accurately.

Finally, fluorophores when excited can transit into a so-called triplet state. Triplet states are due to (forbidden and thus rare) spin flips of the electron in the excited state [21]. As it has now the same spin orientation as the electron in the ground state it cannot return until another rare spin-flip occurs. Therefore, the lifetime of the triplet state can be very long and is for most fluorophores in the µs range. Molecules in different parts of the laser beam are exposed to different excitation intensities and have a different probability to reside in a triplet state. Therefore, different triplet dynamics are expected. However, as shown by Widengren, here as well, a single exponential is most of the time sufficient to fit the experimental data [32] (math box 5.5).

$$1 + \frac{f_T}{1 - f_T} e^{-\tau/\tau_T} \tag{5.46}$$

Here f_T represents the fraction of the molecules that reside in the triplet state and τ_T is the characteristic triplet time (see figures 5.12 and 5.13). Sometimes this is called bunching time as the photons seem to come in bunches when the molecule is in the

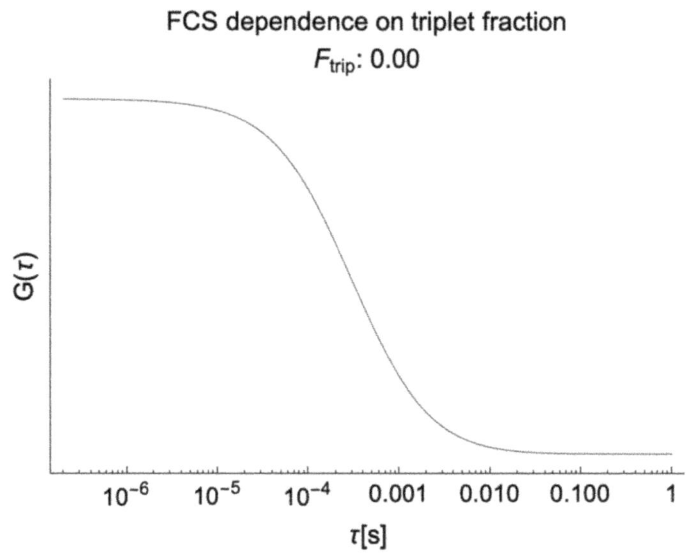

Figure 5.12. 3D ACF for free diffusion containing a triplet contribution. The diffusion time is $\tau_D = 300$ μs, and the triplet time is set to $\tau_D = 5$ μs. The triplet fraction, F_{trip}, varies from 0 to 0.8 in steps of 0.1. Animation available at https://iopscience.iop.org/book/978-0-7503-2080-1.

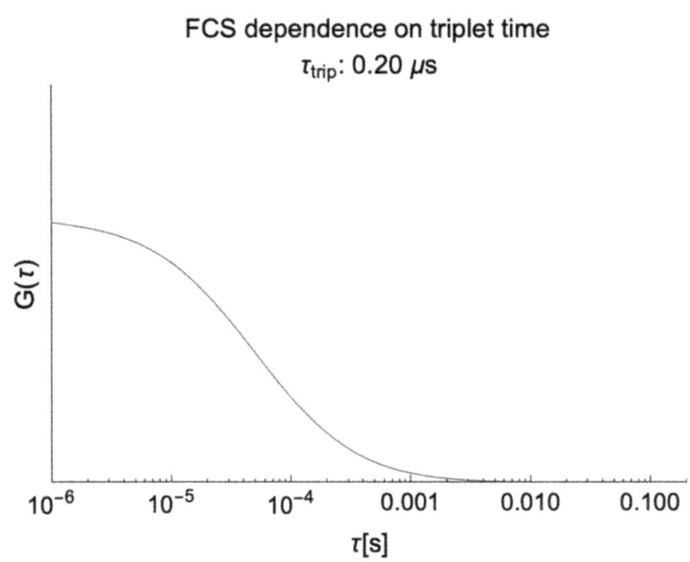

Figure 5.13. 3D ACF for free diffusion containing a triplet contribution. The diffusion time is $\tau_D = 50$ μs. The triplet fraction is set to $F_{trip} = 0.3$. The triplet time varies from 0.2 μs to 5.9 μs in steps of 0.3 μs. Animation available at https://iopscience.iop.org/book/978-0-7503-2080-1.

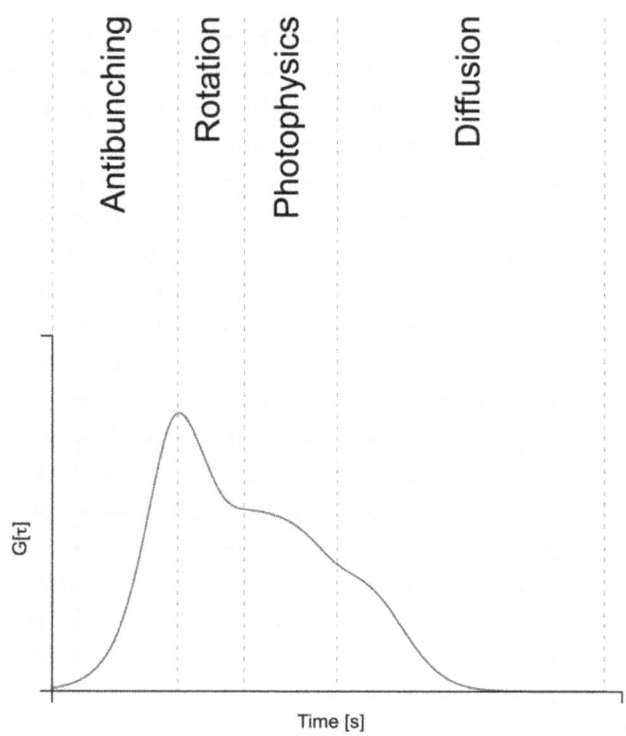

Figure 5.14. ACF according to the model in equation (5.47). The panels on the right show the different processes contributing to the ACF. First, due to antibunching photons will not come in groups at times below the fluorescence lifetime of the fluorophore as it can emit only one photon at a time. Second, as the molecule changes orientation due to rotational diffusion, its absorption dipole will change orientation with respect to the laser polarization. As the absorption changes with orientation, so will the fluorescence emission. Therefore, FCS will detect fluctuations in the emission on the characteristic time scale of the rotation of the fluorophore. Third, transitions to the triplet state will lead to intermittent dark states. Sometimes this is called photon bunching as photons are detected in 'bunches' between the dark states. Finally, diffusion will lead to fluorescence fluctuations. Note that the time axis of the correlation function is plotted on a logarithmic scale to be able to present all processes in one graph but also because the correlation function is calculated on a semi-logarithmic time scale (see chapter 4). Animation available at https://iopscience.iop.org/book/978-0-7503-2080-1.

singlet state with intermittent breaks when the molecule resides in the triplet state and does not emit fluorescence (figure 5.14).

Other processes, including isomerization, charge transfer, or protonation/de-protonation processes can be modelled with similar functions [23], each contributing an exponential term with their own time scales and amplitude factors. A good example showing the application of multiple processes in FCS is the study of GFP by Widengren *et al* [22].

This leads us to the final ACF including all these contributions for a single particle, which is for the 3D case:

$$G(\tau) = \frac{1}{N}\left[\left(1 + \frac{\tau}{\tau_D}\right)^{-1}\left(1 + \frac{\tau}{K^2\tau_D}\right)^{-1/2}\right](1 - e^{-\tau/\tau_A})(1 + Re^{-\tau/\tau_R}) \\ \times\left(1 + \frac{f_T}{1 - f_T}e^{-\tau/\tau_T}\right) + G_\infty \tag{5.47}$$

Note that if there are multiple particles each can have its own lifetime, rotational diffusion, and characteristic triplet times, that need to be attributed to each component. Furthermore, if the particles under investigation carry not one but N_b fluorophores then the equation can be written as

$$G(\tau) = \frac{1}{N}\left[\left(1 + \frac{\tau}{\tau_D}\right)^{-1}\left(1 + \frac{\tau}{K^2\tau_D}\right)^{-1/2}\right]\left(1 - \frac{1}{N_b}e^{-\tau/\tau_A}\right)(1 + Re^{-\tau/\tau_R}) \\ \times\left(1 + \frac{f_T}{N_b(1 - f_T)}e^{-\tau/\tau_T}\right) + G_\infty \tag{5.48}$$

Math box 5.5

The triplet state

For the derivation of the triplet dynamics we use a simple three state system with a ground state S_0, an excited state S_1, and a triplet state T (see figure MB 5.5.1). Excitation from S_0 to S_1 happens with rate k_{12}, which is proportional to the excitation intensity and the absorption cross-section of the fluorophore. Fluorescence emission from S_1 to S_0 happens with rate k_{21}. Finally, we assume that transition to the triplet state (intersystem crossing) happens only from S_1 to T with rate k_{23}, and de-excitation happens only from T to S_0 with rate k_{31}. The last transition represents phosphorescence. We neglect other transitions as they are very unlikely. For example, the transition from T to S_1, leading to delayed fluorescence, is very unlikely as it would require extra energy as the triplet state, T, is lower in energy than the excited singlet state, S_1.

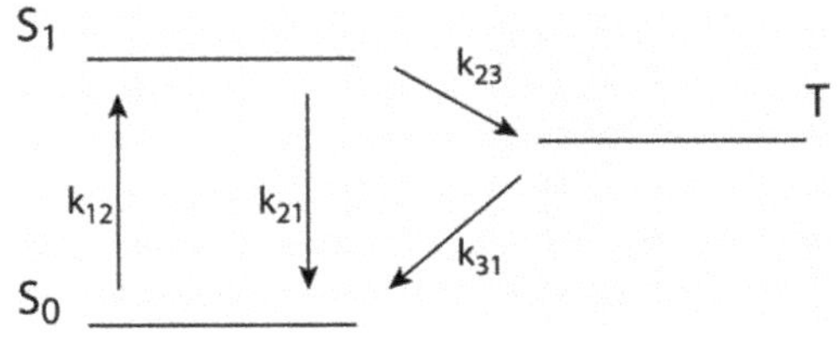

Figure MB 5.5.1. Three state system for the derivation of the triplet state dynamics.

For the three-state system in figure MB 5.5.1 the differential equations describing the change in the state populations can be written as:

$$\frac{dS_0(t)}{dt} = -k_{12}S_0(t) + k_{21}S_1(t) + k_{31}T(t)$$
$$\frac{dS_1(t)}{dt} = k_{12}S_0(t) - (k_{21} + k_{23})S_1(t)$$
$$\frac{dT(t)}{dt} = k_{23}S_1(t) - k_{31}T(t)$$

The system can be written in matrix form:

$$\frac{d}{dt}\begin{pmatrix} S_0(t) \\ S_1(t) \\ T(t) \end{pmatrix} = \begin{pmatrix} -k_{12} & k_{21} & k_{31} \\ k_{12} & -(k_{21}+k_{23}) & 0 \\ 0 & k_{23} & -k_{31} \end{pmatrix}\begin{pmatrix} S_0(t) \\ S_1(t) \\ T(t) \end{pmatrix}$$

The solution of such a system,

$$\vec{x}' = A\vec{x}$$

is given by the following general equation which contains the eigenvalues, λ_1, of the matrix as the factors in the exponential functions, and the eigenvectors, $\vec{\eta}^{(i)}$, and the constants c_i.

$$\vec{x} = c_1 e^{\lambda_1 t}\vec{\eta}^{(1)} + c_2 e^{\lambda_2 t}\vec{\eta}^{(2)} + c_3 e^{\lambda_3 t}\vec{\eta}^{(3)}$$

This system can be solved and the eigenvalues and eigenvectors can be calculated analytically. The values c_i will have to be determined using boundary conditions. The expressions for the eigenvectors and eigenvalues are quite complex, however. We therefore provide the calculations for the eigenvalues and their approximations only. The same calculations and approximations can be performed for the eigenvectors. The exact solutions for the eigenvalues are:

$$\lambda_1 = 0$$
$$\lambda_2 = \frac{1}{2}\left(\begin{array}{c} -k_{12} - k_{21} - k_{23} - k_{31} \\ \hline -\sqrt{(k_{12} + k_{21} + k_{23} + k_{31})^2 - 4(k_{12}k_{23} + k_{12}k_{31} + k_{21}k_{31} + k_{23}k_{31})} \end{array}\right)$$
$$\lambda_3 = \frac{1}{2}\left(\begin{array}{c} -k_{12} - k_{21} - k_{23} - k_{31} \\ \hline +\sqrt{(k_{12} + k_{21} + k_{23} + k_{31})^2 - 4(k_{12}k_{23} + k_{12}k_{31} + k_{21}k_{31} + k_{23}k_{31})} \end{array}\right)$$

The eigenvalues can be simplified assuming that the transition rates from S_1 to T and from T to S_0 are much smaller than the transition rate from S_1 to S_0, i.e. $k_{21} \gg k_{23}, k_{31}$.

$$\lambda_1 = 0$$
$$\lambda_2 = \frac{1}{2}\left(\begin{array}{c} -k_{12} - k_{21} - k_{23} - k_{31} \\ \hline -\sqrt{(k_{12} + k_{21})^2 - 4(k_{12}k_{23} + k_{12}k_{31} + k_{21}k_{31})} \end{array}\right)$$
$$\lambda_3 = \frac{1}{2}\left(\begin{array}{c} -k_{12} - k_{21} - k_{23} - k_{31} \\ \hline +\sqrt{(k_{12} + k_{21})^2 - 4(k_{12}k_{23} + k_{12}k_{31} + k_{21}k_{31})} \end{array}\right)$$

Here, we can replace the square root either by $(k_{12} + k_{21})$ for λ_2 and neglect terms k_{23} and $-k_{31}$ as they are small. Or, for λ_3, we replace the square root by the first two terms of its Taylor series $\sqrt{1+x} \approx 1 + \frac{x}{2}$. We need to develop the square root in this case as all terms k_{12} and k_{21} will cancel and we cannot neglect the other terms.

$$\begin{aligned}
\lambda_1 &= 0\\
\lambda_2 &= \frac{1}{2}(-k_{12} - k_{21} - (k_{12} + k_{21}))\\
\lambda_3 &= \frac{1}{2}\left(-k_{12} - k_{21} - k_{23} - k_{31} + k_{12} + k_{21} - \frac{2(k_{12}k_{23} + k_{12}k_{31} + k_{21}k_{23} + k_{23}k_{31})}{k_{12} + k_{21}}\right)\\
&\approx -\frac{k_{12}k_{23} + k_{12}k_{31} + k_{21}k_{23}}{k_{12} + k_{21}}
\end{aligned}$$

This can then be further simplified to

$$\begin{aligned}
\lambda_1 &= 0\\
\lambda_2 &= -(k_{12} + k_{21})\\
\lambda_3 &= -\left(k_{31} + \frac{k_{12}k_{23}}{k_{12} + k_{21}}\right)
\end{aligned}$$

The first eigenvector has the following exact solution:

$$\vec{\eta}^{(1)} = \begin{pmatrix} \dfrac{(k_{21} + k_{23})k_{31}}{k_{12}k_{23}} \\ \dfrac{k_{31}}{k_{23}} \\ 1 \end{pmatrix}$$

The terms for the second and third eigenvector are again more complex. However, they can be simplified using the same assumption as above, i.e. $k_{21} \gg k_{23}, k_{31}$.

$$\vec{\eta}^{(2)} = \begin{pmatrix} \dfrac{k_{12} + k_{21}}{k_{23}} \\ -\dfrac{k_{12} + k_{21}}{k_{23}} \\ 1 \end{pmatrix}$$

$$\vec{\eta}^{(3)} = \begin{pmatrix} -\dfrac{k_{23} + k_{31}}{2\,k_{23}} \\ -\dfrac{k_{23} - k_{31}}{2\,k_{23}} \\ 1 \end{pmatrix}$$

We can use now the initial condition at time 0 with all molecules in the ground state:

$$\begin{pmatrix} S_0(0) \\ S_1(0) \\ T(0) \end{pmatrix} = \begin{pmatrix} 1 \\ 0 \\ 0 \end{pmatrix}$$

to solve for the coefficients c_i

$$\begin{pmatrix} 1 \\ 0 \\ 0 \end{pmatrix} = c_1\vec{\eta}^{(1)} + c_2\vec{\eta}^{(2)} + c_3\vec{\eta}^{(3)}$$

The final solution is then

$$\begin{pmatrix} S_0(t) \\ S_1(t) \\ T(t) \end{pmatrix} = c_1\vec{\eta}^{(1)} + c_2\vec{\eta}^{(2)}e^{-(k_{12}+k_{21})t} + c_3\vec{\eta}^{(3)}e^{-\left(k_{31}+\frac{k_{12}k_{23}}{k_{12}+k_{21}}\right)t}$$

The solutions for these equations can again be simplified assuming that $k_{21} \gg k_{23}, k_{31}$, and developing terms of the form $\sqrt{1+x}$ by the first terms of the Taylor series of the square root, $1 + \frac{x}{2}$. Then we obtain the following approximation:

$$\begin{pmatrix} S_0(t) \\ S_1(t) \\ T(t) \end{pmatrix} = \begin{pmatrix} \dfrac{k_{21}k_{31}}{k_{21}k_{31} + k_{12}(k_{23} + k_{31})} \\ \dfrac{k_{12}k_{31}}{k_{21}k_{31} + k_{12}(k_{23} + k_{31})} \\ \dfrac{k_{12}k_{23}}{k_{21}k_{31} + k_{12}(k_{23} + k_{31})} \end{pmatrix}$$
$$+ \begin{pmatrix} \dfrac{k_{12}}{k_{12} + k_{21}} \\ -\dfrac{k_{12}}{k_{12} + k_{21}} \\ \dfrac{k_{12}k_{23}^2}{(k_{12} + k_{21})[k_{12}(k_{23} + k_{31}) + k_{21}k_{31}]} \end{pmatrix} e^{-(k_{12}+k_{21})t}$$
$$+ \begin{pmatrix} \dfrac{k_{12}k_{21}k_{23}}{(k_{12} + k_{21})[k_{12}(k_{23} + k_{31}) + k_{21}k_{31}]} \\ \dfrac{k_{12}^2k_{23}}{(k_{12} + k_{21})[k_{12}(k_{23} + k_{31}) + k_{21}k_{31}]} \\ -\dfrac{k_{12}k_{23}}{k_{12}(k_{23} + k_{31}) + k_{21}k_{31}} \end{pmatrix} e^{-\left(k_{31}+\frac{k_{12}k_{23}}{k_{12}+k_{21}}\right)t}$$

Written for the three equations separately, this is

$$\mathrm{S_0}(t) = \frac{k_{21}k_{31}}{k_{21}k_{31} + k_{12}(k_{23} + k_{31})} + \frac{k_{12}}{k_{12} + k_{21}} e^{-(k_{12}+k_{21})t}$$
$$+ \frac{k_{12}k_{21}k_{23}}{(k_{12} + k_{21})[k_{12}(k_{23} + k_{31}) + k_{21}k_{31}]} e^{-(k_{31}+\frac{k_{12}k_{23}}{k_{12}+k_{21}})t}$$

$$\mathrm{S_1}(t) = \frac{k_{12}k_{31}}{k_{21}k_{31} + k_{12}(k_{23} + k_{31})} - \frac{k_{12}}{k_{12} + k_{21}} e^{-(k_{12}+k_{21})t}$$
$$+ \frac{k_{12}^2k_{23}}{(k_{12} + k_{21})[k_{12}(k_{23} + k_{31}) + k_{21}k_{31}]} e^{-\left(k_{31}+\frac{k_{12}k_{23}}{k_{12}+k_{21}}\right)t}$$

$$\mathrm{T}(t) = \frac{k_{12}k_{23}}{k_{21}k_{31} + k_{12}(k_{23} + k_{31})}$$
$$+ \frac{k_{12}k_{23}(k_{23} + k_{31})}{(k_{12} + k_{21})[k_{12}(k_{23} + k_{31}) + k_{21}k_{31}]} e^{-(k_{12}+k_{21})t}$$
$$- \frac{k_{12}k_{23}}{k_{12}(k_{23} + k_{31}) + k_{21}k_{31}} e^{-\left(k_{31}+\frac{k_{12}k_{23}}{k_{12}+k_{21}}\right)t}$$

The first eigenvalue is 0 and represents the steady state values for $t \to \infty$. The second eigenvalue is the antibunching term that describes the exchanges between the $\mathrm{S_0}$ and $\mathrm{S_1}$ states. As this is in the nanosecond range it can be neglected for the discussion of the triplet state whose characteristic time is in the microsecond range. This leads to

$$\mathrm{S_0}(t) = \frac{k_{21}k_{31}}{k_{23}k_{31} + k_{12}(k_{23} + k_{31})} + \frac{k_{12}k_{21}k_{23}}{(k_{12} + k_{21})[k_{12}(k_{23} + k_{31}) + k_{21}k_{31}]} e^{-\left(k_{31}+\frac{k_{12}k_{23}}{k_{12}+k_{21}}\right)t}$$

$$\mathrm{S_1}(t) = \frac{k_{12}k_{31}}{k_{12}(k_{23} + k_{31}) + k_{21}k_{31}} + \frac{k_{12}^2k_{23}}{(k_{12} + k_{21})[k_{12}(k_{23} + k_{31}) + k_{21}k_{31}]} e^{-\left(k_{31}+\frac{k_{12}k_{23}}{k_{12}+k_{21}}\right)t}$$

$$\mathrm{T}(t) = \frac{k_{12}k_{23}}{k_{12}(k_{23} + k_{31}) + k_{21}k_{31}}\left(1 - e^{-(k_{31}+\frac{k_{12}k_{23}}{k_{12}+k_{21}})t}\right)$$

The influence of the triplet state, T, will depend on how it influences the $\mathrm{S_1}$ state, as photon emission, the parameter measured in the correlation functions, happens only from the excited singlet state. Using the equation for $\mathrm{S_1}$ derived for the presence of a triplet state in the derivation of the correlation function, the correlation function is given by [32, 33]:

$$G(\tau) = \frac{1}{N} g_D(\tau) g(\tau) + G_\infty = \frac{1}{N} g_D(\tau) \frac{\mathrm{S_1}(\tau)}{\bar{\mathrm{S}}_1} + G_\infty$$

where $g_D(\tau)$ is the term describing the diffusion dependence of the ACF. For 3D diffusion, for instance, this would be $g_D(\tau) = (1 + \frac{\tau_D}{\tau_D})^{-1}(1 + \frac{\tau_D}{K^2\tau_D})^{-1/2}$. The reader can verify from the equations above that

$$\frac{S_1(t)}{\bar{S}_1} = 1 + \frac{\bar{T}}{1 - \bar{T}} e^{-\lambda_3 t}$$

Leading to

$$G(\tau) = \frac{1}{N}g_D(\tau)[1 + \frac{\bar{T}}{1 - \bar{T}}e^{-\lambda_3\tau}] + G_\infty$$

With $f_T = \bar{T}$ and $\lambda_3 = 1/\tau_T$, this leads to the final equation for the correlation function including a triplet (equation (5.47)):

$$G(\tau) = \frac{1}{N}g_D(\tau)[1 + \frac{f_T}{1 - f_T}e^{-\tau/\tau_T}] + G_\infty$$

Sometimes this equation is written as

$$G(\tau) = \frac{1}{N(1 - f_T)}g_D(\tau)[1 - f_T + f_T e^{-\tau/\tau_T}] + G_\infty = \frac{1}{N'}g_D(\tau)[1 - f_T + f_T e^{-\tau/\tau_T}] + G_\infty$$

Both functions can be used in data fitting but note that the interpretation of the parameter N and N' is different. N is the average number of particles in the observation volume. N' is the average number of particles residing in the single state in the observation volume.

Math box 5.6

Approximation of the full ACF including various processes
The full ACF for various processes is the addition of the correlation functions of these processes:

$$G(\tau) = \frac{1}{N}\left[\left(1 + \frac{\tau}{\tau_D}\right)^{-1}\left(1 + \frac{\tau}{K^2\tau_D}\right)^{-1/2} + \frac{f_T}{(1 - f_T)}e^{-\tau/\tau_T} - e^{-\tau/\tau_A} + Re^{-\tau/\tau_R}\right] + G_\infty$$

To see how this can be simplified, let's take a look first at the combination of diffusion and triplet, two processes which are independent of each other. The ACF in this case is given by

$$G(\tau) = \frac{1}{N}\left[\left(1 + \frac{\tau}{\tau_D}\right)^{-1}\left(1 + \frac{\tau}{K^2\tau_D}\right)^{-1/2} + \frac{f_T}{(1 - f_T)}e^{-\tau/\tau_T}\right] + G_\infty$$

The exponential of the triplet term $e^{-\tau/\tau_T}$ has a significant value above 0 only for $\tau \lesssim \tau_T$. In addition, as $\tau_D \gg \tau_T$, the diffusive term at $\tau \lesssim \tau_T$ is on the order of 1:

$$\left(1 + \frac{\tau_T}{\tau_D}\right)^{-1}\left(1 + \frac{\tau_T}{K^2\tau_D}\right)^{-1/2} \approx 1$$

Therefore, the above ACF can be approximated by

$$G(\tau) = \frac{1}{N}\left[\left(1 + \frac{\tau}{\tau_D}\right)^{-1}\left(1 + \frac{\tau}{K^2\tau_D}\right)^{-1/2}\right]\left(1 + \frac{f_T}{1 - f_T}e^{-\tau/\tau_T}\right) + G_\infty$$

$$= \frac{1}{N}\left[\left(1 + \frac{\tau}{\tau_D}\right)^{-1}\left(1 + \frac{\tau}{K^2\tau_D}\right)^{-1/2} + \left(1 + \frac{\tau}{\tau_D}\right)^{-1}\left(1 + \frac{\tau}{K^2\tau_D}\right)^{-1/2}\frac{f_T}{1 - f_T}e^{-\tau/\tau_T}\right] + G_\infty$$

As we have just shown, either the diffusive term is 1 or the triplet term is 0. Therefore, this is a good approximation for

$$G(\tau) = \frac{1}{N}\left[\left(1 + \frac{\tau}{\tau_D}\right)^{-1}\left(1 + \frac{\tau}{K^2\tau_D}\right)^{-1/2} + \frac{f_T}{(1 - f_T)}e^{-\tau/\tau_T}\right] + G_\infty$$

Extending the same principle to all other processes, which all happen at very different characteristic time scales, we obtain

$$G(\tau) = \frac{1}{N}\left[\left(1 + \frac{\tau}{\tau_D}\right)^{-1}\left(1 + \frac{\tau}{K^2\tau_D}\right)^{-1/2}\right](1 - e^{-\tau/\tau_A})(1 + Re^{-\tau/\tau_R})\left(1 + \frac{f_T}{1 - f_T}e^{-\tau/\tau_T}\right) + G_\infty$$

Both functions can be found in literature [22, 34].

Note that if the particles in question have multiple labels N_B then equation (5.48) needs to be used.

Antibunching, rotation, and other fast processes can be used to extract information on molecular states (see math box 5.6 on a remark on the final form of the ACFs for equations (5.47) and (5.48)). However, as mentioned above, the measurement of fast processes requires longer measurement times. Therefore, in the rest of the book we will concentrate mainly on processes with characteristic times above μs. But we will discuss shortly in chapter 9 how the different processes that lead to fluorophore blinking, can be experimentally separated and can be distinguished from translational diffusion.

5.7 Spatial and spatiotemporal correlation techniques

In the previous sections, we discussed temporal correlation functions, i.e. functions that describe correlations in time. For that purpose, one or more signals were recorded from a single point over time, followed by the calculation of the ACF or CCF. In fluorescence microscopy one often records data in an imaging format, i.e. we have not only temporal but also spatial information. Hence, correlation functions can also be calculated in space [35]. And if the images are recorded with some inherent time resolution, e.g. when images are scanned in a confocal microscope and thus each point is taken at a different time, or when whole images are recorded with a high repetition rate, then one can extract spatial and temporal correlations simultaneously [36]. As we will see, this not only multiplexes measurements but also provides a plethora of new information.

In this chapter, we will discuss how this extra spatial information can be extracted from correlation functions. We will start with spatiotemporal correlations between single points, explain how this can be extended by scanning the observation volume in a sample and finally will end up with camera-based FCS approaches that take in the full spatiotemporal information.

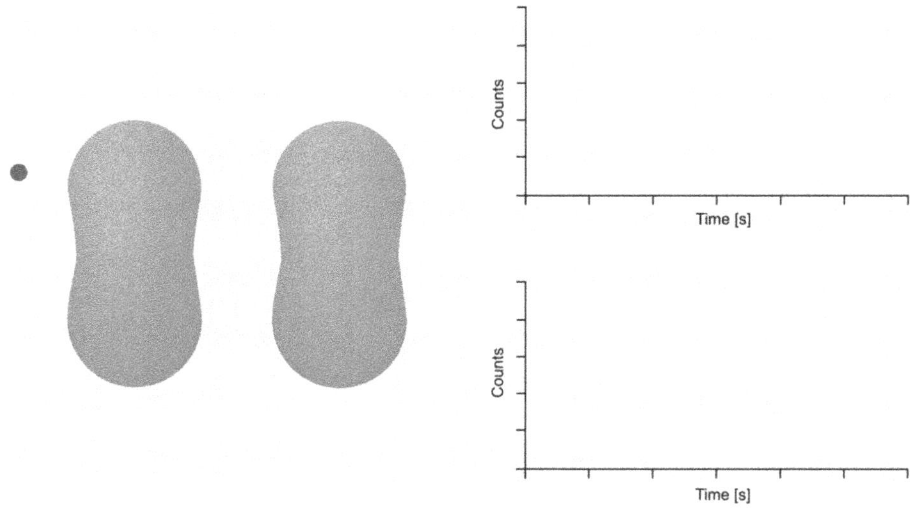

Figure 5.15. Schematic of two-focus FCS. A particle moving through the two focal volumes will give rise to fluorescence signals at different points in time. If the distance between the two observation volumes is known, then the time difference between the fluorescence signals will provide the diffusion coefficient and/or flow speed of the particle. This arrangement gives rise to multiple correlation functions. First there is an ACF for each observation volume. Second there are two CCFs: A CCF from the first to the second volume that will determine the time a particle required to go from the first to the second volume, and vice versa. Animation available at https://iopscience.iop.org/book/978-0-7503-2080-1.

5.7.1 Two-focus FCS

The simplest way to exploit spatial correlations is in a two-focus FCS system in which the intensities from two confocal observation volumes are cross-correlated as shown in figure 5.15 [37]. This requires the creation of two confocal volumes using two separate laser beams and pinholes (see chapter 3). This setup results in ACFs from each confocal volume and in CCFs between the volumes.

For the case of two focal volumes, we need to derive the correlation function anew. For this purpose, we need to first determine which diffusion propagator to use. The probability to find a particle at the location of the second observation volume, $\vec{r}'$, if it was found at the location of the first observation volume, $\vec{r}$, a time, t, earlier, must depend on the distance, $\vec{r} - \vec{r}'$, between the two locations. But this is just the propagator for free Brownian diffusion derived in math box 5.3.

$$\langle C(\vec{r}, t)C(\vec{r}', t + \tau)\rangle = \frac{\langle C\rangle}{(4\pi Dt)^{3/2}} e^{-\frac{(\vec{r}-\vec{r}')^2}{4Dt}} \tag{5.49}$$

The difference in the derivation for two-focus FCS will lie in the CEFs, $W(\vec{r})$, of the volumes that are cross-correlated as they will be displaced from each other by the vector $\vec{\rho}$ and have now different coordinates.

$$G(\tau) = \frac{\iiiint\!\!\iint e^{\frac{-2(x^2+(x'-\rho_x)^2)}{w_0^2}} e^{\frac{-2(y^2+(y'-\rho_y)^2)}{w_0^2}} e^{\frac{-2(z^2+(z'-\rho_z)^2)}{z_0^2}} \frac{e^{-\frac{(x-x')^2+(y-y')^2+(z-z')^2}{4Dt}}}{4\pi Dt} dx dx' dy dy' dz dz'}{\langle C\rangle \left(\iiint e^{\frac{-2x^2}{w_0^2}} e^{\frac{-2y^2}{w_0^2}} e^{\frac{-2z^2}{z_0^2}} dx dy dz\right)\left(\iiint e^{\frac{-2(x'-\rho_x)^2}{w_0^2}} e^{\frac{-2(y'-\rho_y)^2}{w_0^2}} e^{\frac{-2(z'-\rho_z)^2}{z_0^2}} dx' dy' dz'\right)} + G_\infty \tag{5.50}$$

With this we can follow the strategy in math box 5.2 to derive the ACF and arrive at

$$G(\tau) = \frac{1}{N}\left(1 + \frac{\tau}{\tau_D}\right)^{-1}\left(1 + \frac{w_0^2\tau}{z_0^2\tau_D}\right)^{-1/2} e^{-\frac{\rho_x{}^2+\rho_y{}^2}{4Dt+w_0^2}-\frac{\rho_z{}^2}{4Dt+z_0^2}} \tag{5.51}$$

In math box 5.4, we had derived the ACF for diffusion and flow in 2D. The extension to 3D is simple and just contains the extra term for the third dimensions.

$$\langle C(\vec{r}, t)C(\vec{r}\,', t + \tau)\rangle = \frac{\langle C\rangle}{(4\pi Dt)^{3/2}} e^{-\frac{(x-x'-v_x\tau)^2+(y-y'-v_y\tau)^2+(z-z'-v_z\tau)^2}{4Dt}} \tag{5.52}$$

which will then result by integration in the ACF for diffusion and flow in 3D

$$G(\tau) = \frac{1}{N}\left(1 + \frac{4D\tau}{w_0^2}\right)^{-1}\left(1 + \frac{4D\tau}{z_0^2}\right)^{-1/2} e^{-\frac{v_x^2\tau^2+v_y^2\tau^2}{4D\tau+w_0^2}-\frac{v_z^2\tau^2}{4D\tau+z_0^2}} \tag{5.53}$$

Finally, we can combine the two, the two observation volumes and flow. We can again use the propagator for flow but, as before, have to modify the observation volume positions (see equation (5.50)). This results in the ACF for diffusion and flow for two displaced observation volumes

$$G(\tau) = \frac{1}{N}\left(1 + \frac{4D\tau}{w_0^2}\right)^{-1}\left(1 + \frac{4D\tau}{z_0^2}\right)^{-1/2} e^{-\frac{(v_x\tau-\rho_x)^2+(v_y\tau-\rho_y)^2}{4D\tau+w_0^2}-\frac{(v_z\tau-\rho_z)^2}{4D\tau+z_0^2}} \tag{5.54}$$

Depending on the distance between the two confocal volumes, photons excited in laser beam 1 can be detected by the detector 2 and vice versa. This leads to pseudo-autocorrelations (see section 5.7.7) since molecules illuminated by laser beam 1 (or laser beam 2) contribute now to both detectors. Therefore, data needs to be corrected for this cross-talk as we will discuss in the next section.

As we have two observation volumes A and B, the CCF between these volumes can be calculated in two ways, namely either from A to B, CCF_{AB}, or from B to A, CCF_{BA}. In the case of isotropic free diffusion this does not make a difference, as the movement of particles from A to B or from B to A is the same and $\text{CCF}_{\text{AB}} = \text{CCF}_{\text{BA}}$. However, if flow is present, then the movement from one volume to the other will depend on the flow direction and the two CCFs will in general not be the same (figure 5.16). Assuming, without loss of generality, that the flow proceeds from A to B, then CCF_{AB} will contain contribution from the flow from A

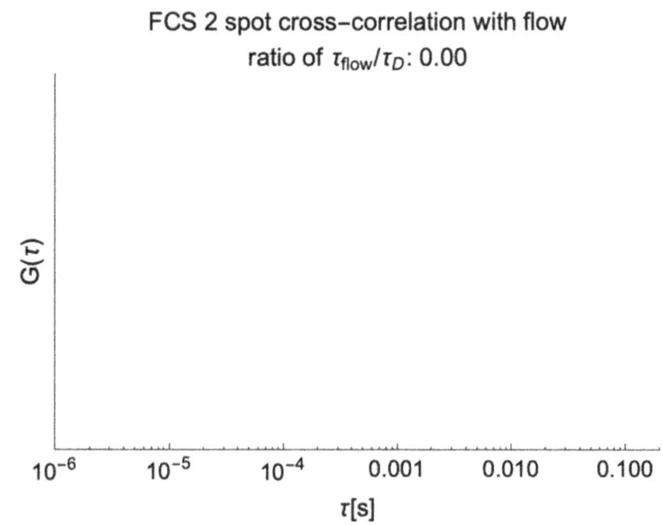

Figure 5.16. CCFs obtained by two-focus FCS. Animation available at https://iopscience.iop.org/book/978-0-7503-2080-1.

to *B* as well as some diffusive component and pseudo-autocorrelations. However, CCF_{BA} will contain only diffusive components and pseudo-autocorrelations. Therefore, by subtracting the two (CCF_{AB} − CCF_{BA}) one can eliminate the pseudo-autocorrelations and obtain a CCF that contains only the flow component (figure 5.16).

If there is no flow and only diffusion is present then $CCF_{AB} = CCF_{BA}$. In this case the CCF will have a peak that indicates how long it takes for a particle to diffuse from volume *A* to *B* and vice versa. If the distance between the volumes is known then together with the peak position this will provide absolute diffusion coefficients without the need for a calibration of focal volumes, as would be required in confocal FCS.

Two-focus FCS has mainly been used to determine flow. However, there exist more advanced systems, in which the two foci are spaced at a fixed, well-defined distance but are overlapping and pass through a single pinhole. Such systems allow very precise absolute diffusion coefficient measurements and the determination of particle sizes with better than nm resolution [38–40]. Finally, by determining the CCF between two points one can also establish whether particles can either travel from point *A* to point *B*, and how long that takes. This can be used to determine the connectivity in a sample and detect the existence of barriers. This was very effectively used in a pair-correlation function approach [41].

5.7.2 Scanning FCS

Scanning FCS is arguably the first and also the most simple way to extract spatiotemporal information from a sample [42]. It records signals not from a single point but moves the point through space at a defined speed in one of various patterns. The most simple pattern would be a line or circular scan, but also random scans or area scans, i.e. recording of an image or region of interest, can be employed. Using this approach we can extract information of molecular processes in time and space.

Linear scanning in a sample at constant speed, is the same as if the sample undergoes uniform flow with the same speed in the opposite direction. So the correlation function is the same as the one for diffusion and flow (equation (5.38)) and the discussion of the ACF properties from section 5.5 also applies here. For simplicity, we assume here a 2D case:

$$G(\tau) = \frac{1}{N}\left(1 + \frac{4D\tau}{w_0^2}\right)^{-1} e^{-\frac{v^2\tau^2}{4D\tau + w_0^2}} + G_\infty \tag{5.55}$$

As Brownian diffusion is isotropic, i.e. the diffusion has the same properties in all directions, scanning in any direction will have the same influence on the ACF.

This method can be used in two very different ways. If the sample is stationary and there is no movement, then scanning over the sample can determine the average number of molecules in the sample, which can be extracted from the amplitude of the ACF [43]. In this case, only the amplitude information is retrieved, as the width of the correlation function reflects mainly the scanning speed used. A similar approach for particle counting is used in image correlation spectroscopy (ICS) by Petersen [35] as will be discussed in section 5.7.4.

If diffusion and flow exist in the sample, then we can extract the *speed* of the flow as can be seen from equation (5.55). However, the laser beam is rotationally symmetric and flow in any direction will have the same effect on the measurement. Therefore, we cannot determine the *velocity* as the direction of the flow cannot be established. By scanning the laser beam with a known velocity, $\vec{v}_{\text{scan}}$, we can break this symmetry and the ACF becomes

$$G(\tau) = \frac{1}{\langle C\rangle \pi w_0^2}\left(1 + \frac{4D\tau}{w_0^2}\right)^{-1} e^{-\frac{(\vec{v}_{\text{flow}} + \vec{v}_{\text{scan}})^2\tau^2}{4D\tau + w_0^2}} + G_\infty \tag{5.56}$$

as the flow and scanning velocities will add up (figure 5.17). If we analyse the net flow, we can see that it depends on the angle between scan and flow.

$$v_{\text{nett}}^2 = (\vec{v}_{\text{flow}} + \vec{v}_{\text{scan}})^2 = v_{\text{flow}}^2 + v_{\text{scan}}^2 + 2v_{\text{flow}}v_{\text{scan}}\cos(\theta) \tag{5.57}$$

However, to determine the angle θ we have to know the flow *speed*. We therefore, can determine the flow *velocity* with two measurements, one without scanning to determine the flow *speed* and a second measurement using scanning to determine the direction of the flow [44–46], the two together providing the flow velocity.

The advantage of scanning FCS is that it can even take account of immobile particles, which will be registered when scanning the beam over them [43]. This can be used for particle counting. It also can reduce photobleaching as particles stay shorter times in the laser beam and it has been used to differentiate free from

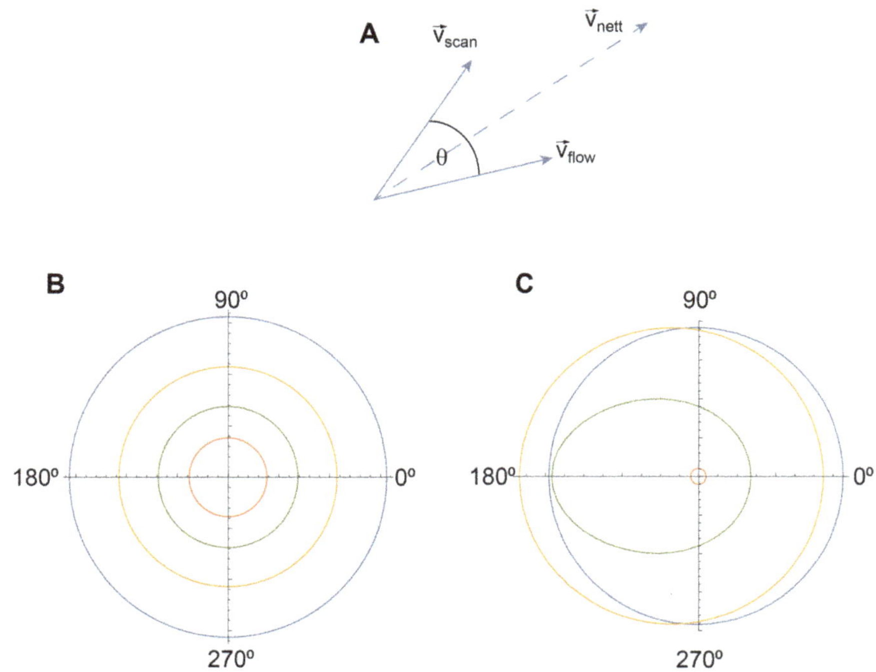

Figure 5.17. Line scanning FCS. (A) Vector diagram showing how the actual flow and laser scan create a net flow. The net flow ($\vec{v}_{net}$) depends on the angle θ between line scan ($\vec{v}_{scan}$) and flow direction ($\vec{v}_{flow}$) and their speeds (equation (5.57)). (B) Polar plot of the lag time at which the ACF has dropped to half its amplitude. In the case of diffusion without flow, an increase in scan rate will reduce the lag time at which the ACF has dropped to its half-amplitude. As there is no flow, scans in all directions result in the same ACF, leading to a circular plot. The red circle has the highest scan rate and the blue circle the lowest. (C) In the case of diffusion with flow, the addition of flow and laser scanning adds up and leads to shorter lag times at which the ACF drops to its half-amplitude, if the two directions are opposite, or longer lag times, if the laser follows the flow and thus tends to keep particles longer in the focus. This leads to non-circular plots. The blue circle has again no flow and is circular; the yellow and green curves have intermediate flow rates leading to stronger deviations from circularity with increasing flow rate. The red circle represents a flow rate that is so high that diffusion and flow are negligible, resulting again in a circular plot.

hindered diffusion [47]. In a slightly more elaborate application, scanning FCS was used to periodically scan the beam along the surface normal of a fluctuating or moving membrane. In the data evaluation only points on the membrane were used for the correlation. This allows measuring even dynamic systems in which it is difficult to keep the sample in the observation volume. While this limits the dynamics that can be measured, as they must be slower than the scanning rate, and could also limit the number of photons per molecule detected, it has been put to very effective use even in organisms [48, 49].

Finally, scanning FCS has resulted in the development of raster image correlation spectroscopy (RICS) that uses the time structure in a scanned image to determine spatiotemporal information of the sample [36]. RICS will be discussed in section 5.7.6. But first let's look at some other scanning and imaging techniques.

Figure 5.18. Principle of circular scanning FCS. A confocal volume is scanned in a circular pattern. A slow moving particle will be scanned over multiple times at a particular position of the scan circle. This leads to the characteristic oscillatory pattern of the correlation function. Note that if the particles did not move and never bleached, then the correlation function would not decay and one would see only the oscillations due to the circular scanning. But as the particles diffuse, the circular scanning FCS correlation function is bounded on the upper side by the ACF for the scan volume. Depending on their path, the particles might also be scanned over again when crossing the scan path a second time. This leads to spatial cross-correlations between points on the scan path, which is the reason why the correlation function is bounded on the lower side by the spatial CCF between two opposing points on the scan path (see also figure 5.19). Animation available at https://iopscience.iop.org/book/978-0-7503-2080-1.

5.7.3 Circular scanning fluorescence correlation spectroscopy

An alternative to linear scanning is circular scanning FCS. As the name indicates in this case the excitation beam is scanned in a circular pattern (figure 5.18). To derive its correlation function, we need to look how the circular scanning changes our intensity pattern and diffusion propagator. Circular scanning can be described from two different points of view. One can assume that the laser beam is scanned and thus use periodically changing laser intensity positions. But one can also assume the lasers are stationary and thus the sample is moving. In this second case, which we will follow here for the 2D case, we modify the diffusion propagator accordingly:

$$\langle C(\vec{r}, t)C(\vec{r}\,', t+\tau)\rangle = \frac{\langle C\rangle}{4\pi Dt} e^{-\frac{((x-\rho\cos(\varphi+\omega t))-(x'-\rho\cos(\varphi+\omega(t+\tau))))^2+((y-\rho\sin(\varphi+\omega t))-(y'-\rho\sin(\varphi+\omega(t+\tau))))^2}{4Dt}} \tag{5.58}$$

Here, the position $\vec{r}$ is a vector with components x and y, ρ is the radius of the scanning circle, ω is the circular frequency, and φ is the phase determined by the time and position the scan started.

Following our by now familiar approach, we use the diffusion propagator in the integrals for FCS (see math box 5.1), leading to the correlation function, that turns out to consist of the correlation function for diffusion and a scanning dependent term.

$$G(\tau) = \frac{1}{\langle C\rangle \pi w_0^2}\left(1 + \frac{4D\tau}{w_0^2}\right)^{-1} e^{-\frac{4\rho^2 \sin(\frac{\omega t}{2})^2 \tau^2}{4D\tau + w_0^2}} + G_\infty \tag{5.59}$$

The correlation function will oscillate due to the $\sin(\omega t/2)^2$ component in the exponential. It will oscillate between two extremes when this component takes its minimum and maximum values, namely 0 and 1. In the first case the correlation function simplifies to the correlation function for diffusion

$$G(\tau) = \frac{1}{\langle C\rangle \pi w_0^2}\left(1 + \frac{4D\tau}{w_0^2}\right)^{-1} + G_\infty \tag{5.60}$$

In the second case it will simplify to

$$G(\tau) = \frac{1}{\langle C\rangle \pi w_0^2}\left(1 + \frac{4D\tau}{w_0^2}\right)^{-1} e^{-\frac{4\rho^2\tau^2}{4D\tau + w_0^2}} + G_\infty \tag{5.61}$$

Comparing this to equation (5.51) shows that this is just the 2D cross-correlation function of two points with a distance 2ρ. Therefore, the correlation function for circular scanning oscillates between an upper envelop which corresponds to the ACF for diffusion and a lower limit which is the CCF for two observation volumes with a distance of the diameter of the scanning circle, 2ρ (figure 5.19).

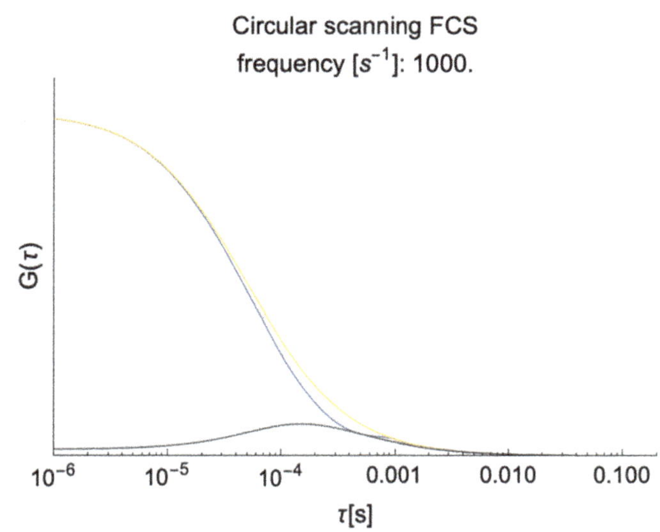

Figure 5.19. Example correlation functions for circular scanning FCS for various scan frequencies. The ACF (blue) shows a characteristic oscillatory pattern determined by the period T ($=f^{-1}$) of the circular scan. The ACF is limited on the lower side by the CCF of two point on opposite sides of the circle (gray) and on the upper side by the ACF of the size of the scanned volume (yellow). Animation available at https://iopscience.iop.org/book/978-0-7503-2080-1.

In 2D experiments on membranes or bilayers, circular scanning FCS has been shown to be insensitive to the position of the laser focus in respect to the membrane measured, which is an important advantage over single-focus conventional FCS (see chapter 7). Furthermore, the diffusion coefficient determined by circular scanning FCS is insensitive to drift or fluctuations of the membrane [50]. Circular scanning FCS has seen various developments and applications over the years and we refer to some examples for the interested reader [50–53].

5.7.4 Image correlation spectroscopy (ICS)

Image correlation spectroscopy (ICS) is the spatial analogue to FCS. In ICS one calculates spatial correlation functions in which the fluctuations in space of an observable are characterized. This was used early on by Nils Peterson to obtain information on the distribution of receptors on cell membranes [54, 55]. In this case one calculates the spatial ACF by

$$G(\xi, \psi) = \frac{\langle f(x, y)f(x + \xi, y + \psi)\rangle}{\langle f(x, y)\rangle^2} \tag{5.62}$$

where x and y are spatial coordinates, and ξ and ψ are the spatial lags, analogous to the temporal lag time τ. The spatial ACF is in this case two-dimensional. But it could also be calculated in one or three dimensions. The ACFs are calculated over one image and can be calculated quite quickly due to the Wiener–Khinchin theorem that relates the Fourier transform $f(\nu)$ of a function $f(x)$ to its autocorrelation $g(\xi)$ by

$$\mathcal{F}_\nu\{|f(\nu)|^2\} = \int_{-\infty}^{\infty} f^*(\xi)f(\xi + x)dx = g(\xi) \tag{5.63}$$

where $f^*(x)$ denotes the complex conjugate of $f(x)$. The function $|f(\nu)|^2$ is called the power spectrum of the signal $f(x)$. The Wiener–Khinchin theorem thus states that the Fourier transform of the power spectrum of a signal is equal to the autocorrelation function of the signal. Using this relation the spatial autocorrelation function can be written as a sequence of Fourier transforms:

$$G(\xi, \psi) = G(x - x', y - y') = \frac{\mathcal{F}^{-1}_{x, y}\{\mathcal{F}_{\nu_x,\nu_y}\{f(x, y)\}\times\mathcal{F}^*_{\nu_x, \nu_y}\{f(x', y')\}\}}{\langle f(x, y)\rangle^2} \tag{5.64}$$

The Fourier transforms are easily calculated as most software packages nowadays provide fast algorithms for Fourier transforms. The result is then divided by the square of the mean intensity for normalization, analogous to FCS. For long distances the correlations vanish and $G(\xi, \psi)$ will converge to 1.

In FCS the width of the ACF was determined by the average time particles required to traverse the observation volume. Thus, the width is determined by the size of the observation volume and the diffusion coefficient. In analogy, in ICS, the width of the spatial correlation function is determined by the size of the observation volume, which is defined by the PSF of the system (assuming pixels much smaller

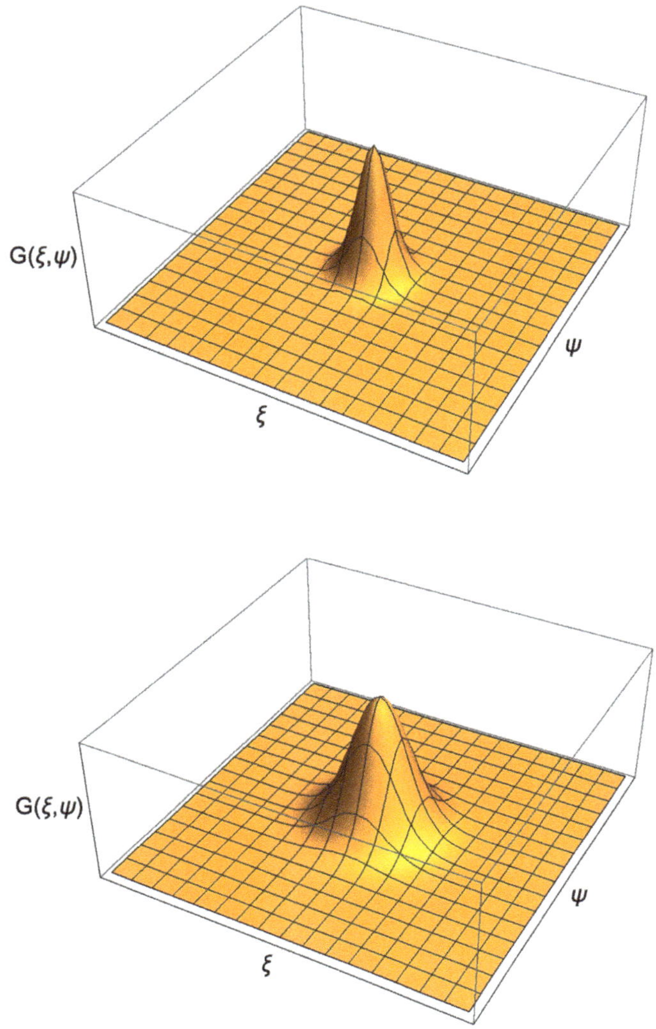

Figure 5.20. Examples of ICS functions. The left pane shows the ICS function limited by the PSF of the microscope. The panel on the right shows broadening of the ICS curve due to structures in the image.

than the PSF) and the size of the structures in the image (figure 5.20). Assuming the structures in the image are below the resolution limit, i.e. they are much smaller than the PSF, the correlation function will have the shape of the PSF, which can be approximated by a 2D Gaussian function:

$$G(\xi, \psi) = G(0{,}0)e^{-\frac{\xi^2+\psi^2}{\omega^2}} + G_\infty \tag{5.65}$$

Here we introduce again G_∞ as a fit parameter as the ideal value of 1 is only reached for infinitely many data points. ω represents the radius of the PSF, and $G(0{,}0)$ is the amplitude of the spatial correlation function. Similar to FCS, the amplitude $G(0,0)$ is inversely proportional to the number of particles in the observation volume/area. This relation can be used to count particles in an image as the total number of particles in an image, N_{image}, is given by

$$N_{\text{image}} = \frac{\text{Image area}}{G(0,0)\pi w_0^2} \tag{5.66}$$

where ω_0 is the radius of the PSF, as determined by fitting equation (5.65) to the spatial ACF [35]. And when performing ICS under varying conditions, one can monitor changes of the number of particles on a cell membrane and determine the number of particles in clusters [56].

In analogy one can also define spatial CCFs

$$G_X(\xi, \psi) = \frac{\langle f(x, y)g(x + \xi, y + \psi)\rangle}{\langle f(x, y)\rangle\langle g(x, y)\rangle} \tag{5.67}$$

In this case the values $f(x, y)$ and $g(x, y)$ might represent data taken from the same locations but at two different wavelengths, similar to our explanation for temporal CCFs. This technique is referred to as image cross-correlation spectroscopy (ICCS) and follows very similar principles to its temporal counterpart, FCCS, and was also used for the measurement for molecular interactions [57].

ICS and ICCS provide complementary information to FCS and FCCS. However, they do not exploit temporal information as FCS/FCCS do not use spatial information. Therefore, in the next sections, we explore techniques that use both temporal and spatial information.

5.7.5 Spatiotemporal image correlation spectroscopy (STICS)

In STICS, the ICS function of equation (5.67) is extended to the time domain by acquiring multiple images in time. As the acquisition of images is typically slower than that of single points in confocal FCS, the time resolution of STICS is in general lower. However, the advantage of being able to obtain simultaneously spatial information often outweighs this disadvantage. The correlation function for STICS is defined by

$$G(\xi, \psi, \tau) = \frac{\langle f(x, y, t)f(x + \xi, y + \psi, t + \tau)\rangle}{\langle f(x, y, t)\rangle^2} \tag{5.68}$$

We already saw the models for the correlation functions for FCS and ICS which can be directly translated to STICS [58]. Assuming a 2D case, e.g. a measurement in a membrane, we obtain

$$G(\xi, \psi, 0) = G(0,0,0)e^{-\frac{\xi^2+\psi^2}{\omega^2}} + G_\infty \tag{5.69a}$$

$$G(0,0, \tau) = G(0,0,0)\left(1 + \frac{4D\tau}{w_0^2}\right)^{-1} + G_\infty \tag{5.69b}$$

where the amplitude, $G(0, 0, 0)$, is inversely proportional to the number of particles N. The first equation describes the ICS result of each frame, where the peak will be centred and the amplitude will contain information about the number of particles. The second equation describes the correlation function along a pixel in time. The solution given here is correct assuming a Gaussian intensity profile similar to what is observed in FCS. We will encounter other solutions in imaging FCS with different illumination profiles (section 5.7.6).

The strength of STICS comes through the possibility to observe the evolution of the spatial correlation in time (figure 5.21). This can include multiple changes. First, the width of the PSF does not stay constant if the particles diffuse. Second, the location of the peak of the spatial correlation will not remain constant but the peak can move, e.g. in the case of flow. In this case the peak will have new coordinates, $x(\tau)$ and $y(\tau)$, depending on the lag time τ. Third, the amplitude of the peak of the spatiotemporal correlation, $G(0, 0, \tau)$, will change with time (see equation (5.69b)). Therefore, the spatiotemporal correlation function is described by the following equation

$$G(\xi, \psi, \tau) = G(0,0, \tau)e^{-\frac{(\xi - x(\tau))^2 + (\psi - y(\tau))^2}{w^2(\tau)}} + G_\infty \tag{5.70}$$

This function can be fitted to the experimental data to obtain diffusion coefficients [59]. The functions $x(\tau)$, $y(\tau)$, and $w^2(\tau)$ take the following values, depending on the situation:

Diffusion: $x(\tau) = y(\tau) = 0$ and $w^2(\tau) = 4D\tau + w_0^2$

Flow: $(\tau) = v_x\tau$, $y(\tau) = v_y\tau$, and $w^2(\tau) = w_0^2$

Diffusion and Flow: $(\tau) = v_x\tau$, $y(\tau) = v_y\tau$, and $w^2(\tau) = 4D\tau + w_0^2$

STICS can be extended to spatiotemporal cross-correlation spectroscopy (STICCS) by collecting frames in two colours and cross-correlating the images of the different colours [60].

5.7.6 Raster image correlation spectroscopy (RICS)

Raster image correlation spectroscopy (RICS) is an extension of ICS that uses the inherent time structure of images recorded on a confocal microscope to recover information on dynamic systems. A confocal microscope scans a beam over a sample line by line and records over (typically) fixed time intervals intensity values. Each value will represent a pixel in the final image. Therefore, each pixel is recorded at a particular time. This can be used to calculate spatiotemporal correlations along and across lines that contain information on different time scales (figure 5.22).

The correlation function for RICS resembles the correlation function for flow or scanning FCS, however, with time scales fixed by the scanning process. In the following, we denote the side length of a square pixel by a, the scanning time between pixels as τ_p, the time to scan a line as τ_l, and the spatial lags in the x and y directions as ξ and ψ. Using these terms the RICS function can be written as

$$G_{\mathrm{RICS}}(\tau) = G(\xi, \psi)e^{-\frac{a^2(|\xi|^2 + |\psi|^2)}{4D(\tau_p|\xi| + \tau_l|\psi|) + w_0^2}} + G_\infty \tag{5.71}$$

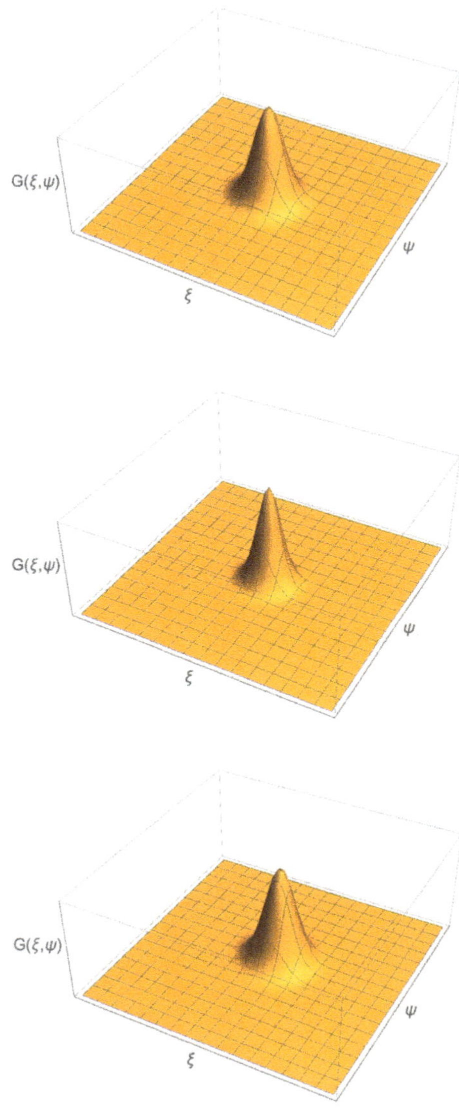

Figure 5.21. Examples of STICS functions. STICS combines spatial and temporal correlations. In the examples here we show movies of how the spatial correlation functions change with lag time. In the case of diffusion, the peak will stay centered but will broaden with lag time due to diffusion (top panel). In the middle panel the STICS for flow is shown. In this case the peak will move along spatial lags depending on the flow direction but it will not broaden. In the bottom panel the STICS function in the presence of flow and diffusion is shown. In this case the peak moves along spatial lags and broadens with lag time. For an interactive version see the accompanying 'Introduction to FCS—interactive exercises.cdf file' (available at https://iopscience.iop.org/book/978-0-7503-2080-1).

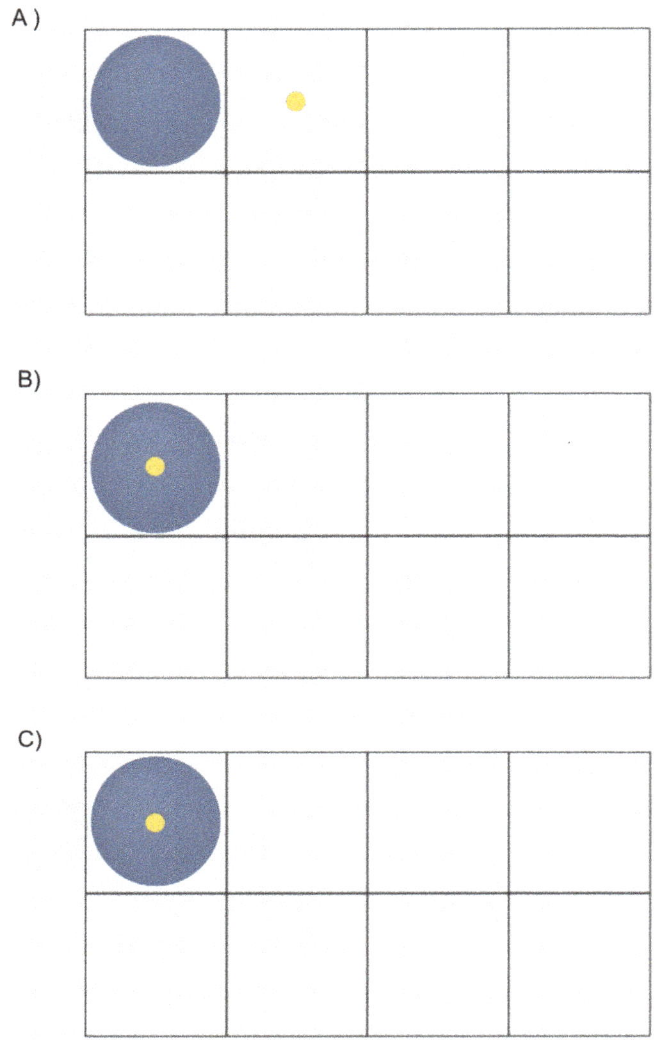

Figure 5.22. RICS principle. In RICS the laser beam (blue circle) is scanned over an image and the intensity per pixel is recorded. Shown are 8 pixels distributed over two lines. The correlations recorded in space and time provide information on mobility and concentration. In panel (A) scanning over stationary particles provides information on the number of particles, similar to ICS. However, if particles move, RICS also provides information on mobility, similar to STICS. In panel (B) a particle moves fast and thus is observed in consecutive pixels in the same line. In panel (C) a slow moving particle is too slow to be observed in consecutive pixels of the same line. However, it is detected in pixels in different lines as the second line is scanned at a later time. As the time delay between different pixels varies, RICS can observe multiple timescales. For an interactive version see the accompanying file 'Introduction to FCS—interactive exercises. cdf' (available at https://iopscience.iop.org/book/978-0-7503-2080-1).

with

$$G(\xi, \psi) = \frac{1}{N}\left(1 + \frac{4D(\tau_p|\xi| + \tau_l|\psi|)}{w_0^2}\right)^{-1}\left(1 + \frac{4D(\tau_p|\xi| + \tau_l|\psi|)}{w_z^2}\right)^{-1/2} \tag{5.72}$$

$G(\xi, \psi)$ describes the decay of the correlation with time due to diffusion. The times in $G_{\mathrm{RICS}}(\tau)$ are given by the difference in recording time between two pixels as the product of the spatial lags times the characteristic scanning times for pixels and lines, $\tau_p|\xi| + \tau_l|\psi|$. The exponential term describes the correlation of the PSF. As the PSF is broadened by diffusion it is also time dependent and changes with τ_p and τ_l.

This leads to an interesting effect. Scanning in a confocal microscope is performed line by line. Therefore, pixels that are neighbours in the same line are recorded with very small time difference, which is the pixel dwell time (τ_p). However, neighbouring pixels that are not in the same line but belong to two different lines, have a much larger difference of recording time, namely the time it takes to scan a whole line (τ_l). Therefore, the RICS correlation function tests two different time scales in the x and y directions. Norris *et al* used that to differentiate fast and slow diffusion and the occurrence of transient binding, which slows down movement [61].

If the diffusion is slow, then the fast and slow axes show a different profile. The short axis will show a profile similar to the PSF. In essence the scanning process is faster than the particle can move and the correlation function will be similar to that for stationary particles. The slow axis will show a profile that is broadened by diffusion. As the scanning to the neighbouring pixel in the next line takes considerably longer compared to scanning to the next pixel in the same line, the particle will have had time to move and thus will not be seen at the same point. This leads to the broadening of the profile. If the diffusion is fast, the profile along the fast axis will be already broadened. But the profile along the slow axis will be flat as the beam will not be fast enough to catch the particle in the next line and the correlation function is flat. If there exists slow diffusion and binding in a system, then the correlation along both directions will reflect the PSF shape similar to the stationary case, as particles do not move appreciably during the time of a scan. Finally, if there exists fast diffusion and binding, then the correlations for both spatial directions will be broadened due to the fast diffusion but eventual binding will not allow the particle to diffuse very far and the scan in the second line can still catch the particle. So in this case, despite the fast diffusion, the correlation across lines is not flat. Which of the cases is seen will depend sensitively on the values of the diffusion coefficient and the on and off rates of the binding event. But the possibility of RICS to measure these differences can be a powerful tool to determine the nature of the molecular processes underlying diffusion in complex networks. We have added an interactive example in the file called 'Introduction to FCS—interactive exercises.cdf' (available at https://iopscience.iop.org/book/978-0-7503-2080-1).

RICS has been extended in several ways. RICS can be performed as a cross-correlation approach (RICCS) when recording multiple colours simultaneously. Other improvements concerned the scanning of arbitrary regions [62]. An exciting extension has been the use of stimulated emission depletion with RICS (STED-RICS) that provides resolution of dynamics with super-resolution capabilities [62, 63].

5.7.7 Imaging fluorescence correlation spectroscopy (Imaging FCS)

Technological advances have driven new developments in FCS over the years. In fact, the basic theory on fluctuations had already been around for decades before in the 1970s we finally had the necessary technology available to perform FCS measurements. At that time, it was the development of lasers and sensitive detectors that made FCS possible. In the last decade, we had the appearance of highly sensitive and fast cameras that made multiplexing of FCS possible. Although ICS was already performed since the 1980s, the collection of frames was not sufficiently fast to allow correlations in time and thus to extract information on molecular processes for each pixel. With the advent of electron multiplying charge coupled device (EMCCD) cameras, that had very high sensitivity and that could record data with a time resolution on the millisecond time range it became possible to extract spatiotemporal information from image stacks as has been demonstrated by STICS [58]. This was followed by the extraction of FCS functions for each pixel using first a spinning disk approach [64] and then TIRF and light sheet illumination [65–67] (figure 5.23).

The idea in imaging FCS is that one records the fluorescence data on a fast and sensitive camera or array detector and determines the temporal correlations for all pixels or between pixels or pixel areas on the camera. This mode of performing FCS has the advantage that it can measure thousands of temporal correlation functions

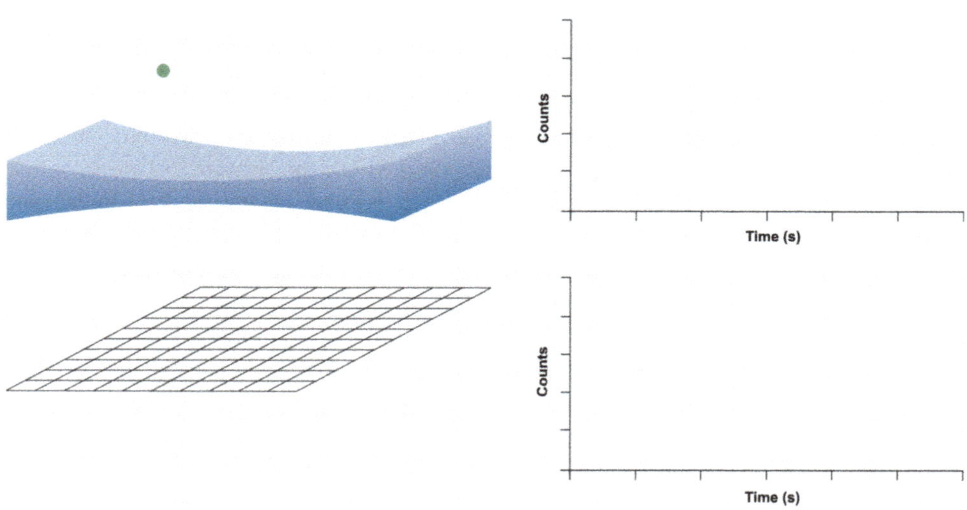

Figure 5.23. Imaging FCS principle. A thin cross-section of a sample is illuminated. Fluorescently labelled particles moving through the illuminated region emit fluorescence that can be projected onto a camera, recording the whole sample cross-section. Each particle on the camera records fluorescence with each particle moving through its observation volume. If the fluorescence intensity is recorded sufficiently fast and with sufficient duration, correlation functions can be calculated for each and between each pixel. Animation available at https://iopscience.iop.org/book/978-0-7503-2080-1.

simultaneously, reducing experimental time enormously and providing FCS data in an imaging format. Cameras are still much slower than the point detectors typically used in confocal FCS. Thus, they have been mainly used for membrane probes or slow processes. However, point detector arrays and cameras have been shown to be fast enough to measure molecules in the cell cytoplasm or in solution [67, 68].

In this case, confocal microscopy cannot be used as an illumination scheme, as one needs to illuminate many pixels simultaneously with as little cross-talk between pixels as possible. In confocal microscopy a strongly focused laser is used for illumination. This laser illuminates a large part of the sample and a confocal pinhole is required to reduce signal from all regions outside the focal plane. If one uses two laser beams and pinholes simultaneously to create two confocal detection areas, then the out-of-focus light of the second laser cannot be completely filtered out by the first pinhole and vice versa. This leads to cross-talk of the signal between the two laser beams, increases background, and causes the signal-to-noise ratio and thus the ACFs to deteriorate. So there needs to be a minimum separation between confocal spots if they are to be recorded simultaneously, to minimize the crosstalk between pinholes. This effect is well known for spinning disk confocal microscopes, leading, on one hand, to patterns with well separated confocal pinholes, and on the other hand to the recommendation not to use spinning disk confocal microscopes with thick samples (figure 5.24). This effect has also been shown in camera-based FCS [69]. A distance of at least 10 pinhole diameters between confocal detection areas has been suggested to ensure that the cross-talk is sufficiently low so that diffusion times are independent between pixels. However, even then, the ACF

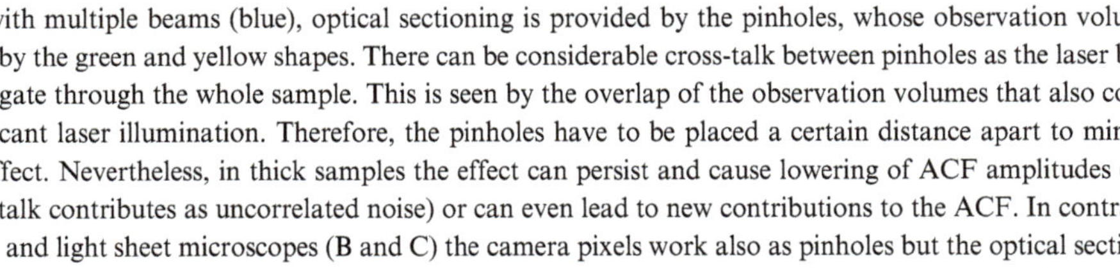

Figure 5.24. Cross-talk from out-of-focus-light in different illumination schemes. (A) In a confocal illumination with multiple beams (blue), optical sectioning is provided by the pinholes, whose observation volume is given by the green and yellow shapes. There can be considerable cross-talk between pinholes as the laser beams propagate through the whole sample. This is seen by the overlap of the observation volumes that also contain significant laser illumination. Therefore, the pinholes have to be placed a certain distance apart to minimize this effect. Nevertheless, in thick samples the effect can persist and cause lowering of ACF amplitudes (if the cross-talk contributes as uncorrelated noise) or can even lead to new contributions to the ACF. In contrast, in TIRF and light sheet microscopes (B and C) the camera pixels work also as pinholes but the optical sectioning is provided mainly by the illumination, relaxing the requirement of placing pixels far apart to avoid cross-talk. (B) In a TIRF microscope cross-talk is minimized as the evanescent wave extends only ~100 nm into the sample. In addition, when observing lipid bilayers or plasma membranes, the signal comes from a very narrow region (<10 nm). Therefore, cross-talk is largely negligible (note that because of the lateral extend of the PSF neighboring pixels experience cross-talk from in-focus-light and are still correlated as they see partially the same sample, see figure 5.25). (C) In a light sheet microscope, the optical sectioning is achieved by the extension of the illuminating light sheet which can be as thin as 1 μm. Nevertheless, in this configuration cross-talk from out-of-focus-light between camera pixels persists and requires correction of the FCS model (see remarks in math box 5.7).

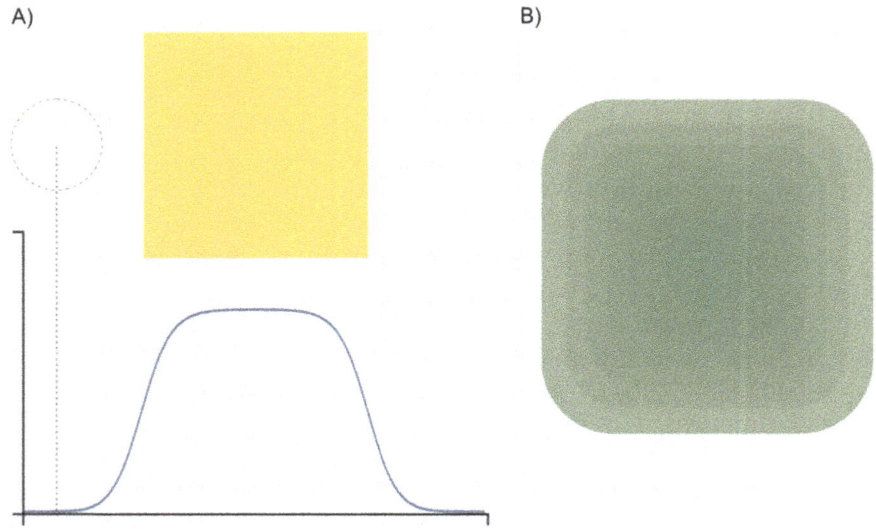

Figure 5.25. The actual observation area of a pixel is larger than its actual size (actual pixel size: yellow; observation area: green). (A) A point particle, depicted as the dashed circle, is imaged in a microscope according to the PSF of the system. As soon as a signal from the particle image falls onto the pixel, it is detected. The animation shows a particle moving across a pixel and the signal (bottom graph) that is detected. (B) The observation volume of the pixel is thus the convolution of pixel and PSF. Animation available at https://iopscience.iop.org/book/978-0-7503-2080-1.

amplitude is still influenced by the increased background due to cross-talk, and concentrations are systematically overestimated. The difference in the influence of the background on correlation function width and amplitude is easily explained. For a particle to contribute to a correlation and thus influence the ACF width, one needs to capture at least two photons from the same particle, otherwise no correlation exists. However, every photon contributes to the denominator of the ACF. So even if only one photon from a particle is detected, it will lead to a decrease of the ACF amplitude. Thus, there is a minimum distance where the likelihood to detect two photons from a particle in a neighboring confocal volume becomes negligible and the width of the ACF is only defined by one confocal volume. However, single photons from particles from the neighboring volume are still detected and influence the amplitude of the ACF. One has to place the neighboring confocal volumes significantly further apart to make this cross-talk negligible.

To avoid these complications and allow a contiguous detection of areas in a sample, different illumination schemes need to be employed.

The cross-talk in confocal microscopy originates from out of focal planes. Therefore, in an ideal system, if fluorophores can be illuminated only within the focal plane, then cross-talk can be avoided. At the moment, there are two different illumination schemes available that illuminate thin sections across a sample, total internal reflection (TIR) and single plane illumination microscopy (SPIM) (chapter 4). Both illumination schemes provide illumination of a thin section of the sample but achieve this in very different ways.

TIR illuminates a thin section adjacent to the cover slide. Because of TIR the light cannot travel into the sample but instead creates an exponentially decaying, non-propagating, electromagnetic field, called evanescent field, at the cover slide surface with a penetration depth of ~30–300 nm, depending on the wavelength and the angle of TIR.

$$\theta_c = \arcsin\left(\frac{n_2}{n_1}\right) \tag{5.73a}$$

$$d = \frac{\lambda}{4\pi}\left(n_1^2 \sin^2\theta - n_2^2\right)^{-\frac{1}{2}} \tag{5.73b}$$

Here, θ_c represents the critical angle above which light is totally internal reflected, and d is the penetration depths (decay length of the exponential) of the evanescent wave, for $\theta \geqslant \theta_c$. Within this evanescent wave, fluorophores can still be excited. This was first used for single spot FCS with derivation of the appropriate ACF models [3, 70–72]. In imaging FCS this approach is extended to a whole camera, where every pixel is a detection element and works as a pinhole. This method is referred to as imaging TIR-FCS or ITIR-FCS.

Let's start by deriving the ACF for the 2D case in the xy-plane. Because of the different illumination intensities compared to confocal FCS or ICS we have to start by defining again the four characteristics that define the correlation function:

i) The size of the sample $S(\vec{r})$: here we assume that the sample is confined to 2D but within the xy-plane the sample can be infinitely large. This situation corresponds to the case when diffusion in a lipid membrane is measured.
ii) The intensity distribution of the excitation light $I(\vec{r})$: in TIR this can be considered constant as the distance over which the intensity varies (>10 μm) is typically much larger than the size of the pixel in the sample (< 1 μm).
iii) The collection efficiency of the optical system $CEF(\vec{r})$: as detection elements we use square pixels of a camera. However, the fluorophores, which we can consider point particles, will create a diffraction pattern in the shape of the PSF on the camera. So the effective detection area will be the convolution of the pixel area with the PSF of the optical system.
iv) The fluctuation causing process: here we consider Brownian diffusion and we can use the propagators we have derived earlier (math box 5.3).

With these assumptions we can derive the ACF model for camera-based FCS. In math box 5.7, we calculate the convolution of the pixel with the PSF to arrive at the effective observation area for a pixel [73]:

$$A_{\text{eff}} = a^2 \otimes \text{PSF} = \frac{a^2}{\left(\frac{w_0}{a\sqrt{\pi}}\left(e^{\frac{a^2}{w_0^2}} - 1\right) + \text{erf}\left(\frac{a}{w_0}\right)\right)^2} \tag{5.74}$$

We can use this detection area together with the assumption that the excitation intensity is constant over the observation area to determine the ACF model for Brownian diffusion (math box 5.7):

Math box 5.7

The correlation function and observation area/volume for imaging FCS
First, we will derive the ACF for the TIR case. As in the case for FCS we can start by defining the elements needed to derive the ACF.

(a) The intensity profile of the exciting light source $I(\vec{r}, t)$. We assume here that the light source is constant in space and time and can be described by a constant value I_0.
(b) The CEF of the optical detection system CEF($\vec{r}$); this is determined by the PSF of the microscope and the pixel size and shape. In the case of confocal FCS, we approximated the combined action of intensity profile and CEF($\vec{r}$) by a 3D Gaussian. Here, we approximate the PSF as a 2D Gaussian function. The pixel size and shape we will take account of by integrating over a pixel, which is assumed to have uniform detection efficiency over its whole area. Here, we will assume square pixels, as that is the typical case on cameras. For round pixels the equations have been derived [72].
(c) The product of the fluorophore absorption cross-section, its quantum yield, and the overall detection efficiency (this accounts for the detector quantum efficiency, and any losses within the system) κ
(d) The extent of the sample $S(\vec{r})$, which we assume to extend over that whole plane, neglecting any extension in the z-direction, thus $S(\vec{r}) = 1$.
(e) The fluorophore concentration fluctuations $\delta C(\vec{r}, t)$

Using the information above, we can write the intensity fluctuations

$$\delta F(t) = \kappa \int\int I(\vec{r})\mathrm{CEF}(\vec{r})S(\vec{r})\delta C(\vec{r}, t)d\vec{r} = \kappa I_0 \int\int e^{\frac{-2x^2}{w_0^2}} e^{\frac{-2y^2}{w_0^2}} \delta C(\vec{r}, t)dxdy$$

From math box 5.3, we already know the diffusion propagator for 2D diffusion:

$$\langle \delta C(\vec{r}, t)\delta C(\vec{r}', t + \tau)\rangle = \frac{\langle C\rangle}{4\pi Dt} e^{-\frac{(x-x')^2+(y-y')^2}{4Dt}}$$

For the ACF, we therefore obtain

$$G(\tau) = \frac{\kappa^2 \int\int\int\int I(\vec{r})\mathrm{CEF}(\vec{r})S(\vec{r})\langle \delta C(\vec{r}, t)\delta C(\vec{r}', t + \tau)\rangle dxdx'dydy'}{\left(\int\int \kappa I(\vec{r})\langle C\rangle \mathrm{CEF}(\vec{r})dxdy\right)^2}$$

$$= \frac{\iint_0^a \iint_0^a \iint_{-\infty}^{\infty} \iint_{-\infty}^{\infty} e^{\frac{-2(x^2+x_0^2)}{w_0^2}} e^{\frac{-2(y^2+y_0^2)}{w_0^2}} e^{\frac{-2(x'^2+x'^2_0)}{w_0^2}} e^{\frac{-2(y'^2+y'^2_0)}{w_0^2}} \frac{e^{-\frac{(x-x')^2+(y-y')^2}{4Dt}}}{4\pi Dt} dxdx'dydy'dx_0dx_0'dy_0dy_0'}{\left(\int\int e^{\frac{-2x^2}{w_0^2}} e^{\frac{-2y^2}{w_0^2}} dxdy\right)^2}$$

where the common factors $\kappa^2 I_0^2\langle C\rangle$ in numerator and denominator were cancelled out. Note that we integrate over whole space ($-\infty$ to ∞) to obtain the contribution of all fluorescent particles. Then we integrate over the pixel area (0 to a) to obtain the detected fluorescence. In confocal FCS these integrations over the pixel/pinhole area, were subsumed in the approximation of $W(\vec{r}) = I(\vec{r})\mathrm{CEF}(\vec{r})$ as a 3D Gaussian.

The integrals can be solved and lead to the final solution

$$G(\tau) = \frac{1}{\langle C\rangle a^2}\left[\frac{\sqrt{4D\tau + w_0^2}}{a\sqrt{\pi}}\left(e^{\frac{a^2}{4D\tau + w_0^2}} - 1\right) + \mathrm{erf}\left(\frac{a}{\sqrt{4D\tau + w_0^2}}\right)\right]^2$$

For SPIM-FCS, the third dimension has to be included. This is determined by the light sheet profile, which is a Gaussian function (see section 5.7.7). Thus, the solution in the z-direction is the same as for confocal FCS and we obtain for the SPIM-FCS model

$$G(\tau) = \frac{1}{\sqrt{\pi}\, w_z \langle C\rangle a^2}\left[\frac{\sqrt{4D\tau + w_0^2}}{a\sqrt{\pi}}\left(e^{\frac{a^2}{4D\tau + w_0^2}} - 1\right) + \mathrm{erf}\left(\frac{a}{\sqrt{4D\tau + w_0^2}}\right)\right]^2\left(1 + \frac{4D\tau}{w_z^2}\right)^{-\frac{1}{2}}$$

The extent of the observation volume in the z-direction, w_z, is determined by the Gaussian profile of the light sheet, w_{ill}, and the spatial filtering effect of the pixel, that acts as a pinhole. The pinhole effect of the pixel can also be modeled by a Gaussian with a radius w_{det}. The value of w_z is then given by

$$\frac{1}{w_z^2} = \frac{1}{w_{ill}^2} + \frac{1}{w_{det}^2}$$

In the case that the pixel is large, $1/w_{det}^2$ can be neglected and $w_z^2 = w_{ill}^2$.

The actual mathematical derivations of the imaging FCS effective observation area (ITIR-FCS) and volume (SPIM-FCS) are more intricate and have been published before [73]. Here we demonstrate a simpler way to determine the observation area/volume for imaging FCS by demanding that the amplitude of the ACF is equal to the inverse of the number of particles N:

$$G(0) = \frac{1}{N}$$

Using the just derived correlation function:

$$G(\tau) = \frac{1}{\langle C\rangle a^2}\left[\frac{\sqrt{4D\tau + w_0^2}}{a\sqrt{\pi}}\left(e^{\frac{a^2}{4D\tau + w_0^2}} - 1\right) + \mathrm{erf}\left(\frac{a}{\sqrt{4D\tau + w_0^2}}\right)\right]^2$$

We can see that

$$G(0) = \frac{A_{\mathrm{eff}}}{\langle C\rangle a^2 A_{\mathrm{eff}}}\left[\frac{w_0}{a\sqrt{\pi}}\left(e^{\frac{a^2}{w_0^2}} - 1\right) + \mathrm{erf}\left(\frac{a}{w_0}\right)\right]^2$$

where we multiplied the right-hand side by 1 ($=A_{\mathrm{eff}}/A_{\mathrm{eff}}$). Using the definition that $N = \langle C\rangle A_{\mathrm{eff}}$ we can write

$$G(0) = \frac{A_{\mathrm{eff}}}{Na^2}\left[\frac{w_0}{a\sqrt{\pi}}\left(e^{\frac{a^2}{w_0^2}} - 1\right) + \mathrm{erf}\left(\frac{a}{w_0}\right)\right]^2$$

Therefore, for $G(0) = 1/N$, the following has to hold:

$$A_{\text{eff}} = \frac{a^2}{\left[\frac{w_0}{a\sqrt{\pi}}\left(e^{\frac{a^2}{w_0^2}} - 1\right) + \operatorname{erf}\left(\frac{a}{w_0}\right)\right]^2}$$

For SPIM-FCS the same approach leads to

$$V_{\text{eff}} = \frac{\sqrt{\pi}\, w_z a^2}{\left[\frac{w_0}{a\sqrt{\pi}}\left(e^{\frac{a^2}{w_0^2}} - 1\right) + \operatorname{erf}\left(\frac{a}{w_0}\right)\right]^2}$$

With these definitions we arrive at the final ACF models for TIR- and SPIM-based FCS:

$$\text{ITIR} - \text{FCS}: G(\tau) = \frac{1}{N}\frac{\left[\frac{\sqrt{4D\tau + w_0^2}}{a\sqrt{\pi}}\left(e^{\frac{a^2}{4D\tau + w_0^2}} - 1\right) + \operatorname{erf}\left(\frac{a}{\sqrt{4D\tau + w_0^2}}\right)\right]^2}{\left[\frac{w_0}{a\sqrt{\pi}}\left(e^{\frac{a^2}{w_0^2}} - 1\right) + \operatorname{erf}\left(\frac{a}{w_0}\right)\right]^2}$$

$$\text{SPIM} - \text{FCS}: G(\tau) = \frac{1}{N}\frac{\left[\frac{\sqrt{4D\tau + w_0^2}}{a\sqrt{\pi}}\left(e^{\frac{a^2}{4D\tau + w_0^2}} - 1\right) + \operatorname{erf}\left(\frac{a}{\sqrt{4D\tau + w_0^2}}\right)\right]^2\left(1 + \frac{4D\tau}{w_z^2}\right)^{-\frac{1}{2}}}{\left[\frac{w_0}{a\sqrt{\pi}}\left(e^{\frac{a^2}{w_0^2}} - 1\right) + \operatorname{erf}\left(\frac{a}{w_0}\right)\right]^2}$$

We would like to note here, that this SPIM-FCS function is an approximation, as it assumes that the observation volume of a pixel is limited in the xy-directions even above and below the focal plane (figure 5.24). In confocal FCS this assumption is justifiable as the laser beam falls off in all the xy-dimensions as a Gaussian and thus particles further away than the beam radius contribute very little to the signal independent of their position along the z-axis. However, in SPIM-FCS the intensity profile in the xy-dimensions is constant. Therefore, particles that are out of the focal plane still contribute to the signal even if they are laterally displaced. This function here neglects cross-talk. It can still fit the experimental correlation function but will be somewhat biased (we estimate that the diffusion coefficient is underestimated by about 30% but this will depend somewhat on the exact experimental conditions). Including cross-talk in the derivation will lead to equations that cannot be integrated and numerical solutions need to be used.

$$G(\tau) = \frac{1}{N}\left[\frac{\frac{\sqrt{4Dt + w_0^2}}{a\sqrt{\pi}}\left(e^{\frac{a^2}{4Dt + w_0^2}} - 1\right) + \operatorname{erf}\left(\frac{a}{\sqrt{4Dt + w_0^2}}\right)}{\frac{w_0}{a\sqrt{\pi}}\left(e^{\frac{a^2}{w_0^2}} - 1\right) + \operatorname{erf}\left(\frac{a}{w_0}\right)}\right]^2 \tag{5.75}$$

This function is very different from the ACF for confocal FCS and the ICS cases. It therefore has also somewhat different properties. The first to note is that the half-amplitude of the ACF no longer relates directly to the diffusion coefficient, although the half amplitude is of course still a measure of the diffusion coefficient. Wider correlation functions correspond to slower diffusing particles. We also no longer define a diffusion time τ_D. If the pixel size a and the PSF radius w_0 are known, we can directly use the diffusion coefficient D as a fit parameter. At first this seems not to make much difference. But unlike in confocal FCS, this means one does not have to calibrate the system on a daily basis to determine the observation volume. The pixel size a is precisely known from the camera specifications, and w_0 needs to be determined for an optical system only once, as it will not change over time. Therefore, in principle absolute diffusion coefficients can be measured by camera-based FCS, assuming that the underlying molecular processes are known. However, TIR is restricted to an observation plane directly adjacent to the coverslip. While this is an ideal geometry for the observation of bilayers and membranes an extension to 3D is desirable. For TIRF the FCS models for diffusion in the axial z-direction have been derived [3, 71]. But this is still restricted to samples at the coverslip surface. Here, we introduce another illumination scheme, that can illuminate a cross section in a 3D sample, single plane illumination microscopy (SPIM).

In SPIM, the illuminated plane is created by focusing a laser beam by a cylindrical lens. As the cylindrical lens focuses the laser beam only in one lateral direction one obtains a strongly elliptical laser beam or a light sheet. The functional form of the intensity distribution is given by a Gaussian Beam with

$$\begin{aligned} I(x, y, z) &= I_0\left[\frac{w_{0x}}{w_x[z]}\right]\left[\frac{w_{0y}}{w_y[z]}\right]e^{-\frac{2x^2}{w_x^2[z]}}\,e^{-\frac{2y^2}{w_y^2[z]}} \\ &= \frac{I_0}{\sqrt{1+(z/z_{0x})^2}\sqrt{1+(z/z_{0y})^2}}e^{-2\left(\frac{x^2}{w_x^2[z]}+\frac{y^2}{w_y^2[z]}\right)} \end{aligned} \tag{5.76}$$

and

$$w_i[z] = w_{0i}\sqrt{1+(z/z_{0i})^2} \tag{5.77a}$$

$$z_{0i} = \frac{\pi w_{0i}^2}{\lambda} \tag{5.77b}$$

and $i = x$ or y. The parameter z_0 is called the Rayleigh range and describes the distance at which the size of the beam has increased by $\sqrt{2}$. The depth of focus is just twice the Rayleigh range, $2 \times z_{0i}$, and is a measure over which the beam stays within $\sqrt{2} \times w_{0i}$, and thus has only a moderate change of thickness (figure 5.26). Equation (5.77) describes one of the dilemmas of light sheet microscopy. It shows that the Rayleigh range depends on the minimum width of the light sheet. In FCS we would like to have a thin light sheet to minimize the observation volume. But then the

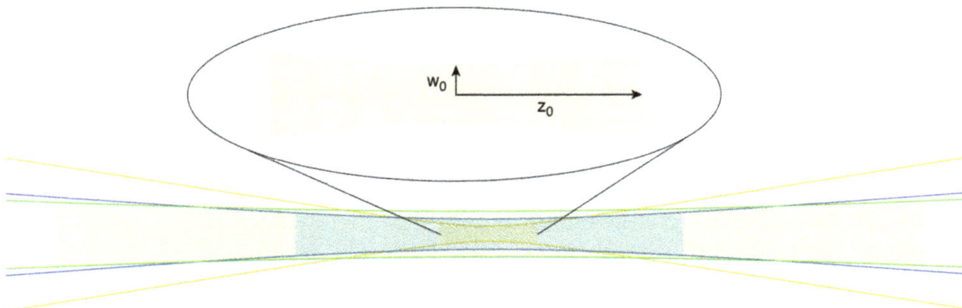

Figure 5.26. Light sheets for three different thicknesses of 1, 2, and 3 μm, assuming a wavelength of 500 nm, depicted in yellow, blue, and green (for values see table 5.1). The shaded areas depict the region within the depth of focus, i.e. between $-z_0$ and z_0. The figure illustrates how the thickness of the light sheet influences the length of the light sheet where the variation in thickness is below a factor $\sqrt{2}$ (equation (5.77a)). As the thickness influences the measured diffusion time, one either needs to measure in a region where the thickness does not vary strongly or needs to correct the measured diffusion coefficient for the varying thickness over the image. On the other hand, in FCS one aims to use thin light sheets to minimize the observation volume to be able to observe the fluctuations from single molecules.

Table 5.1. Relation between light sheet thickness (w_0) and length as measured by the depth of focus ($2 \times z_0$), assuming a wavelength of $\lambda = 500$ nm.

w_0[μm]	1	1.2	1.5	2	3	5	10
$2 \times z_0$[μm]	13	18	28	50	113	314	1257

usable light sheet length, as measured by the depth of focus, is small. This problem is exacerbated as the depth of focus depends on the square of the width.

Table 5.1 shows how the usable depth of focus changes with beam width. In SPIM the beam is asymmetric and the Rayleigh ranges for the x and y dimensions of the elliptical Gaussian beam will also be different.

The light sheet can be used to illuminate any plane in a sample. Using a detection objective orthogonal to the light sheet, the light sheet is brought to the focal plane of the detection objective (figure 3.6). Light sheets with a thickness of ~ 1 μm and a length of ~ 10 μm have been used for FCS [74]. As now fluorescence is only excited in the focal plane, cross-talk between pixels is minimized, although not completely eliminated, and essentially the whole illuminated area can be recorded on a camera.

Without loss of generality, we assume that the cylindrical lens focuses only the y-direction. In that case, the thickness of the light sheet will be given by w_{0y} and the usable length of the light sheet by $2 \times z_{0y}$. In the x-direction, the light sheet width is only limited by the size of the objective and can be as large as a millimetre or more. Therefore, the intensity variation in the xz-plane is very slow and over the size of a pixel the intensity can be regarded as constant. In the y-dimension, the intensity changes as a Gaussian function. And therefore the FCS fitting function has to take account of both, the constant intensity in the xz-plane and the Gaussian intensity change in the y-dimensions, to describe the full shape of the ACF. Thus, the SPIM effective observation volume and the ACF show the same dependence as TIR for the xz-plane, while in the y-dimension, the Gaussian shape of the intensity profile leads to a similar ACF as in confocal FCS (math box 5.7):

$$V_{\text{eff}} = \frac{\sqrt{\pi} w_z a^2}{\left[\frac{w_{xy}}{a\sqrt{\pi}}\left(e^{-\frac{a^2}{w_{xy}^2}} - 1\right) + \text{erf}\left(\frac{a}{w_{xy}}\right)\right]^2} \tag{5.78}$$

$$g(\tau) = \frac{1}{N} \frac{\left[\frac{\sqrt{4D\tau + w_{xy}^2}}{a\sqrt{\pi}}\left(e^{-\frac{a^2}{4D\tau + w_{xy}^2}} - 1\right) + \text{erf}\left(\frac{a}{\sqrt{4D\tau + w_{xy}^2}}\right)\right]^2}{\left[\frac{w_{xy}}{a\sqrt{\pi}}\left(e^{-\frac{a^2}{w_{xy}^2}} - 1\right) + \text{erf}\left(\frac{a}{w_{xy}}\right)\right]^2} \left(1 + \frac{4D\tau}{w_z^2}\right)^{-\frac{1}{2}} \tag{5.79}$$

It should be noted that this is only an approximation. In confocal FCS the illumination is limited in the focal plane due to the focused laser beam. But in SPIM-FCS the sheet illumination is extended over the whole focal plane. Therefore, particles that are outside the focal plane but still within the light sheet can contribute to the signal in the pixel that would not be seen in the confocal system [75]. However, for this case there is no analytic solution and numerical approaches have to be used to account for this additional signal.

We should pause here to make some estimations of the time a molecule takes to diffuse across a pixel and across an image in relation to the speed with which the camera records images. The mean squared displacement (MSD) of a molecule for Brownian diffusion in two dimensions is given by

$$\langle x^2 \rangle = 4Dt \tag{5.80}$$

First, we need to estimate some typical value ranges for these parameters. Typical values for the diffusion coefficient range from about 0.1 $\mu m^2\ s^{-1}$ for a protein in a membrane to about 400 $\mu m^2\ s^{-1}$ for a small organic dye in solution. The fastest recording time per frame, t, that is reachable by cameras today can be as low as 1–10 μs and as high as 1 ms. These ranges lead to the furthest and smallest distance a molecule will diffuse while being recorded. The furthest distance a molecule can move during the time a frame is acquired is given by the MSD for the fastest diffusion coefficient and the slowest recoding time per frame:

$$\sqrt{\langle x^2 \rangle} = \sqrt{4 \times 400\ \mu m^2\ s^{-1} \times 10^{-3}\ s} \approx 1.3\ \mu m$$

The shortest distance a molecule can move during the recording of a frame will be determined by the slowest expected diffusion coefficient and the fastest recording time (assuming 10 μs):

$$\sqrt{\langle x^2 \rangle} = \sqrt{4 \times 0.1\ \mu m^2\ s^{-1} \times 10^{-5}\ s} = 2\ nm$$

These two values set limits to the correlations that can be measured. In the first case, for instance, a particle will move by about 1.3 μm each frame. Assuming that pixel sizes between 100–400 nm in object space are used, this corresponds typically to 3–13 pixels. Therefore, the molecule will not be detected in the same pixel in subsequent frames and the correlation cannot be measured. A solution to that can be either the use of a faster camera or the use of larger pixels. The camera speeds are limited by the available technology, but with a recording speed of 0.01 ms one can reach a limit that still allows recording of fast diffusing fluorophores in solution, if the pixels are not too small:

$$\sqrt{\langle x^2 \rangle} = \sqrt{4 \times 400\ \mu\text{m}^2\ \text{s}^{-1} \times 10^{-5}\ \text{s}} \approx 126\ \text{nm}$$

If a camera cannot record that fast, then pixel sizes larger than 130 nm can be used. However, note that the pixel size will be limited as a larger pixel means a larger observation area and thus smaller amplitudes due to the $1/N$ proportionality of the ACF, thus limiting the pixel size.

At the lower end, slowly diffusing particles cannot be detected if they do not move sufficiently far to ultimately leave the observation area of a pixel. In that case one might choose to lower the acquisition speed of the camera and possibly the illumination intensity. Using a slower speed increases the distance a particle moves during acquisition and with sufficiently small pixels correlations can be recorded:

$$\sqrt{\langle x^2 \rangle} = \sqrt{4 \times 0.1\ \mu\text{m}^2\ \text{s}^{-1} \times 10^{-3}\text{s}} = 20\ \text{nm}$$

In this case the pixel observation area is limited by the PSF of the optical system and using too small pixels will reduce the signal intensity and decrease the signal-to-noise ratio.

5.7.8 The FCS diffusion law

We have seen in section 5.4 that for free diffusion the mean squared displacement is a linear function of time (figure 5.7). In other words, the time it takes a particle to diffuse through an area is just proportional to that area. For anomalous diffusion however, this is not the case and we expect a nonlinear behavior of the time a particle spends in an observed area as a function of that area. Therefore, if we can measure the diffusion time τ_D as a function of the area A observed we can draw conclusions about the diffusive mode of the particle.

There are two ways to achieve this in FCS. Either we calculate the ACF and from each point in the ACF we can extract the MSD. Or we measure FCS at various observation areas.

5.7.8.1 Extracting the mean squared displacement form the ACF

We discuss here the 2D case for simplicity but this can be extended to 3D [76]. Using equations (5.6) and (5.8) we can write the 2D ACF as

$$G(\tau) = \frac{1}{N}\left(1 + \frac{4D\tau}{w_0^2}\right)^{-1} + G_\infty \tag{5.81}$$

The mean squared displacement is given by $\langle x \rangle^2 = 4D\tau$, and thus

$$G(\tau) = \frac{1}{N}\left(1 + \frac{\langle x \rangle^2}{w_0^2}\right)^{-1} + G_\infty \tag{5.82}$$

This can be rearranged to obtain

$$\langle x \rangle^2 = \left(\frac{1}{N(G(\tau) - G_\infty)} - 1\right)w_0^2 \tag{5.83}$$

This means that for each experimentally measured value $G(\tau)$ we can extract the MSD. Deviations from linearity then can indicate deviations from free diffusion. This has been used to characterize anomalous diffusion and identify non-Gaussian processes in variable-lengthscale FCS [11, 12].

5.7.8.2 Measuring FCS at various observation areas: the FCS diffusion law

If we perform FCS on a sample with different observation areas then we can plot the diffusion time versus the area, the so-called diffusion law plot. Again, derivations from linearity indicate different diffusive modes. This was exploited for the first time in FRAP [77, 78], and was later adapted for FCS [79].

Let's contrast three different cases (figure 5.27). The first case is free diffusion. As already mentioned, one expects a linear plot of the diffusion time versus the observation area with the line crossing the y-axis at a 0 intercept, independent of the length scale or size of the observation area. As the second case, consider a particle that diffuses within a meshwork. Within the meshes the particle diffuses freely and thus we have a linear behavior of the diffusion law plot for small observation areas with a 0 y-intercept. However, once the particle encounters a barrier in the meshwork it requires a certain energy to cross that barrier and thus will have a certain probability whether it can hop over the barrier or stay within the current mesh. This mode of diffusion is therefore called hop diffusion. The hopping over the barriers will slow down diffusion. Assuming a constant average mesh size, this means that for observation areas larger than the mesh size we will observe a slower average diffusion coefficient which is determined by the meshwork. A slower diffusion coefficient means a change in slope of the diffusion law plot. As a matter of fact, the slope will increase as it will take now more time for a particle to diffuse farther. If that second part of the diffusion law plot, at large lengths scales, is interpolated to 0, it will show a negative y-intercept. This diffusive mode is well understood and a theoretical treatment has been given [80]. A typical example is proteins in the cellular plasma membrane which can show hop diffusion if they are hindered by the cytoskeleton.

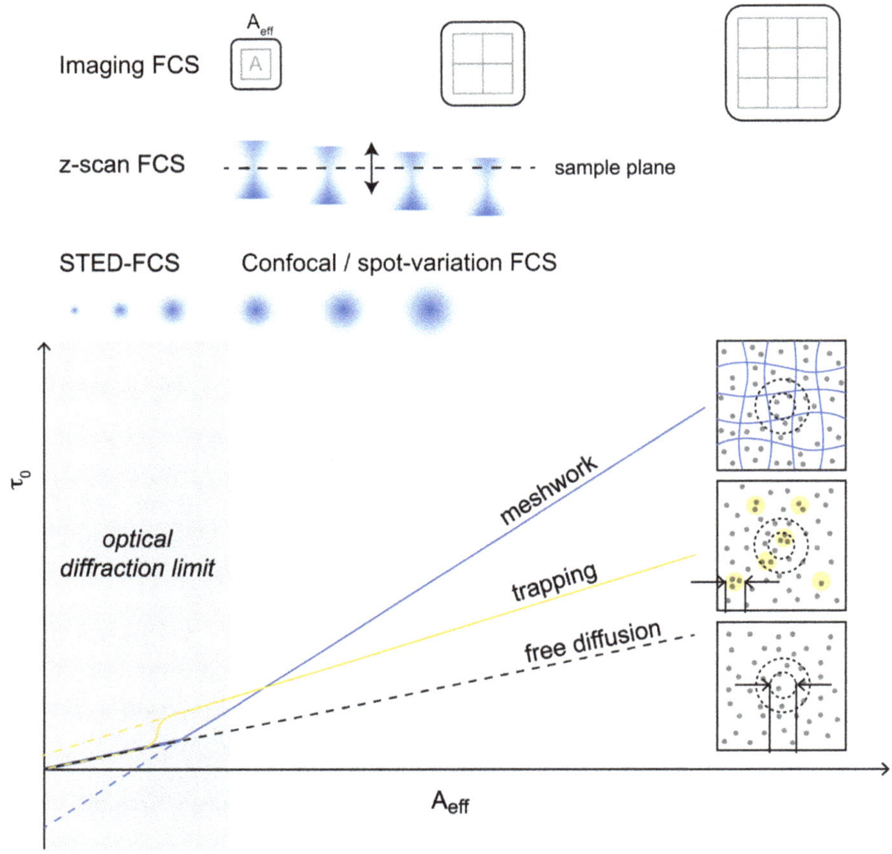

Figure 5.27. The FCS diffusion law. The upper part of the graph shows the various ways to produce observation areas of different size. This is achieved in different ways. In confocal or spot-variation FCS, the spot size in the sample plane can be varied by changing the size of the laser beam before the objective, with a smaller beam diameter resulting in a larger spot size. In STED-FCS this can be tuned by the intensity of the depletion laser beam. In z-scan FCS the laser beam is scanned orthogonally across the sample plane and thus creates spots of different size on the membrane. Finally, in imaging FCS, one can bin multiple pixels to obtain different observation areas, with the advantage that this can be done on one single data set, while the other methods require multiple measurements, for each spot size at least one. The lower part of the graph shows the different diffusion modes and their diffusion law plots as discussed in the text.

In the third case, we consider particles that are transiently trapped at trapping sites distributed throughout the plane of the sample. Again, when measuring at small length scales, much smaller than the free distances between trapping sites, one expects free diffusion. But when the size of the observation area is on the order of the inter-trapping site distances, then diffusion becomes complex. Depending on the likelihood to be trapped and the time length of trapping, the diffusion law plot can change its slope in a more complex behavior. The slope will increase, as trapping slows down diffusion. But at larger observation areas with sizes much larger than the inter-trapping site distances, i.e. when the observation areas contain many trapping sites, we obtain again an average diffusion coefficient. This diffusion coefficient is expected to be lower than the one for free diffusion, thus having a larger slope. Researchers have found mainly positive intercepts for these cases [81, 82]. However, in some cases negative intercepts were found in lipid bilayers despite the hindrance being conveyed by trapping in lipid domains [83, 84]. In these latter cases, typically domain densities were high, close to the percolation threshold, indicating that in these systems the domains act similar to barriers for diffusion. Simulations for various cases of transient binding and nanodomain confinement demonstrate that they result in positive intercepts that can be further analyzed to obtain detailed information on the interaction processes [85].

With the exception of STED-FCS, other FCS modalities are typically constrained to diffraction limited observation areas of ~200 nm radius or more. Thus, one normally cannot observe the first parts of the diffusion law plot but see only the behavior at length scales that are larger than the mesh works and the trapping site dimensions. This is particularly true when measuring diffusion on cell membranes as the cytoskeleton and lipid domains are assumed to have dimensions less than 100 nm. So typically, we cannot observe the full diffusion law plot. Therefore, we extrapolate the data at large length scales to 0 area and use the y-intercept to distinguish different diffusive modes. In general, this is done with an equation of the following form:

$$\tau_D(A_{\mathrm{eff}}) = \tau_0 + \frac{A_{\mathrm{eff}}}{4D_{\mathrm{eff}}} \tag{5.84}$$

Here, A_{eff} represents the effective observation area, which can be given by the confocal spot size in confocal or STED-FCS or the observation area of a camera pixel in imaging FCS. The y-intercept, τ_0, is used to interpret the diffusion mode based on its magnitude and whether it is positive, negative or 0 as discussed above.

Lastly, it is possible that multiple diffusive modes coexist and thus the actual intercept of the diffusion law plot at large areas is a weighted average between the intercepts of the individual diffusive modes. To differentiate between these possibilities, it is therefore necessary to measure diffusion in the sample before and after disruption of the diffusion obstacles, be it trapping sites or diffusion barriers. It is then the change of the intercept that provides information on the diffusive mode [81, 86].

As shown in figure 5.27 the diffusion law plot can be created by various means. One can change the confocal spot by changing the beam size (see chapter 3 and the accompanying CDF file—https://iopscience.iop.org/book/978-0-7503-2080-1) and measure at different areas. One can scan the laser beam in the z-direction orthogonal to the sample plane, so-called z-scan FCS [87, 88]. At different positions different area sizes are illuminated due to the diverging nature of the focused laser beam. Or one can use STED-FCS which provides various beam sizes depending on the depletion laser power [89], giving access to dimensions below the diffraction limit. Or one can use the data in imaging FCS. By calculating the ACF for different $n \times n$ pixel binning, one obtains the diffusion plot from a single measurement [90, 91]. Finally, a similar evaluation can be performed with STICS, deriving a mean-squared displacement plot versus time delay from the evolution of the spatial correlation function in time [92]. The diffusion law has been used to determine the diffusion modes of lipids and proteins in artificial bilayers, cells and organisms (see, e.g. [93–95]).

5.8 Other FCS modalities

Many other FCS modalities have been implemented over the years using various illumination schemes changing the illumination pattern in space and time, as well as changing the detection modalities. While we cannot discuss all of them in this book, we hope that the FCS basics from this chapter will enable the reader to understand the literature and we therefore provide a short, albeit incomplete, overview of several FCS techniques developed over the decades that address particular weak points in the FCS scheme, to provide starting points for the interested reader to delve deeper into the potentials of FCS.

As we have seen in this chapter, FCS and its various modalities rely on the fact that we possess a good model of the actual observation volume. In confocal FCS this is, for instance, a three-dimensional Gaussian function (math box 5.2) that is defined by the Gaussian beam characteristics of the focused laser and the pinhole. There have been a number of approaches that either characterize the observation volume experimentally to improve data fitting and parameter recovery or that engineer new illumination geometries in various experimental situations. Culbertson *et al* measured the confocal 3D PSF and used this PSF to simulate the experimental situation with different parameters until they obtained a best fit [96]. This is a powerful approach as it no longer relies on a fit model but uses the best determined experimental situation and then uses our understanding of different molecular processes to find the best process and parameters that fit the situation. A disadvantage of this method is that simulating different situations takes a much longer time than data fitting. However, with computers getting faster and more efficient, this strategy becomes more realistic. Another limitation in FCS is the diffraction limit of the observation volume. This has been addressed by zero-mode waveguide FCS [97] and stimulated emission depletion (STED) FCS [89], both of which reduce the observation volume of FCS considerably. Due to the universal applicability of STED, it has been combined with a variety of other FCS modalities, including STED-RICS [63], TIRF-STED-FCS with almost isotropic resolution of the observation volume of ~50 nm [98], Line-scan STED-FCS reducing problems of photobleaching [99], STED-FLCS [100], and it has been implemented with adaptive optics to extend STED-FCS to 3D [101]. The necessity of probe labelling has been circumvented by inverse FCS, in which the solution is filled with a fluorophore and the fluctuations caused by unlabelled particles diffusing through the observation volume are recorded and correlated [102]. The problem of the fluorescence background and its removal has been addressed by FLCS, and by time-gated FCS [103], which has also been shown to be able to resolve chemical reaction and conformational fluctuations. The concentration range of FCS, which is limited by the signal-to-noise ratio and the correlated noise of the system, mainly caused by the stability of the lasers, can be considerably extended, reaching 10s of μM or more, when using stabilized lasers and high count rate detectors [104]. Scanning FCS has been used to correct measurements for sample movement and characterize ligand–receptor interactions *in vivo* [49, 50]. And the need for calibration in confocal FCS has been circumvented by dual-focus FCS [39], which provides absolute diffusion coefficients and precise particle sizes. Its extension to dual-colour dual-focus line-scanning FCS [105], is particularly well adapted to live cell measurements by eliminating sample artefacts and providing absolute concentrations and diffusion coefficients. For further information on FCS modalities we refer the reader to regular published reviews, which often aim at specific scientific areas in which FCS has been used as well as articles reviewing particular FCS modalities.

5.9 Synopsis

- Theoretical correlation functions are calculated by taking account of the extent of the sample, the excitation profile, the detection efficiency of the system (the latter two define the observation profile), and a function that describes the probability of the system to be in two particular states at two times that differ by a lagtime τ (e.g. for diffusion this is the diffusion propagator).
- Various processes can cause fluctuations, including translational and rotational diffusion or photophysical processes, for which the theoretical correlation function can be derived. The different processes can be differentiated by varying experimental conditions (observation volume extent, laser power, viscosity, pH, etc).
- Fluorescence cross-correlation spectroscopy (FCCS) between signals of different color allows quantification of interactions.
- Image correlation spectroscopy (ICS) exploits the spatial correlations in an image to characterize number and size of clusters.
- There are various forms of spatiotemporal FCCS that exploit not only the correlation in time but also in space.
- Scanning FCS can be used to characterize flow and transport
- Raster image correlation spectroscopy (RICS) exploits the inherent time structure in laser scanning confocal images to obtain information on diffusion and concentrations.
- Spatiotemporal ICS (STICS) and imaging FCS use the simultaneous recording of all pixels in an image to calculate ICS and FCS curves at each pixel. They not only multiplex FCS but also provide new information by including spatial correlations.
- The dependence of the diffusion coefficient on the length scale over which it is measured, the so-called FCS diffusion law plot, can provide information on diffusive modes, even if that is determined by structures well below the optical diffraction limit.

Table 5.2 provides a summary of the ACF models discussed in this chapter. In the right column, we provide definitions of the different parameters used. To avoid

Table 5.2. Summary of the ACF models.

FCS	ACF	Parameter definitions
2D free diffusion (symmetric)	$G(\tau) = \frac{1}{N}\left(1 + \frac{\tau}{\tau_D}\right)^{-1} + G_\infty$	$\tau_D = \frac{w_0^2}{4D}$ $w_0 = w_{xy}$: e^{-2} radius in the focal (xy-) plane D: diffusion coefficient $N = \langle C \rangle \pi^{3/2} w_0^2 z_0 G_\infty$: convergence value for $t \to \infty$
2D free diffusion (asymmetric)	$G(\tau) = \frac{1}{N}\left(1 + \frac{\tau}{\tau_D}\right)^{-1/2}\left(1 + \frac{\tau}{K^2\tau_D}\right)^{-1/2} + G_\infty$	w_0, w_1: e^{-2} radii in two orthogonal directions (any of x, y, z)$K = \frac{w_1}{w_0}$
Free diffusion in 3D	$G(\tau) = \frac{1}{N}\left(1 + \frac{\tau}{\tau_D}\right)^{-1}\left(1 + \frac{\tau}{K^2\tau_D}\right)^{-1/2} + G_\infty$	$w_z = z_0$: e^{-2} radius along the optical axis $K = \frac{z_0}{w_0}$
Free diffusion including rotation and photophysics	$G(\tau) = \frac{1}{N}\left[\left(1 + \frac{\tau}{\tau_D}\right)^{-1}\left(1 + \frac{\tau}{K^2\tau_D}\right)^{-1/2}\right]\left(1 - \frac{1}{N_b}e^{-\tau/\tau_A}\right)(1 + Re^{-\tau/\tau_R})$ $\times\left(1 + \frac{f_T}{N_b(1-f_T)}e^{-\tau/\tau_T}\right) + G_\infty$	τ_A: antibunching; fluorescence lifetime N_b: labels per molecule $\tau_R = \frac{1}{6D_R}$; $D_R = \frac{kT}{8\pi\eta r_h^3}$ R: normalization factor; to be fitted τ_T: triplet time f_T: triplet fraction
Anomalous diffusion	$G(\tau) = \frac{1}{\langle C \rangle \pi^{3/2} K w_0^3}\left(1 + \left(\frac{\tau}{\tau_D}\right)^\alpha\right)^{-1}\left(1 + \frac{1}{K^2}\left(\frac{\tau}{\tau_D}\right)^\alpha\right)^{-1/2} + 1$	$D_t(t) = \frac{1}{2m}\frac{\partial \mathrm{MSD}}{\partial t} = \frac{\alpha}{2m}\Gamma t^{\alpha-1}\tau_D = \frac{\alpha w_0^2}{4D_t} = \frac{w_0^2}{2/3\Gamma t^{\alpha-1}}$
Diffusion and flow	$G(\tau) = \frac{1}{N}\left(1 + \frac{\tau}{\tau_D}\right)^{-1}\left(1 + \frac{\tau}{K^2\tau_D}\right)^{-1/2} e^{-\frac{v_x^2\tau^2 + v_y^2\tau^2}{4D\tau + w_0^2} - \frac{v_z^2\tau^2}{4D\tau + K^2 w_0^2}}$	v_i: flow speed in the ith direction
2 focus FCS for diffusion and flow	$G(\tau) = \frac{1}{N}\left(1 + \frac{\tau}{\tau_D}\right)^{-1}\left(1 + \frac{\tau}{K^2\tau_D}\right)^{-1/2} e^{-\frac{(v_x\tau - \rho_x)^2 + (v_y\tau - \rho_y)^2}{4D\tau + w_0^2} - \frac{(v_z\tau - \rho_z)^2}{4D\tau + z_0^2}}$	ρ_i: displacement between foci in i^{th} direction
Flow and scan in 2D	$G(\tau) = \frac{1}{N}\left(1 + \frac{\tau}{\tau_D}\right)^{-1} e^{-\frac{(\vec{v}_{\text{flow}} + \vec{v}_{\text{scan}})^2\tau^2}{4D\tau + w_0^2}} + G_\infty$	$N = \langle C \rangle \pi^{3/2} w_0^2 \vec{v}_{\text{flow}}$: flow velocity $\vec{v}_{scan}$: scan velocity $v_{\text{nett}}^2 = (\vec{v}_{\text{flow}} + \vec{v}_{\text{scan}})^2 = v_{\text{flow}}^2 + v_{\text{scan}}^2 + 2v_{\text{flow}}v_{\text{scan}}\cos(\theta)$ θ: angle between $\vec{v}_{\text{flow}}$ and $\vec{v}_{\text{scan}}$

Table 5.2. *(Continued)*

FCS	ACF	Parameter definitions
Circular scanning in 2D	$G(\tau) = \frac{1}{N}\left(1 + \frac{\tau}{\tau_D}\right)^{-1} e^{-\frac{4\rho^2 \sin(\frac{\omega t}{2})^2 \tau^2}{4D\tau + w_0^2}} + G_\infty$	$N = \langle C \rangle \pi^{3/2} w_0^2 \omega$: angular scan frequency
RICS	$G_{\text{RICS}}(\tau) = G(\xi, \psi) e^{-\frac{a^2(\lvert\xi\rvert^2 + \lvert\psi\rvert^2)}{4D(\tau_p\lvert\xi\rvert + \tau_l\lvert\psi\rvert) + w_0^2}} + G_\infty G(\xi, \psi) = \frac{1}{N}\left(1 + \frac{4D(\tau_p \lvert \xi \rvert + \tau_l \lvert \psi \rvert)}{w_0^2}\right)^{-1} \left(1 + \frac{4D(\tau_p \lvert \xi \rvert + \tau_l \lvert \psi \rvert)}{z_0^2}\right)^{-1/2}$	a: pixel side length (assuming square pixels) τ_p: pixel dwell time τ_l: line scan time ξ and ψ are the spatial lags in the x and y direction
STICS	$G(\xi, \psi, \tau) = G(0,0, \tau) e^{-\frac{(\xi - x(\tau))^2 + (\psi - y(\tau))^2}{w^2(\tau)}} + G_\infty$	$G(0,0, \tau)$: time dependent peakDiffusion: $x(\tau) = y(\tau) = 0$, $w^2(\tau) = 4D\tau + w_0^2$Flow: $x(\tau) = v_x\tau$, $y(\tau) = v_y\tau$, $w^2(\tau) = w_0^2$Diffusion and Flow: $x(\tau) = v_x\tau$, $y(\tau) = v_y\tau$, $w^2(\tau) = 4D\tau + w_0^2$
Imaging FCS/ ITIR-FCS	$G(\tau) = \frac{1}{N}\left[\frac{\frac{\sqrt{4Dt + w_0^2}}{a\sqrt{\pi}}\left(e^{\frac{a^2}{4Dt + w_0^2}} - 1\right) + \text{erf}\left(\frac{a}{\sqrt{4Dt + w_0^2}}\right)}{\frac{w_0}{a\sqrt{\pi}}\left(e^{\frac{a^2}{w_0^2}} - 1\right) + \text{erf}\left(\frac{a}{w_0}\right)}\right]^2 + G_\infty$	
SPIM-FCS	There is no analytic solution due to cross-talk contributions. The solution has to be approximated by numerical integration. For the ACF the following function can be used approximately but will lead to underestimation of the real D by up to 30%: $G(\tau) = \frac{1}{N}\frac{\left[\frac{\sqrt{4D\tau + w_0^2}}{a\sqrt{\pi}}\left(e^{\frac{a^2}{4D\tau + w_0^2}} - 1\right) + \text{erf}\left(\frac{a}{\sqrt{4D\tau + w_0^2}}\right)\right]^2}{\left[\frac{w_0}{a\sqrt{\pi}}\left(e^{\frac{a^2}{w_{xy}^2}} - 1\right) + \text{erf}\left(\frac{a}{w_0}\right)\right]^2}\left(1 + \frac{4D\tau}{z_0^2}\right)^{-\frac{1}{2}} + G_\infty$	$w_0 = w_{xy}$: e^{-2} radius in the focal plane$w_z = z_0$: e^{-2} radius along the optical axis
Multiple components	$G(\tau) = \frac{\sum_{i=1}^{n} \kappa_i^2 f_i g(\tau)}{N(\sum_{i=1}^{n} \kappa_i f_i)^2} + G_\infty$	$g(\tau)$: any model described above

repetition we define in each row only new parameters not used in any of the previous rows and assume that all previous definitions are valid.

5.10 Exercises

Note that for exercises 4-11, you will need to have the computable document format player which is freely available at https://www.wolfram.com/cdf/. With the CDF Player you can open the 'Introduction to FCS—interactive exercises.cdf' file available at https://iopscience.iop.org/book/978-0-7503-2080-1.

1. Verify the transition from 5.3 to 5.10.

2. Extend the multi-particle ACF to n particles.

3. Equation (5.21) describes the dependence of the diffusion coefficient on the mass and hydrodynamic radius of the particle under investigation.
   a. What is the expected value of D if the particle changes its hydrodynamic radius by a factor 2?
   b. What is the expected value of D if the particle changes mass by a factor 2, assuming the particle remains spherical?
   c. If the particle changes mass, e.g. as would happen in dimerization, do you expect a change in D smaller or greater than what you calculated in b?

4. In the 'Confocal FCS' section of the 'Introduction to FCS—interactive exercises.cdf' (available at https://iopscience.iop.org/book/978-0-7503-2080-1), choose G2D1P and G3D1P for model and reference function, respectively for comparison.
   a. Keeping N, τ_D and G_∞ constant, change the structure factor K and observe the differences. At what value for K do the functions become visibly indistinguishable?
   b. What happens for $K = 1$.
   c. Using equation (5.16), explain how the differences between the 2D and 3D model develop for small and large K.

5. In the 'Confocal FCS' section of the 'Introduction to FCS—interactive exercises.cdf' (available at https://iopscience.iop.org/book/978-0-7503-2080-1), choose G3D2PQ for both, model and reference, functions. Keep the default values of the parameters, but set $\tau_{D2} = 0.004$ (s) for both functions.
   a. Change the q value (relative brightness, see equation (5.28)) to 2 in the model function. Which other parameter(s) do you need to change in the model function so that the reference and model function overlap again?
   b. Can you explain why these two parameters need to be changed but the other parameters stay constant? Verify this by changing the q value to other values (e.g. 3 or 4).

6. Choose G3D1P and G3D1P1T models for comparison. Keep common parameters equal for both models. And keep the triplet fraction at $F_{\text{trip}} = 0.5$.
   a. At what triplet time τ_{trip} is there no obvious transition anymore between the triplet and diffusive contribution (no two 'humps')?
   b. Does this change with the value of F_{trip}?

7. Investigate the influence of the different parameters on the correlation function. For this purpose, go to the section 'Noise and Background' of the 'Introduction to FCS—interactive exercises.cdf' (available at https://iopscience.iop.org/book/978-0-7503-2080-1). Start with $N = 1$, $cps = 1000$, a background of 100 counts per second, and 20 s measurement time.
   a. Change the measurement time to see the increase in noise. Does the amplitude change?
   b. Change the background to see the changes in amplitude. Does the noise change?
   c. If you start with the default values as stated above and change the number of particles, how does the difference between the ACFs with and without background change? Explain why.

8. Start as in question one with the default values but choose $cps = 100$, and a background of 1000 counts per second. Now change N from 1 to 30. How does the amplitude change? Why do you see this deviation from the background free case, where the amplitude always decreases with N?

9. Using the 'STICS and RICS.cdf' verify how the STICS function (first graph) varies with different parameters. In particular compare the influence of beam radius and diffusion coefficient on the peak shape, and observe how the peak position depends on the flow velocities in the x- and y-directions.

10. Select the 'Imaging FCS' section of the 'Introduction to FCS—interactive exercises.cdf' file (available at https://iopscience.iop.org/book/978-0-7503-2080-1).
    a. Compare ITIR-FCS and SPIM-FCS models for varying parameters; in particular investigate how the z-resolution (σ_z) influences SPIM FCS and how SPIM-FCS and ITIR-FCS converge fore large values of σ_z.
    b. Observe how spatial cross-talk leads to a non-zero amplitude at distances up to about 3 pixels in ITIR-FCS.

References

[1] Gennerich A and Schild D 2000 Fluorescence correlation spectroscopy in small cytosolic compartments depends critically on the diffusion model used *Biophys. J.* **79** 3294–306

[2] Thompson N L 1991 Fluorescence correlation spectroscopy *Topics in Fluorescence Spectroscopy* (Techniques) vol 1 ed J R Lakowicz (New York: Plenum Press) pp 337–78
[3] Hassler K, Leutenegger M, Rigler P, Rao R, Rigler R, Gösch M and Lasser T 2005 Total internal reflection fluorescence correlation spectroscopy (TIR-FCS) with low background and high count-rate per molecule *Opt. Express* **13** 7415–23
[4] Weidemann T, Wachsmuth M, Tewes M, Rippe K and Langowski J 2002 Analysis of ligand binding by two-colour fluorescence cross-correlation spectroscopy *Single Molecules* **3** 49–61
[5] Berg H C 1993 *Random Walks in Biology* (Princeton, NJ: Princeton University Press)
[6] Elson E and Magde D 1974 Fluorescence correlation spectroscopy. I. Conceptual basis and theory *Biopolymers* **13** 1–27
[7] Schmitz K S 2017 Nonequilibrium thermodynamics *Physical Chemistry* ed K S Schmitz (Boston: Elsevier) ch 14 pp 659–93
[8] Saxton M J 2007 Modeling 2D and 3D diffusion *Methods Mol. Biol* **400** 295–321
[9] Wu J and Berland K M 2008 Propagators and time-dependent diffusion coefficients for anomalous diffusion *Biophys. J.* **95** 2049–52
[10] Shusterman R, Alon S, Gavrinyov T and Krichevsky O 2004 Monomer dynamics in double- and single-stranded DNA polymers *Phys. Rev. Lett.* **92** 048303
[11] Banks D S, Tressler C, Peters R D, Fling F H X and Fradin C X C 2016 Characterizing anomalous diffusion in crowded polymer solutions and gels over five decades in time with variable-lengthscale fluorescence correlation spectroscopy *Soft Matter.* **12** 4190–203
[12] Stolle M D N and Fradin C 2019 Anomalous diffusion in inverted variable-lengthscale fluorescence correlation spectroscopy *Biophys. J.* **116** 791–806
[13] Almeida P and Vaz W 1995 Lateral diffusion in membranes *Handbook of Biological Physics* vol 1 (Amsterdam: Elsevier) 305–57
[14] Saxton M J 1999 Lateral diffusion of lipids and proteins *Membrane Permeability – 100 Years since Ernest Overton* (Amsterdam: Elsevier) pp 229–82
[15] Metzler R and Klafter J 2000 The random walk's guide to anomalous diffusion: a fractional dynamics approach *Phys. Rep.* **339** 1–77
[16] Klafter J and Sokolov I M 2011 *First Steps in Random Walks: From Tools to Applications* 1st edn (Oxford: Oxford University Press)
[17] Höfling F and Franosch T 2013 Anomalous transport in the crowded world of biological cells *Rep. Prog. Phys.* **76** 046602
[18] Krapf D 2015 *Mechanisms Underlying Anomalous Diffusion in the Plasma Membrane* (Amsterdam: Elsevier)
[19] Milon S, Hovius R, Vogel H and Wohland T 2003 Factors influencing fluorescence correlation spectroscopy measurements on membranes: simulations and experiments *Chem. Phys.* **288** 171–86
[20] Kalwarczyk T, Kwapiszewska K, Szczepanski K, Sozański K, Szymanski J, Michalska B, Patalas-Krawczyk P, Duszynski J and Holyst R 2017 Apparent anomalous diffusion in the cytoplasm of human cells: the effect of probes' polydispersity *J. Phys. Chem.* B **121** 9831–37
[21] Widengren J and Rigler R 1994 Triplet-state monitoring by fluorescence correlation spectroscopy *J Fluoresc.* **4** 255–58
[22] Widengren J, Mets Ü and Rigler R 1999 Photodynamic properties of green fluorescent proteins investigated by fluorescence correlation spectroscopy *Chem. Phys.* **250** 171–86

[23] Widengren J and Seidel C A M 2000 Manipulation and characterization of photo-induced transient states of Merocyanine 540 by fluorescence correlation spectroscopy *Phys. Chem. Chem. Phys.* **2** 3435–41
[24] Kask P, Piksarv P, Pooga M, Mets Ü and Lippmaa E 1989 Separation of the rotational contribution in fluorescence correlation experiments *Biophys. J.* **55** 213–20
[25] Vester M, Staut T, Enderlein J and Jung G 2015 Photon antibunching in a cyclic chemical reaction scheme *J. Phys. Chem. Lett.* **6–7** 1149–54
[26] Sharma A, Enderlein J and Kumbhakar M 2016 Photon antibunching in complex intermolecular fluorescence quenching kinetics *J. Phys. Chem. Lett.* **7** 3137–41
[27] Ta H, Kiel A, Wahl M and Herten D-P 2010 Experimental approach to extend the range for counting fluorescent molecules based on photon-antibunching *Phys. Chem. Chem. Phys.* **12** 10295
[28] Nettels D, Hoffmann A and Schuler B 2008 Unfolded protein and peptide dynamics investigated with single-molecule FRET and correlation spectroscopy from picoseconds to seconds *J. Phys. Chem.* B **112** 6137–46
[29] Aragon S and Pecora R 1975 Fluorescence correlation spectroscopy and Brownian rotational diffusion *Biopolymers* **14** 119–37
[30] Ehrenberg M and Rigler R 1974 Rotational Brownian motion and fluorescence intensify fluctuations *Chem. Phys.* **4** 390–401
[31] Ehrenberg M and Rigler R 1976 Fluorescence correlation spectroscopy applied to rotational diffusion of macromolecules *Quart. Rev. Biophys.* **9** 69–81
[32] Widengren J, Mets U and Rigler R 1995 Fluorescence correlation spectroscopy of triplet states in solution: a theoretical and experimental study *J. Phys. Chem.* **99** 13368–79
[33] Widengren J 2001 Photophysical aspects of FCS measurements *Fluorescence Correlation Spectroscopy: Theory and Applications* ed R Rigler and E S Elson (Berlin: Springer) pp 276–01
[34] Felekyan S, Kühnemuth R, Kudryavtsev V, Sandhagen C, Becker W and Seidel C A M 2005 Full correlation from picoseconds to seconds by time-resolved and time-correlated single photon detection *Rev. Sci. Instrum.* **76** 083104
[35] Petersen N O, Höddelius P L, Wiseman P W, Seger O and Magnusson K E 1993 Quantitation of membrane receptor distributions by image correlation spectroscopy: concept and application *Biophys. J.* **65** 1135–46
[36] Brown C M, Dalal R B, Hebert B, Digman M A, Horwitz A R and Gratton E 2008 Raster image correlation spectroscopy (RICS) for measuring fast protein dynamics and concentrations with a commercial laser scanning confocal microscope *J. Microsc.* **229** 78–91
[37] Brinkmeier M, Dörre K, Stephan J and Eigen M 1999 Two-beam cross-correlation: a method to characterize transport phenomena in micrometer-sized structures *Anal. Chem.* **71** 609–16
[38] Dertinger T, Pacheco V, Von Der Hocht I, Hartmann R, Gregor I and Enderlein J 2007 Two-Focus fluorescence correlation spectroscopy: a new tool for accurate and absolute diffusion measurements *Chem. Phys. Chem.* **8** 433–43
[39] Müller C B, Loman A, Richtering W and Enderlein J 2008 Dual-focus fluorescence correlation spectroscopy of colloidal solutions: influence of particle size *J. Phys. Chem.* B **112** 8236–40
[40] Dertinger T, Loman A, Ewers B, Müller C B, Krämer B and Enderlein J 2008 The optics and performance of dual-focus fluorescence correlation spectroscopy *Opt. Express* **16** 14353–68

[41] Digman M A and Gratton E 2009 Imaging barriers to diffusion by pair correlation functions *Biophys. J.* **97** 665–73

[42] Magde D, Webb W and Elson E 1978 Fluorescence correlation spectroscopy. III *Unif. Transl. Lam. Flow Biopolym.* **17** 361–76

[43] Bieschke J, Giese A, Schulz-Schaeffer W, Zerr I, Poser S, Eigen M and Kretzschmar H 2000 Ultrasensitive detection of pathological prion protein aggregates by dual-color scanning for intensely fluorescent targets *Proc. Nat. Acad. Sci.* **97** 5468

[44] Gösch M, Blom H, Holm J, Heino T and Rigler R 2000 Hydrodynamic flow profiling in microchannel structures by single molecule fluorescence correlation spectroscopy *Anal. Chem.* **72** 3260–65

[45] Pan X, Yu H, Shi X, Korzh V and Wohland T 2007 Characterization of flow direction in microchannels and zebrafish blood vessels by scanning fluorescence correlation spectroscopy *J. Biomed. Opt.* **12** 014034

[46] Pan X, Shi X, Korzh V, Yu H and Wohland T 2009 Line scan fluorescence correlation spectroscopy for three-dimensional microfluidic flow velocity measurements *J. Biomed. Opt.* **14** 024049

[47] Schneider F, Waithe D, Lagerholm B C, Shrestha D, Sezgin E, Eggeling C and Fritzsche M 2018 Statistical analysis of scanning fluorescence correlation spectroscopy data differentiates free from hindered diffusion *ACS Nano* **12** 8540–46

[48] Ries J and Schwille P 2006 Studying slow membrane dynamics with continuous wave scanning fluorescence correlation spectroscopy *Biophys. J.* **91** 1915–24

[49] Ries J, Yu S R, Burkhardt M, Brand M and Schwille P 2009 Modular scanning FCS quantifies receptor-ligand interactions in living multicellular organisms *Nat Meth.* **6** 643–45

[50] Petrášek Z, Derenko S and Schwille P 2011 Circular scanning fluorescence correlation spectroscopy on membranes *Opt. Express* **19** 25006–21

[51] Ruan Q 2004 Spatial-temporal studies of membrane dynamics: scanning fluorescence correlation spectroscopy (SFCS) *Biophys. J.* **87** 1260–67

[52] Skinner J P, Chen Y and Müller J D 2005 Position-sensitive scanning fluorescence correlation spectroscopy *Biophys J.* **89** 1288–301

[53] Petrášek Z and Schwille P 2008 Precise measurement of diffusion coefficients using scanning fluorescence correlation spectroscopy *Biophys. J.* **94** 1437–48

[54] St-Pierre P R and Petersen N O 1992 Average density and size of microclusters of epidermal growth factor receptors on A431 cells *Biochemistry* **31** 2459–63

[55] Nohe A and Petersen N O 2004 Analyzing protein-protein interactions in cell membranes *Bioessays* **26** 196–203

[56] Nohe A and Petersen N O 2007 *Image Correlation Spectroscopy Sci. STKE* **2007** l7

[57] Comeau J, Kolin D and Wiseman P 2008 Accurate measurements of protein interactions in cells via improved spatial image cross-correlation spectroscopy *Mol. Biosyst.* **4** 672–85

[58] Hebert B, Costantino S and Wiseman P W 2005 Spatiotemporal image correlation spectroscopy (STICS) theory, verification, and application to protein velocity mapping in living CHO cells *Biophys. J.* **88** 3601–14

[59] Wiseman P W 2013 *Image Correlation Spectroscopy: Mapping Correlations in Space, Time, and Reciprocal Space* 1st edn (Amsterdam: Elsevier)

[60] Toplak T, Pandzic E, Chen L, Vicente-Manzanares M, Horwitz A R and Wiseman P W 2012 STICCS reveals matrix-dependent adhesion slipping and gripping in migrating cells *Biophys. J.* **103** 1672–82

[61] Norris S C P, Humpolíčková J, Amler E, Huranová M, Buzgo M, Macháň R, Lukáš D and Hof M 2011 Acta biomaterialia *Acta Biomater.* **7** 4195–203
[62] Hendrix J, Dekens T, Schrimpf W and Lamb D C 2016 Arbitrary-region raster image correlation spectroscopy *Biophys. J.* **111** 1785–96
[63] Hedde P N, rlich R E M D O, Blomley R, Gradl D, Oppong E, Cato A C B and Nienhaus G U 2013 Stimulated emission depletion-based rasterimage correlation spectroscopy revealsbiomolecular dynamics in live cells *Nat. Commun.* **4** 1–8
[64] Sisan D R, Arevalo R, Graves C, Mcallister R and Urbach J S 2006 Spatially resolved fluorescence correlation spectroscopy using a spinning disk confocal microscope *Biophys. J.* **91** 4241–52
[65] Kannan B, Guo L, Sudhaharan T, Ahmed S, Maruyama I and Wohland T 2007 Spatially resolved total internal reflection fluorescence correlation microscopy using an electron multiplying charge-coupled device camera *Anal. Chem.* **79** 4463–70
[66] Wohland T, Shi X, Sankaran J and Stelzer E 2010 Single plane illumination fluorescence correlation spectroscopy (SPIM-FCS) probes inhomogeneous three-dimensional environments *Opt. Express* **18** 10627–41
[67] Krieger J W, Singh A P, Bag N, Garbe C S, Saunders T E, Langowski J and Wohland T 2015 Imaging fluorescence (cross-) correlation spectroscopy in live cells and organisms *Nat. Protoc.* **10** 1948–74
[68] Struntz P and Weiss M 2015 Multiplexed measurement of protein diffusion in Caenorhabditis elegans embryos with SPIM-FCS *J. Phys. D: Appl. Phys.* **49** 044002
[69] Kannan B, Har J, Liu P, Maruyama I, Ding J and Wohland T 2006 Electron multiplying charge-coupled device camera based fluorescence correlation spectroscopy *Anal. Chem.* **78** 3444–51
[70] Hassler K, Anhut T, Rigler R, Gösch M and Lasser T 2005 High count rates with total internal reflection fluorescence correlation spectroscopy *Biophys. J.* **88** L01–3
[71] Thompson N L and Steele B L 2007 Total internal reflection with fluorescence correlation spectroscopy *Nat. Protoc.* **2** 878–90
[72] Ries J, Petrov E P and Schwille P 2008 Total internal reflection fluorescence correlation spectroscopy: effects of lateral diffusion and surface-generated fluorescence *Biophys. J.* **95** 390–99
[73] Sankaran J, Bag N, Kraut R S and Wohland T 2013 Accuracy and precision in camera-based fluorescence correlation spectroscopy measurements *Anal. Chem.* **85** 3948–54
[74] Singh A P, Krieger J W, Buchholz J, Charbon E, Langowski J and Wohland T 2013 The performance of 2D array detectors for light sheet based fluorescence correlation spectroscopy *Opt. Express* **21** 8652
[75] Hess S and Webb W 2002 Focal volume optics and experimental artifacts in confocal fluorescence correlation spectroscopy *Biophys. J.* **83** 2300–17
[76] Shusterman R, Gavrinyov T and Krichevsky O 2008 Internal dynamics of superhelical DNA *Phys. Rev. Lett.* **100** 098102
[77] Yechiel E and Edidin M 1987 Micrometer-scale domains in fibroblast plasma membranes *J. Cell Biol.* **105** 755–60
[78] Edidin M and Stroynowski I 1991 Differences between the lateral organization of conventional and inositol phospholipid-anchored membrane proteins. A further definition of micrometer scale membrane domains *J. Cell Biol.* **112** 1143–50

[79] Wawrezinieck L, Rigneault H, Marguet D and Lenne P-F 2005 Fluorescence correlation spectroscopy diffusion laws to probe the submicron cell membrane organization *Biophys. J.* **89** 4029–42

[80] Destainville N 2008 Theory of fluorescence correlation spectroscopy at variable observation area for two-dimensional diffusion on a meshgrid *Soft Matter.* **4** 1288

[81] Lenne P-F, Wawrezinieck L, Conchonaud F, Wurtz O, Boned A, Guo X-J, Rigneault H, He H-T and Marguet D 2006 Dynamic molecular confinement in the plasma membrane by microdomains and the cytoskeleton meshwork *EMBO J.* **25** 3245–56

[82] Huang S, Lim S Y, Gupta A, Bag N and Wohland T 2017 Plasma membrane organization and dynamics is probe and cell line dependent *BBA – Biomembranes* **1859** 1483–92

[83] Favard C, Wenger J, Lenne P-F and Rigneault H 2011 FCS Diffusion laws in two-phase lipid membranes: determination of domain mean size by experiments and Monte Carlo simulations *Biophys. J.* **100** 1242–51

[84] Bag N, Yap D H X and Wohland T 2013 *Biochimica et Biophysica Acta. BBA – Biomembranes.* **1838** 802–13

[85] Ruprecht V, Wieser S, Marguet D and Schütz G J 2011 Spot variation fluorescence correlation spectroscopy allows for superresolution chronoscopy of confinement times in membranes *Biophys. J.* **100** 2839–45

[86] Veerapathiran S and Wohland T 2018 The imaging FCS diffusion law in the presence of multiple diffusive modes *Methods* **140–141** 140–50

[87] Sorscher S and Klein M 1980 Profile of a focussed collimated laser beam near the focal minimum characterized by fluorescence correlation spectroscopy *Rev. Sci. Instrum.* **51** 98–102

[88] Benda A, Beneš M, Mareček V, Lhotský A, Hermens W T and Hof M 2003 How to determine diffusion coefficients in planar phospholipid systems by confocal fluorescence correlation spectroscopy *Langmuir* **19** 4120–26

[89] Eggeling C *et al* 2009 Direct observation of the nanoscale dynamics of membrane lipids in a living cell *Nature* **457** 1159–62

[90] Huang H, Simsek M F, Jin W and Pralle A 2015 Effect of receptor dimerization on membrane lipid raft structure continuously quantified on single cells by camera based fluorescence correlation spectroscopy *PLoS One* **10** e0121777

[91] Ng X W, Bag N and Wohland T 2015 Characterization of lipid and cell membrane organization by the fluorescence correlation spectroscopy diffusion law *CHIMIA* **69** 112–19

[92] Di Rienzo C, Gratton E, Beltram F and Cardarelli F 2013 Fast spatiotemporal correlation spectroscopy to determine protein lateral diffusion laws in live cell membranes *Proc. Natl Acad. Sci.* **110** 12307–12

[93] Marguet D, Lenne P-F, Rigneault H and He H-T 2006 Dynamics in the plasma membrane: how to combine fluidity and order *EMBO J.* **25** 3446–57

[94] Ganguly S and Chattopadhyay A 2010 Cholesterol depletion mimics the effect of cytoskeletal destabilization on membrane dynamics of the serotonin *Biophys. J.* **99** 1397–407

[95] Ng X W, Teh C, Korzh V and Wohland T 2016 The secreted signaling protein Wnt3 is associated with membrane domains *in vivo*: a SPIM-FCS study *Biophys. J.* **111** 418–29

[96] Culbertson M J, Williams J T B, Cheng W W L, Stults D A, Wiebracht E R, Kasianowicz J J and Burden D L 2007 Numerical fluorescence correlation spectroscopy for the analysis of molecular dynamics under nonstandard conditions *Anal. Chem.* **79** 4031–39

[97] Levene M J 2003 Zero-mode waveguides for single-molecule analysis at high concentrations *Science* **299** 682–86

[98] Leutenegger M, Ringemann C, Lasser T, Hell S W and Eggeling C 2012 Fluorescence correlation spectroscopy with a total internal reflection fluorescence STED microscope (TIRF-STED-FCS) *Opt. Express* **20** 5243–63

[99] Benda A, Ma Y and Gaus K 2015 Self-calibrated line-scan STED-FCS to quantify lipid dynamics in model and cell membranes *Biophys. J.* **108** 596–609

[100] Vicidomini G *et al* 2015 STED-FLCS: An advanced tool to reveal spatiotemporal heterogeneity of molecular membrane dynamics *Nano Lett.* **15** 5912–18

[101] Barbotin A, Galiani S, Urbančič I, Eggeling C and Booth M J 2019 Adaptive optics allows STED-FCS measurements in the cytoplasm of living cells *Opt. Express* **27** 23378–18

[102] Wennmalm S and Widengren J 2010 Inverse-fluorescence cross-correlation spectroscopy *Anal. Chem.* **82** 5646–51

[103] Lamb D C, Schenk A, Röcker C, Scalfi-Happ C and Nienhaus G U 2000 Sensitivity enhancement in fluorescence correlation spectroscopy of multiple species using time-gated detection *Biophys. J.* **79** 1129–38

[104] Laurence T A, Ly S, Bourguet F, Fischer N O and Coleman M A 2014 Fluorescence correlation spectroscopy at micromolar concentrations without optical nanoconfinement *J. Phys. Chem.* B **118** 9662–67

[105] Dörlich R M, Chen Q, Niklas Hedde P, Schuster V, Hippler M, Wesslowski J, Davidson G and Nienhaus G U 2015 Dual-color dual-focus line-scanning FCS for quantitative analysis of receptor-ligand interactions in living specimens *Sci. Rep.* **5** 1–11

Thorsten Wohland, Sudipta Maiti and Radek Macháň

Chapter 6

Theoretical fluorescence cross-correlation spectroscopy (FCCS) models

FCCS analyses cross-correlations between fluorescence intensity time traces originating from distinct sources. In the most common FCCS modality, dual-clour FCCS (DC-FCCS), the two intensity traces are signals from two fluorophore species with distinct emission spectra. Positive cross-correlation, then, indicates co-diffusion of fluorophores of different species, which can be used as an indication of complex formation between particles labelled by the distinct fluorophore species. Here we present theoretical models that allow quantitative characterisation of the complex formation based on FCCS results, mainly the correlation function amplitudes. First, we develop the theory for an ideal DC-FCCS experiment and then we gradually include corrections needed for a realistic description of most real-life FCCS experiments. Concurrently, we discuss practical aspects of estimating the additional variables that enter the models with those corrections. Thereafter, we present some alternative FCCS modalities which eliminate or minimize some sources of artefacts typical for DC-FCCS. We will include the use of statistical filtering (outlined in chapter 4) in FCCS among those alternative approaches and discuss how it extends the possible applications of FCCS to the study of equilibrium reaction kinetics; a corresponding theoretical model will be derived in that section. After this chapter the reader should have a solid understanding of the theoretical models used in analysing DC-FCCS data, their relation to FCS models derived in chapter 5 and the ways how corrections for artefacts of real FCCS experiments can be incorporated into the models. This knowledge will allow the reader to select and apply an appropriate model to interpret the results of a real FCCS experiment.

6.1 Introduction

FCCS measures and analyses cross-correlations between pairs of distinct fluorescence intensity time traces $F_A(t)$ and $F_B(t)$, typically (though not necessarily) recorded by different detectors and, most importantly, supposed to be signals

doi:10.1088/978-0-7503-2080-1ch6

originating from distinct sources, such as different fluorophore populations or different locations within the sample (spatial FCCS). In general, the CCF between the two fluorescence time traces can be defined as (see section 2.6):

$$G_{AB}(\tau) = \frac{\langle F_A(t)F_B(t+\tau)\rangle}{\langle F_A(t)\rangle\langle F_B(t)\rangle} \tag{6.1}$$

and calculated in practice according to the equation (4.9) with symmetric normalization (see section 4.1.4). Symmetric normalization does not mean that the roles of the signals $F_A(t)$ and $F_B(t)$ in equation (6.1) are identical, since $F_B(t)$ is shifted in time with respect to $F_A(t)$. Analogously, we could define $G_{BA}(\tau)$, which is in general different from $G_{AB}(\tau)$. Systematic differences between the two CCFs arise from an asymmetric relationship (such as causality) between the processes underlying the fluctuations in $F_A(t)$ and $F_B(t)$. This is discussed in section 2.6 and illustrated on an example of scoring goals and fans cheering during a football match. Examples more relevant to the present topic could be found in spatial FCCS either in the presence of oriented movement, e.g. flow or active transport, of molecules between the observation volumes from which $F_A(t)$ and $F_B(t)$ are collected or when the two observation volumes are separated by a semipermeable barrier, which the molecules can cross only in one direction. Another example, FCCS with one of the two fluorophores undergoing photobleaching, is discussed in section 6.2.5.4. In the absence of any such asymmetry and when the studied processes are strictly in equilibrium, no systematic difference between the two CCFs exists. It is then useful to define a symmetrized CCF $G_X(\tau)$ as the average of $G_{AB}(\tau)$ and $G_{BA}(\tau)$ (equation (4.10)). In the following text we will mean $G_X(\tau)$ when talking about CCF, unless stated otherwise.

When the fluctuations in the two intensity time traces $F_A(t)$ and $F_B(t)$ are independent of each other, $G_X(\tau)$ is, within the margin of its random errors, equal to its asymptotic value G_∞ (ideally 1) for all time lags τ. On the other hand, $G_X(\tau)$ larger than G_∞ can be measured only when there exists some form of correlation between the fluctuations in $F_A(t)$ and $F_B(t)$. This, in turn, implies a connection between the processes underlying those fluctuations, the quantitative characterisation of which is the aim of FCCS experiments. However, $G_X(\tau) = G_\infty$ is true only when the signals originating from the two sources are unambiguously assigned to the respective detection channels. When this condition is not satisfied and signal from a single source contributes to both channels, the CCF will be artificially increased by what is in fact the ACF of the signal detected in both channels (see section 6.2.3). The sorting of the signals can be based on various principles (see, for example, the introduction to FLCS in section 4.4); however, the best known and most widespread form of FCCS is dual-colour FCCS (DC-FCCS) in which the separation is based on differences in fluorescence emission spectra. Being frequently regarded as practically synonymous to FCCS, DC-FCCS and the modalities derived from it will occupy most of this chapter; only in the end will we extend the theory to other forms of FCCS.

6.2 Dual-colour FCCS (DC-FCCS)

As already mentioned, DC-FCCS separates the fluorescence time traces originating from two distinct fluorophore populations based on differences in emission spectra.

We will refer to the two populations as a and b, with the fluorophore a emitting at shorter wavelength than fluorophore b. In a typical DC-FCCS experiment fluorophores of both populations are simultaneously excited by two laser lines chosen according to the excitation spectra of the individual populations. Photons belonging to two distinct non-overlapping spectral bands are then selected from the fluorescence collected in the effective observation volume and the photons belonging to the individual spectral bands are detected by separate detectors. The spectral bands are chosen to match the emission spectra of the respective fluorophore populations so that in the first approximation we can assume that each of the two detectors receives photons originating only from the members of one of the fluorophore populations. We will therefore refer to the two detection channels as A and B and denote the intensity traces recorded by the respective detectors as $F_A(t)$ and $F_B(t)$. By correlating the time traces recorded by the respective detectors, we obtain the ACFs $G_A(\tau)$ and $G_B(\tau)$; cross-correlating the time traces gives the CCF $G_X(\tau)$.

Let us introduce a convention at this point that all quantities belonging to the detection channels will be distinguished by uppercase subscripts A and B. On the other hand, all quantities pertaining to the fluorophore populations will be marked by lower case subscripts a and b. This distinction may seem redundant if we only consider the ideal case where all photons are unambiguously assigned to the correct fluorophore populations; however, it will be needed to discuss the non-ideal situation encountered in real experiments.

Let us start first with the basic theory of an ideal DC-FCCS experiment before we proceed to develop the theory further to describe more realistic, non-ideal situations. The key idealization lies in the assumption of the absence of crosstalk between the detection channels. This means that photons emitted by fluorophores a are exclusively detected in detection channel A and photons emitted by fluorophores b in channel B only. Unrealistic as it is, we find it useful as a first introduction to the main concepts of DC-FCCS.

As already mentioned earlier in this chapter, $G_X(\tau)$ larger than G_∞ is observed only if the fluctuations in $F_A(t)$ and in $F_B(t)$ are correlated with each other. If molecular diffusion is the source of the fluctuations, then $G_X(\tau) > G_\infty$ signifies the presence of co-diffusing fluorophores a and b resulting in simultaneous bursts of intensity in both detection channels A and B (see figure 6.1). Co-diffusion of fluorophores of different species is not expected to occur unless both fluorophores are parts of the same diffusing particles (molecules, complexes or aggregates). Therefore, $G_X(\tau) > G_\infty$ is an indication of the presence of particles containing fluorophores of both species (at least one from each species). Let us assume here that it is exactly one fluorophore of each species; we will discuss other stoichiometries later in section 6.2.7. Quantities referring to such double-labelled particles will be identified by the subscript ab.

Using this notation, the CCF is defined by equation (6.1). The theoretical model for the CCF in confocal DC-FCCS can be derived by the same procedure which was used in math box 5.1 for the ACF derivation. Intensity $F_A(t)$ in channel A is proportional to the total concentration C_a^{tot} of fluorophores a, which, in the case discussed here, can be written as the sum of concentrations of the double-labelled

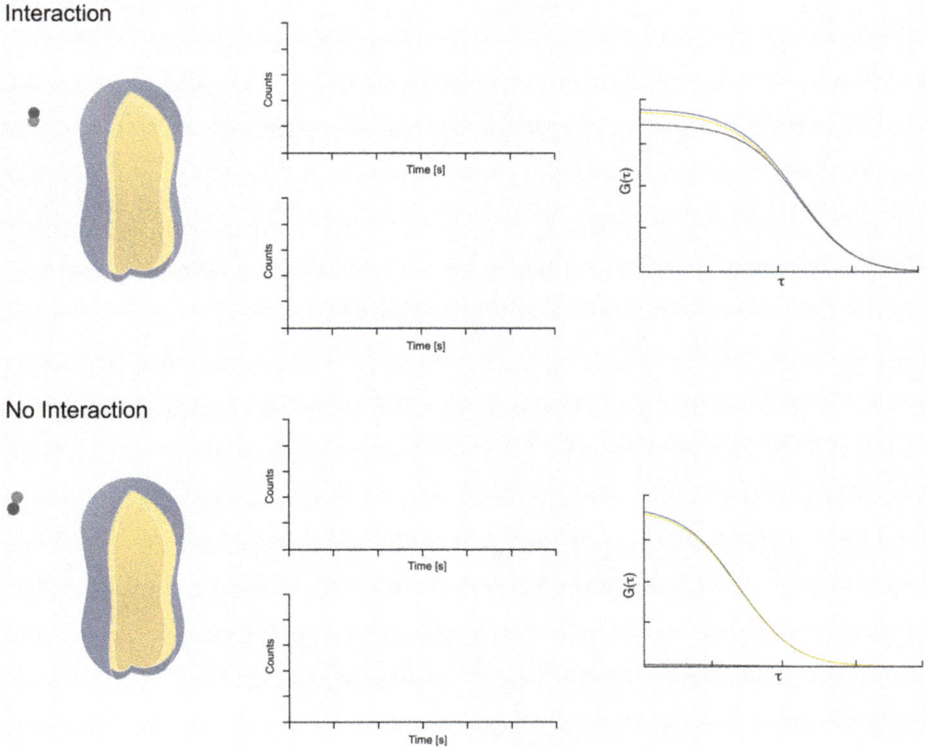

Figure 6.1. Principle of fluorescence cross-correlation spectroscopy (FCCS). If two particles that carry spectrally different fluorescent labels bind to each other (upper panel), then the signal detected in the two wavelength channels will be correlated as it changes in similar ways when bound particles diffuse through the observation volume. This will lead then not only to two ACFs (shown in blue and yellow) that contain information about the particle concentrations and mobility, but also to a CCF (shown in grey) that reflects the concentration and mobility of the bound particles. If the particles move independently and do not interact, then only the ACFs will be measurable while the CCF will be flat. Animation available at https://iopscience.iop.org/book/978-0-7503-2080-1.

particles and particles labelled by fluorophore a only, $C_a^{tot} = C_a + C_{ab}$, and analogously $F_B(t)$ is proportional to $C_b^{tot} = C_b + C_{ab}$. Therefore, the numerator will depend on:

$$\begin{aligned} &\langle \delta C_a^{tot}(\vec{r}, t)\delta C_b^{tot}(\vec{r}', t+\tau)\rangle \\ &= \langle (\delta C_a(\vec{r}, t) + \delta C_{ab}(\vec{r}, t))(\delta C_b(\vec{r}', t+\tau) + \delta C_{ab}(\vec{r}', t+\tau))\rangle \\ &= \langle \delta C_{ab}(\vec{r}, t)\delta C_{ab}(\vec{r}', t+\tau)\rangle \end{aligned} \tag{6.2}$$

Equation (6.2) is based on the assumption that movements of individual single-labelled particles are independent of each other and, thus, only the double-labelled particles contribute to the cross-correlation. The denominator, on the other hand, depends on the product of the average values $\langle \delta C_a^{tot}(\vec{r}, t)\rangle\langle \delta C_b^{tot}(\vec{r}', t+\tau)\rangle = \langle \delta C_a(\vec{r}, t) + \delta C_{ab}(\vec{r}, t)\rangle\langle \delta C_b(\vec{r}, t) + \delta C_{ab}(\vec{r}, t)\rangle$. This finally leads to the following model analogous to equation (5.5) (refer to math box 6.1 for more details on the derivation):

$$G_X(\tau) = \frac{N_{ab}}{(N_a + N_{ab})(N_b + N_{ab})}\left(1 + \frac{\tau}{\tau_{D_{ab}}}\right)^{-1}\left(1 + \frac{\tau}{K^2\tau_{D_{ab}}}\right)^{-1/2} + G_\infty \qquad (6.3a)$$

The temporal decay of $G_X(\tau)$ is determined by the diffusion of the double-labelled particles. The characteristics of their diffusion (typically their diffusion coefficient D_{ab}, which is related to the diffusion time $\tau_{D_{ab}}$ through equation (5.6)) can be extracted from $G_X(\tau)$ by fitting it with an appropriate theoretical model such as equation (6.3), which differs from equation (5.5) only in the expression for the amplitude, the temporal decays of both models having identical functional forms. The ACFs of the individual detection channels contain contributions from the double-labelled particles and single-labelled particles of the species corresponding to the given detection channel. Following equation (5.25), we can write:

$$G_A(\tau) = \frac{1}{(N_a + N_{ab})}\left[f_a\left(1 + \frac{\tau}{\tau_{D_a}}\right)^{-1}\left(1 + \frac{\tau}{K^2\tau_{D_a}}\right)^{-1/2} + (1 - f_a)\left(1 + \frac{\tau}{\tau_{D_{ab}}}\right)^{-1}\left(1 + \frac{\tau}{K^2\tau_{D_{ab}}}\right)^{-1/2}\right] + G_\infty \qquad (6.3b)$$

$$G_B(\tau) = \frac{1}{(N_b + N_{ab})}\left[f_b\left(1 + \frac{\tau}{\tau_{D_b}}\right)^{-1}\left(1 + \frac{\tau}{K^2\tau_{D_b}}\right)^{-1/2} + (1 - f_b)\left(1 + \frac{\tau}{\tau_{D_{ab}}}\right)^{-1}\left(1 + \frac{\tau}{K^2\tau_{D_{ab}}}\right)^{-1/2}\right] + G_\infty \qquad (6.3c)$$

where τ_{D_a} and τ_{D_b} are the diffusion times and f_a and f_b are the fractions of the respective single-labelled particles. If fluctuations caused by fluorophores transiting to their triplet states (or undergoing other photophysical processes) are detected in the respective detection channels, the ACFs would need to account for such processes (see equation (5.47)). However, for the purpose of DC-FCCS analysis, we are only interested in the diffusional parts of the ACFs (as given by equations (6.3b) and (6.3c)) and when talking about ACF amplitudes in this chapter, we will always assume them to be the amplitudes of the diffusional parts of the ACFs (see equation (6.5)). The temporal decays of the ACFs contain in general two terms corresponding to different diffusion coefficients, unless the respective fraction of single-labelled particles is zero. At the same time, we can notice that all three correlation functions contain a term corresponding to the diffusion coefficient D_{ab} of the double-labelled particles. Although in the following text we will concentrate mainly on the correlation function amplitudes which carry the information typically extracted from DC-FCCS experiments, we would like to highlight here the importance of analysing correlation function temporal decays in DC-FCCS experiments. The absence of a term corresponding to the diffusion coefficient D_{ab} of the double-labelled particles in any of the ACFs contradicts the presence of double-labelled particles in the system and any positive cross-correlation observed in such cases must originate from other, possibly artificial, sources.

Just like equation (5.5), equations (6.3a)–(6.3c) are applicable to data measured by a confocal microscope (Gaussian observation volume) in a sample where the fluorescent particles undergo Brownian movement in three dimensions. Analogously, we could use other models derived in chapter 5 for fitting temporal decays of CCFs measured with different geometries of observation volumes (e.g. TIRF) or samples (e.g. diffusion confined to two dimensions in membranes) or to account for non-Brownian diffusion of fluorescent particles (e.g. anomalous diffusion). Note however, that the CCF will not contain any contribution from processes which involve only one of the fluorophores bound to the double-labelled particle. Examples of such processes are fluorescence blinking caused by transition to triplet or other dark states, photo-isomerisation and other excited state reactions; such processes will be manifested only in the ACFs of the corresponding detection channels. The CCF reflects only processes involving fluorophores of both species bound to the same particle; in most cases, the movement of the whole particle is the only such process as illustrated in figure 6.1.

Math box 6.1

The CCF in confocal DC-FCCS

The model for the CCF in confocal DC-FCCS is derived analogously to the model for the ACF in confocal FCS presented in math box 5.1. We will, therefore, concentrate here on the differences and avoid going into details of what is the same in derivation of both models. As in the case of math box 5.1, we assume the extent of the sample to be much larger than the observation volume, which assumption is expressed by setting $S_A = S_B = 1$. We can then write for the CCF (using the notation introduced in math box 5.1)

$$G_{AB}(\tau) = \frac{\kappa_a\kappa_b \int\int W_A(\vec{r})W_B(\vec{r}')\langle\delta C_a^{tot}(\vec{r}, t)\delta C_b^{tot}(\vec{r}', t+\tau)\rangle d\vec{r}d\vec{r}'}{\kappa_a\kappa_b \int W_A(\vec{r})\langle C_a^{tot}(\vec{r}, t)\rangle d\vec{r} \int W_B(\vec{r}')\langle C_b^{tot}(\vec{r}', t)\rangle d\vec{r}'} + G_\infty$$

$$= \frac{\int\int W_A(\vec{r})W_B(\vec{r}')\langle\delta C_{ab}(\vec{r}, t)\delta C_{ab}(\vec{r}', t+\tau)\rangle d\vec{r}d\vec{r}'}{\int W_A(\vec{r})\langle C_a(\vec{r}, t) + C_{ab}(\vec{r}, t)\rangle d\vec{r} \int W_B(\vec{r}')\langle C_b(\vec{r}', t) + C_{ab}(\vec{r}', t)\rangle d\vec{r}'} + G_\infty$$

Similar to equation (6.2), we have assumed that movements of individual single-labelled particles are independent of each other and, thus, only the double-labelled particles contribute to the concentration fluctuation correlation. On the other hand, the average intensities in the individual detection channels used for the CCF normalisation are proportional to the total concentrations of the respective fluorophores.

The concentration fluctuation correlation is described by the diffusion propagator:

$$\langle\delta C_{ab}(\vec{r}, t)\delta C_{ab}(\vec{r}', t+\tau)\rangle = \langle C_{ab}\rangle \frac{e^{-\frac{(\vec{r}-\vec{r}')^2}{4D\tau}}}{8(\pi D_{ab}\tau)^{3/2}}$$

W_A and W_B can be written as

$$W_L(\vec{r}) = \frac{2P_L}{\pi w_0} e^{\frac{-2x^2}{w_0^2}} e^{\frac{-2y^2}{w_0^2}} e^{\frac{-2z^2}{z_0^2}}$$

where the subscript L stands either for A or B and P_L is the power of the respective excitation laser.

The integration in the expression for $G_{AB}(\tau)$ is calculated, like in math box 5.1, according to the formula:

$$\int_{-\infty}^{\infty} e^{-ax^2+\beta x} dx = \sqrt{\frac{\pi}{a}} e^{\frac{\beta^2}{4a}}, \quad \text{for } \alpha > 0$$

which leads to an equation for $G_{AB}(\tau)$ identical in form to equation (6.3a). Assuming no asymmetric relationship between the intensity fluctuations in channels A and B, we will obtain the same equation also for $G_{BA}(\tau)$ and, thus, also for $G_X(\tau)$.

DC-FCCS experiments are usually conducted to characterise (preferably quantitatively) the formation of molecular complexes which then appear as double-labelled particles. We will show now that the correlation function amplitudes can be used for that purpose. As we already know from section 5.2, the ACF amplitude is the reciprocal value of the average number of independently diffusing particles in the observation volume. Note that by the amplitude we mean here the difference between the maximum of the correlation function at $\tau = 0$ and its asymptotic value G_∞ as defined in equation (5.13). Let us look now at the CCF amplitude G_{0_X}; from equation (6.3a) we get:

$$G_{0_X} = \frac{N_{ab}}{(N_a + N_{ab})(N_b + N_{ab})} \tag{6.4}$$

It is obvious that based on G_{0_X} alone we can distinguish whether there are any double-labelled particles in the sample or not (in the latter case the amplitude being zero), but we cannot make any conclusions concerning the relative abundance of the complex (the ratio of the double-labelled to the single-labelled particles). Let us, therefore, include the information from the amplitudes of the two ACFs, G_{0_A} and G_{0_B}, obtained by correlating the signal in the respective detection channels. Following what we said in the previous paragraph, the amplitudes in this case are the reciprocal values of the numbers of all particles carrying the respective fluorophores, as we can readily see from equations (6.3a) and (6.3b):

$$G_{0_A} = \frac{1}{N_a + N_{ab}} \tag{6.5a}$$

$$G_{0_B} = \frac{1}{N_b + N_{ab}} \tag{6.5b}$$

Assuming identical observation volumes V for both channels, we can rewrite equations (6.4) and (6.5a) and (6.5b) in terms of the average concentrations:

$$G_{0_A} = \frac{1}{N_a + N_{ab}} = \frac{1}{V(\langle C_a \rangle + \langle C_{ab} \rangle)} \tag{6.6a}$$

$$G_{0_B} = \frac{1}{N_b + N_{ab}} = \frac{1}{V(\langle C_b \rangle + \langle C_{ab} \rangle)} \tag{6.6b}$$

$$\begin{aligned} G_{0_X} &= \frac{N_{ab}}{(N_a + N_{ab})(N_b + N_{ab})} \\ &= \frac{\langle C_{ab} \rangle}{V(\langle C_a \rangle + \langle C_{ab} \rangle)(\langle C_a \rangle + \langle C_{ab} \rangle)} \end{aligned} \tag{6.6c}$$

Let us consider, for example, a simple complex formation following the reaction scheme:

$$M_a + M_b \rightleftarrows M_a M_b \tag{6.7}$$

To characterise this process by DC-FCCS, we assume that the molecules M_a and M_b are labelled by fluorophore a or b, respectively. Assuming perfect labelling (there are no unlabelled molecules M_a or M_b in the system and each molecule M_a or M_b is labelled by a single fluorophore a or b, respectively), the concentrations of M_a, M_b and M_aM_b are equal to C_a, C_b and C_{ab}, respectively.

The stability of the complex is quantified by its dissociation constant K_D which can be written as:

$$K_D = \frac{[M_a][M_b]}{[M_a M_b]} = \frac{\langle C_a \rangle \langle C_b \rangle}{N_A \langle C_{ab} \rangle} \tag{6.8}$$

where the square brackets denote equilibrium molar concentrations and N_A is the Avogadro constant needed to convert the number concentrations C used throughout this book to molar concentrations used in the definition of K_D ($\langle C_x \rangle = [X] N_A$, where X, x stand for M_a, a or M_b, b or M_aM_b, ab).

Equations (6.6a)—(6.6c) and (6.8) show that the correlation amplitudes obtained in a DC-FCCS experiment contain sufficient information to determine the K_D of the complex (provided V is known). Since FCS is compatible with measurements in samples such as living cells or embryos, it follows that DC-FCCS has the potential to quantify stability of molecular complexes directly in such systems [1, 2]. However, to reach this 'Holy Grail' of molecular interaction studies we must abandon the idealized assumptions we have used in deriving the simple model for the correlation function amplitudes (equations (6.6a)–(6.6c)). Those assumptions are never fully satisfied in real DC-FCCS experiments, rendering the simple model too inaccurate to give a quantitative estimate of the complex stability. In the following sections, we will discuss phenomena that have been disregarded in the discussion of an ideal experiment and introduce their description into the model for correlation amplitudes. The resulting models and the situations under which they are applicable are summarised in table 6.1. However, before doing so, let us introduce the cross-correlation amount Q as another parameter that can be used to characterise, albeit

Table 6.1. This table provides a summary of the CCF amplitudes discussed in this chapter. In the right column we provide definitions of the different parameters used. To avoid repetition we define in each row only new parameters not used in any of the previous rows and assume that all previous definitions are valid. In the cases where in the second column is written 'Use appropriate ACF and CCF formula from above with definitions on the right' please use the new definitions for the concentrations in the right column and plug them into previous equations given in the second column.

FCCS Condition	ACFs and CCF	Parameter definitions
Ideal conditions	$G_{0_A} = \frac{1}{N_a + N_{ab}} = \frac{1}{V(\langle C_a\rangle + \langle C_{ab}\rangle)}$ $G_{0_B} = \frac{1}{N_b + N_{ab}} = \frac{1}{V(\langle C_b\rangle + \langle C_{ab}\rangle)}$ $G_{0_X} = \frac{N_{ab}}{(N_a + N_{ab})(N_b + N_{ab})}$ $= \frac{\langle C_{ab}\rangle}{V(\langle C_a\rangle + \langle C_{ab}\rangle)(\langle C_a\rangle + \langle C_{ab}\rangle)}$	$Q = \max\left[\frac{G_{0_X}}{G_{0_B}}; \frac{G_{0_X}}{G_{0_A}}\right]$
Unequal/non-aligned observation volumes	$G_{0_A} = \frac{1}{V_A(\langle C_a\rangle + \langle C_{ab}\rangle)}$ $G_{0_B} = \frac{1}{V_B(\langle C_b\rangle + \langle C_{ab}\rangle)}$ $G_{0_X} = \frac{\langle C_{ab}\rangle}{V_X(\langle C_a\rangle + \langle C_{ab}\rangle)(\langle C_b\rangle + \langle C_{ab}\rangle)}\exp\left(-\frac{d_x^2 + d_y^2}{w_{0_X}^2} - \frac{d_z^2}{z_{0_X}^2}\right)$ or $G_{0_X} = \frac{\langle C_{ab}\rangle}{V_X^{\mathrm{eff}}(\langle C_a\rangle + \langle C_{ab}\rangle)(\langle C_b\rangle + \langle C_{ab}\rangle)}$	$V_A = \pi^{3/2} w_A^2 z_A$ $V_B = \pi^{3/2} w_B^2 z_B$ $V_X = \pi^{3/2}\left(\frac{w_A^2 + w_B^2}{2}\right)\left(\frac{z_A^2 + z_B^2}{2}\right)^{1/2}$ $V_X^{\mathrm{eff}} = \frac{\exp\left(-\frac{d_x^2 + d_y^2}{w_{0_X}^2} - \frac{d_z^2}{z_{0_X}^2}\right)}{V_X}$
Spectral crosstalk	$G_{0_A} = \frac{(\kappa_{aA} + \kappa_{bA})^2\langle C_{ab}\rangle + \kappa_{aA}{}^2\langle C_a\rangle + \kappa_{bA}{}^2\langle C_b\rangle}{V_A[\kappa_{aA}\langle C_a\rangle + \kappa_{bA}\langle C_b\rangle + (\kappa_{aA} + \kappa_{bA})\langle C_{ab}\rangle]^2}$ $G_{0_B} = \frac{(\kappa_{aB} + \kappa_{bB})^2\langle C_{ab}\rangle + \kappa_{aB}{}^2\langle C_a\rangle + \kappa_{bB}{}^2\langle C_b\rangle}{V_B[\kappa_{aB}\langle C_a\rangle + \kappa_{bB}\langle C_b\rangle + (\kappa_{aB} + \kappa_{bB})\langle C_{ab}\rangle]^2}$ $G_{0_X} = \frac{(\kappa_{aA} + \kappa_{bA})(\kappa_{aB} + \kappa_{bB})\langle C_{ab}\rangle + \kappa_{aA}\kappa_{aB}\langle C_a\rangle + \kappa_{bA}\kappa_{bB}\langle C_b\rangle}{V_X^{\mathrm{eff}}[\kappa_{aA}\langle C_a\rangle + \kappa_{bA}\langle C_b\rangle + (\kappa_{aA} + \kappa_{bA})\langle C_{ab}\rangle]}$ $\times[\kappa_{aB}\langle C_a\rangle + \kappa_{bB}\langle C_b\rangle + (\kappa_{aB} + \kappa_{bB})\langle C_{ab}\rangle]^{-1}$	κ_{kL}: brightness of fluorophore k in channel L
Non-correlated background	$G_{0_A} = \frac{(\kappa_{aA} + \kappa_{bA})^2\langle C_{ab}\rangle + \kappa_{aA}{}^2\langle C_a\rangle + \kappa_{bA}{}^2\langle C_b\rangle}{V_A\left[\kappa_{aA}\langle C_a\rangle + \kappa_{bA}\langle C_b\rangle + (\kappa_{aA} + \kappa_{bA})\langle C_{ab}\rangle + \frac{\langle B_A\rangle}{V_A}\right]^2}$ $G_{0_B} = \frac{(\kappa_{aB} + \kappa_{bB})^2\langle C_{ab}\rangle + \kappa_{aB}{}^2\langle C_a\rangle + \kappa_{bB}{}^2\langle C_b\rangle}{V_B\left[\kappa_{aB}\langle C_a\rangle + \kappa_{bB}\langle C_b\rangle + (\kappa_{aB} + \kappa_{bB})\langle C_{ab}\rangle + \frac{\langle B_B\rangle}{V_B}\right]^2}$ $G_{0_X} = \frac{(\kappa_{aA} + \kappa_{bA})(\kappa_{aB} + \kappa_{bB})\langle C_{ab}\rangle + \kappa_{aA}\kappa_{aB}\langle C_a\rangle + \kappa_{bA}\kappa_{bB}\langle C_b\rangle}{V_X^{\mathrm{eff}}\left[\kappa_{aA}\langle C_a\rangle + \kappa_{bA}\langle C_b\rangle + (\kappa_{aA} + \kappa_{bA})\langle C_{ab}\rangle + \frac{\langle B_A\rangle}{V_X^{\mathrm{eff}}}\right]}$ $\times\left[\ldots \frac{\langle B_B\rangle}{\ldots}\right]^{-1}$	B_L: background counts in channel L

Table 6.1. (*Continued*)

FCCS Condition	ACFs and CCF	Parameter definitions
Non-fluorescent fluorophores	Use appropriate ACF and CCF formula from above with definitions on the right	$\langle C_a^{app}\rangle = p_a[M_a]N_A + p_a(1 - p_b)[M_aM_b]N_A$ $\langle C_b^{app}\rangle = p_b[M_b]N_A + (1 - p_a)p_b[M_aM_b]N_A$ $\langle C_{ab}^{app}\rangle = p_ap_b[M_aM_b]N_A$
Non-labeled proteins	Use appropriate ACF and CCF formula from above with definitions on the right	$\langle C_a^{app}\rangle = p_a[M_a^{(f)}]N_A + p_a(1 - p_b)\left[M_a^{(f)}M_b^{(f)}\right]N_A + p_a\left[M_a^{(f)}M_b^{(e)}\right]N_A$ $\langle C_b^{app}\rangle = p_b\left[M_b^{(f)}\right]N_A + (1 - p_a)p_b\left[M_a^{(f)}M_b^{(f)}\right]N_A + p_b\left[M_a^{(e)}M_b^{(f)}\right]N_A$ $\langle C_{ab}^{app}\rangle = p_ap_b\left[M_a^{(f)}M_b^{(f)}\right]N_A$
Free fluorophore a	Use appropriate ACF and CCF formula from above with definitions on the right	$\langle C_a^{app}\rangle = \langle C_a^{free}\rangle + p_a[M_a]N_A + p_a(1 - p_b)[M_aM_b]N_A$
Quenching and FRET		q_k: quenching (<1) or enhancement factor (>1) of fluorophore k due to quenching or energy transfer

	$G_{0_A} = \frac{(q_a\kappa_{aA} + q_b\kappa_{bA})^2\langle C_{ab}^{app}\rangle + \kappa_{aA}{}^2\langle C_a^{app}\rangle + \kappa_{bA}{}^2\langle C_b^{app}\rangle}{V_A\left[\kappa_{aA}\langle C_a^{app}\rangle + \kappa_{bA}\langle C_b^{app}\rangle + (q_a\kappa_{aA} + q_b\kappa_{bA})\langle C_{ab}^{app}\rangle + \frac{\langle B_A\rangle}{V_A}\right]^2}$ $G_{0_B} = \frac{(q_a\kappa_{aB} + q_b\kappa_{bB})^2\langle C_{ab}^{app}\rangle + \kappa_{aB}{}^2\langle C_a^{app}\rangle + \kappa_{bB}{}^2\langle C_b^{app}\rangle}{V_B\left[\kappa_{aB}\langle C_a^{app}\rangle + \kappa_{bB}\langle C_b^{app}\rangle + (q_a\kappa_{aB} + q_b\kappa_{bB})\langle C_{ab}^{app}\rangle + \frac{\langle B_B\rangle}{V_B}\right]^2}$ $G_{0_X} = \frac{(q_a\kappa_{aA} + q_b\kappa_{aB})(q_a\kappa_{aB} + q_b\kappa_{bB})\langle C_{ab}^{app}\rangle + \kappa_{aA}\kappa_{aB}\langle C_a^{app}\rangle + \kappa_{bA}\kappa_{bB}\langle C_b^{app}\rangle}{V_X^{\text{eff}}\left[\kappa_{aA}\langle C_a^{app}\rangle + \kappa_{bA}\langle C_b^{app}\rangle + (q_a\kappa_{aA} + q_b\kappa_{bA})\langle C_{ab}^{app}\rangle + \frac{\langle B_A\rangle}{V_X^{\text{eff}}}\right]}$ $\times\left[\kappa_{aB}\langle C_a^{app}\rangle + \kappa_{bB}\langle C_b^{app}\rangle + (q_a\kappa_{aB} + q_b\kappa_{bB})\langle C_{ab}^{app}\rangle + \frac{\langle B_B\rangle}{V_X^{\text{eff}}}\right]^{-1}$	
Complex of arbitrary stoichiometry $M_{A_\alpha}M_{B_\beta}$	$G_{0_\Lambda} = \frac{\sum_{\rho=0}^{\alpha}\sum_{\sigma=0}^{\beta}(\rho\kappa_{gL_1} + \sigma\kappa_{rL_1})(\rho\kappa_{gL_2} + \sigma\kappa_{rL_2})\langle C_{\rho g\sigma r}^{app}\rangle}{V_\Lambda^{\text{eff}}\left[\sum_{\rho=0}^{\alpha}\sum_{\sigma=0}^{\beta}(\rho\kappa_{gL_1} + \sigma\kappa_{rL_1})\langle C_{\rho g\sigma r}^{app}\rangle + \frac{\langle B_{L_1}\rangle}{V_\Lambda^{\text{eff}}}\right]}$ $\times\left[\sum_{\rho=0}^{\alpha}\sum_{\sigma=0}^{\beta}(\rho\kappa_{gL_2} + \sigma\kappa_{rL_2})\langle C_{\rho g\sigma r}^{app}\rangle + \frac{\langle B_{L_2}\rangle}{V_\Lambda^{\text{eff}}}\right]^{-1}$	$\langle C_{\rho a\sigma b}^{app}\rangle = \binom{\alpha}{\rho}\binom{\beta}{\sigma}p_a^\rho(1 - p_a)^{\alpha-\rho}p_b^\sigma(1 - p_b)^{\beta-\sigma}[M_{a_\alpha}M_{b_\beta}]N_A$ $\langle C_{1a0b}^{app}\rangle = p_a[M_a]N_A$ $+\binom{\alpha}{1}p_a(1 - p_a)^{\alpha-1}(1 - p_b)^\beta[M_{a_\alpha}M_{b_\beta}]N_A$ $\langle C_{0a1b}^{app}\rangle = p_b[M_b]N_A$ $+\binom{\beta}{1}(1 - p_a)^\alpha p_b(1 - p_b)^{\beta-1}[M_{a_\alpha}M_{b_\beta}]N_A$
Oligomerisaton	Use appropriate ACF and CCF formula from above with definitions on the right	$\langle C_{\rho a\sigma b}^{app}\rangle = \binom{\alpha}{\rho}\binom{\alpha-\rho}{\sigma}p_a^\rho p_b^\sigma(1 - p_a - p_b)^{\alpha-\rho-\sigma}[M_\alpha]N_A$ $\langle C_{1a0b}^{app}\rangle = p_a[M]N_A$ $+\binom{\alpha}{1}p_a(1 - p_a - p_b)^{\alpha-1}[M_\alpha]N_A$ $\langle C_{0a1b}^{app}\rangle = p_b[M]N_A$ $+\binom{\alpha}{1}p_b(1 - p_a - p_b)^{\alpha-1}[M_\alpha]N_A$

only semi-quantitively at best, the relative abundance of double-labelled particles based on DC-FCCS results [3].

6.2.1 Cross-correlation amount

Combining equations (6.4) and (6.5a) and (6.5b), we can write:

$$\frac{G_{0_X}}{G_{0_B}} = \frac{N_{ab}}{N_a + N_{ab}} = \frac{\text{complexed particles}}{\text{total particles carrying } a} \tag{6.9a}$$

and

$$\frac{G_{0_X}}{G_{0_A}} = \frac{N_{ab}}{N_b + N_{ab}} = \frac{\text{complexed particles}}{\text{total particles carrying } b} \tag{6.9b}$$

In other words, the ratio of the CCF amplitude to the ACF amplitude in channel A represents the ratio of the number of complexed particles to the total number of particles carrying fluorophore b and vice versa. We would like to highlight here the channel and fluorophore indicators as the relationship could seem confusing at first.

Under equilibrium, the number of complexed particles formed is limited by the total number of molecules of the less abundant of the two species. N_{ab} can assume values between zero (no cross-correlation) and the total number of particles carrying fluorophore b $(N_b + N_{ab})$ or a $(N_a + N_{ab})$, whichever is lower (complete cross-correlation). Therefore, the ratio of N_{ab} to the smaller of the two total particle numbers $(N_b + N_{ab})$ or $(N_a + N_{ab})$ is indicative of the complex stability (the affinity of the two species of molecules to form a complex). According to equations (6.9a) and (6.9b)

$$\begin{aligned} \frac{N_{ab}}{min[(N_a + N_{ab});(N_b + N_{ab})]} &= \max\left[\frac{N_{ab}}{N_a + N_{ab}}; \frac{N_{gr}}{N_b + N_{ab}}\right] \\ &= \max\left[\frac{G_{0_X}}{G_{0_B}}; \frac{G_{0_X}}{G_{0_A}}\right] \end{aligned} \tag{6.10}$$

In this equation, the functions min$[x;y]$ and max$[x;y]$, provide the minimum and maximum values of x and y, respectively. We will now define the cross-correlation amount Q as:

$$Q = \max\left[\frac{G_{0_X}}{G_{0_B}}; \frac{G_{0_X}}{G_{0_A}}\right] \tag{6.11}$$

In an ideal DC-FCCS experiment, we have discussed so far, the meaning of Q follows from equation (6.10). In the case of a real experiment, the cross-correlation amount Q defined by equation (6.11) is no longer exactly equal to the particle number ratios in equation (6.10). Nevertheless, it can still be used, and indeed often is used, as a semi-quantitative indicator of the complex stability. The main reason for its popularity as an output parameter of DC-FCCS experiments is its straightforward calculation according to equation (6.11) which compensates for the lack of direct quantitative interpretation. As you will see in the following sections, a rigorous description of a real DC-FCCS experiment requires several additional parameters to be included in the model. The experimental determination of those parameters could be problematic and introduce a large relative error to the final output values, such as the K_D of the complex. In such cases it is often a more practical approach to forgo aspirations at a fully quantitative interpretation of the DC-FCCS data and use a simple output such as the cross-correlation amount Q for a relative comparison between samples or between a sample and controls [3–5].

6.2.2 Unequal and non-aligned observation volumes

The actual observation volumes for both detection channels depend strongly on the type of microscope setup used for the FCCS experiment as well as on the geometry of the sample; however, in most cases the observation volume is in some way related to the microscope PSF. The PSF dimensions scale with wavelength and its position is also wavelength dependent in the case of objective chromatic aberrations. Chromatic aberration can lead to an imperfect overlap of the excitation laser beams, further complicating the interpretation.

In a confocal microscope, which is certainly the most widespread instrument for DC-FCCS, the observation volume can be approximated as a 3D Gaussian profile with the e^{-2} radii in the lateral and axial direction equal to w_0 and z_0, respectively (see section 5.1). In the first approximation, both w_0 and z_0 are proportional to the excitation wavelength λ_{ex} through the diffraction limited focusing of the excitation laser beam (see appendix 3.1). As derived in math box 5.2, the effective confocal observation volume is proportional to $w_0^2 z_0$, which means it is proportional to λ_{ex}^3. For a common pair of excitation wavelengths 488 nm and 561 nm this results in observation volumes different by a factor larger than 1.5. One possibility of reducing this difference comes from the fact that the excitation laser beam is focused to a diffraction limited spot in the focal plane only when the back aperture of the objective is fully filled by the beam. If the beam diameter d_{beam} is smaller than the diameter of the objective back aperture d_{obj} (the beam fills only a part of the aperture), the focal spot size increases; the smaller the part of the aperture that is filled, the larger the focal spot becomes (the same idea is sometimes used to study FCS diffusion laws; see section 5.7.8). Under-filling the objective back aperture results in an effective excitation numerical aperture NA_{ex} of the objective being smaller than the objective's native numerical aperture NA [6].

$$\frac{\text{NA}_{ex}}{\text{NA}} = \frac{d_{\text{beam}}}{d_{\text{obj}}} \tag{6.12}$$

Objective under-filling has been used by some authors to expand the observation volume for channel A to match the channel B observation volume [7]. However, since this approach requires non-trivial modifications to commercially available confocal setups and even then is not likely to guarantee a perfect overlap of the observation volumes (for reasons discussed further), it is not widely used in DC-FCCS.

Besides the differences in their dimensions, the two observation volumes may be shifted with respect to each other in both the lateral and the axial direction

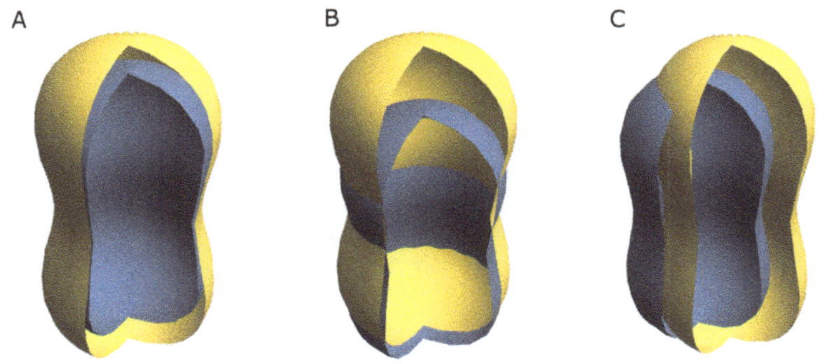

Figure 6.2. Illustration of unequal or non-aligned observation volumes commonly encountered in confocal DC-FCCS. Because of the longer excitation and emission wavelengths, the observation volume of channel *B* (shown in yellow) is larger than the observation volume of channel *A* (shown in blue). Besides the difference in size, the two volumes can be shifted with respect to each other in both the axial (B) and lateral (C) directions.

(see figure 6.2). We will describe this shift by a displacement vector $\vec{d}$, defined as the difference in position vectors of the centres of the observation volumes of channels *B* and *A*, $\vec{d} = \vec{r}_B - \vec{r}_A$. The axial displacement of the centres of the two observation volumes, d_z, is usually caused by the axial chromatic aberration of the microscope objective. This is an intrinsic property of the objective, which can be minimised by the selection of an objective with good quality of chromatic aberration correction, such as an apochromatic objective [8]. Alternatively, the axial chromatic aberration can be reduced by slightly changing the collimation of one of the excitation laser beams entering the objective, so that the centre of the excitation intensity distribution is moved up or down along the optical axis. However, besides not being readily possible with many commercial microscopes, this will not entirely remove the axial chromatic aberration as the detection efficiency for the different fluorescence wavelengths will still be subject to the same aberration.

The lateral chromatic aberration of microscope objectives can be considered negligible, especially for beams close to the optical axis. Thus, the lateral displacement of the observation volume centres, d_x and d_y, if significantly present, is more likely to be caused predominantly by an imperfect overlap of the two laser beams entering the objective. This effect can be entirely eliminated if the two excitation laser lines are emitted by the same laser (for example, different lines of an Argon ion or white light laser) or, when two different lasers need to be used, if the beams from the two lasers enter the microscope via a single optical fibre. This ensures that both laser lines enter the objective following exactly the same path. Note that neither of these scenarios is compatible with expanding channel *A* observation volume by objective back aperture under-filling or with reducing the axial chromatic aberration by changing the beam collimation discussed in the previous paragraphs.

The phenomena causing imperfect observation volume overlap, which we have discussed here, can be incorporated into the theoretical model for $G_X(\tau)$. For free diffusion in all three dimensions, equation (6.3a) becomes (see math box 6.2 for derivation):

$$G_X(\tau) = \frac{\langle C_{ab}\rangle}{V_X(\langle C_a\rangle + \langle C_{ab}\rangle)(\langle C_b\rangle + \langle C_{ab}\rangle)} \times \left(1 + \frac{4D_{ab}\tau}{w_{0_X}^2}\right)^{-1}\left(1 + \frac{4D_{ab}\tau}{z_{0_X}^2}\right)^{-1/2} \times \exp\left(-\frac{d_x^2 + d_y^2}{4D_{ab}\tau + w_{0_X}^2} - \frac{d_z^2}{4D_{ab}\tau + z_{0_X}^2}\right) + G_\infty \tag{6.13}$$

Note that instead of the diffusion time used in equation (6.3a), we characterise the diffusion of the double-labelled particles by their diffusion coefficient D_{ab}. The reason for this is the dependence of the diffusion time on the observation volume, which means that the double-labelled particles will appear in the individual ACFs, $G_A(\tau)$ and $G_B(\tau)$, with different diffusion times and in the CCF with yet a different diffusion time. However, the diffusion coefficient is the same for all the three correlation functions. The apparent e^{-2} radii of the cross-correlation observation volume are:

$$w_{0_X}^2 = \frac{1}{2}\left(w_{0_A}^2 + w_{0_B}^2\right) \tag{6.14a}$$

$$z_{0_X}^2 = \frac{1}{2}\left(z_{0_A}^2 + z_{0_B}^2\right) \tag{6.14b}$$

and the apparent observation volume is:

$$V_X = (\pi/2)^{\frac{3}{2}} w_{0_X}^2 z_{0_X} \tag{6.15}$$

Comparing equations (6.13) and (6.3a)–(6.3c), we can notice that the imperfect observation volume overlap results in a slower temporal decay of the CCF. This can be understood quite intuitively; the average time spent by the double-labelled particles in the observation volume of channel A is equal to the corresponding diffusion time, which determines the temporal decay of the $G_A(\tau)$. If the particle happens to leave the observation volume of channel A in a direction having a component parallel to the displacement vector $\vec{d}$, its trajectory will continue for some more time to be inside the observation volume of channel B, for which time the particle will continue contributing to $G_{AB}(\tau)$, making its decay slower. Analogously, particles entering the observation volume of channel A in a direction with a component antiparallel to $\vec{d}$ prolong the decay of $G_{BA}(\tau)$. Therefore, the decay of the symmetrized CCF $G_X(\tau)$ is also slower.

Math box 6.2

The CCF with displaced observation volumes

We will introduce the displacement vector $\vec{d}$ in the equation for the CCF, leading to:

$$G_{AB}(\tau) = \frac{\kappa_a\kappa_b \int\int W_A(\vec{r})W_B(\vec{r}\,' - \vec{d})\langle\delta C_{ab}(\vec{r}, t)\delta C_{ab}(\vec{r}\,', t + \tau)\rangle d\vec{r}d\vec{r}\,'}{\kappa_a\kappa_b \int W_A(\vec{r})\langle C_a(\vec{r}, t) + C_{ab}(\vec{r}, t)\rangle d\vec{r} \int W_B(\vec{r}\,' - \vec{d})\langle C_b(\vec{r}\,', t) + C_{ab}(\vec{r}\,', t)\rangle d\vec{r}\,'} + G_\infty$$

As in the case of math box 6.1, we assume that $S_A = S_B = 1$ and that the concentration fluctuation correlation depends only on the concentration of the double-labelled particles C_{ab}. The diffusion propagator characterizing the concentration fluctuation correlation is not affected by the displacement and, therefore, described by the same formula as in math box 6.1.

W_A and W_B can be written as:

$$W_A(\vec{r}) = \frac{2P_A}{\pi w_{0_A}} e^{\frac{-2x^2}{w_{0_A}^2}} e^{\frac{-2y^2}{w_{0_A}^2}} e^{\frac{-2z^2}{z_{0_A}^2}}$$

$$W_b(\vec{r} - \vec{d}) = \frac{2P_B}{\pi w_{0_B}} e^{\frac{-2(x-d_x)^2}{w_{0_B}^2}} e^{\frac{-2(y-d_y)^2}{w_{0_B}^2}} e^{\frac{-2(z-d_z)^2}{z_{0_B}^2}}$$

Next, we introduce the expressions for W_A and W_B to the original formula, use the diffusion propagator to express the concentration fluctuation correlation and move constant factors outside the integrals. We can perform the integration separately over each spatial dimension; note that the constant factors contribute to each dimension by their cubic roots. For the x spatial dimension, we get:

$$(G_{AB}(\tau) - G_\infty)^{(x)} = \frac{\frac{\langle C_{ab}\rangle^{1/3}}{2\pi D_{ab}\tau}\int\int e^{\frac{-2x^2}{w_{0_A}^2}} e^{\frac{-2(x'-d_x)^2}{w_{0_B}^2}} e^{-\frac{(x-x')^2}{4D_{ab}\tau}} dxdx'}{(\langle C_a\rangle + \langle C_{ab}\rangle)^{1/3}(\langle C_b\rangle + \langle C_{ab}\rangle)^{1/3}\int e^{\frac{-2x^2}{w_{0_A}^2}} dx \int e^{\frac{-2(x'-d_x)^2}{w_{0_B}^2}} dx'}$$

$$= \left[\frac{\langle C_{ab}\rangle}{(\langle C_a\rangle + \langle C_{ab}\rangle)(\langle C_b\rangle + \langle C_{ab}\rangle)}\right]^{1/3} \frac{w_{0_A}\left(8D_{ab}\tau + w_{0_A}^2\right)^{-1/2}}{\frac{\pi}{2}w_{0_A}w_{0_B}} e^{-\frac{2\,d_x^2}{w_{0_B}^2}}$$

$$\times \int e^{-\frac{2\left(8D\tau + w_{0_A}^2 + w_{0_B}^2\right)}{w_{0_B}^2\left(8D_{ab}\tau + w_{0_A}^2\right)}x'^2 + \frac{4d_x}{w_{0_B}^2}x'} dx'$$

$$= \left[\frac{\langle C_{ab}\rangle}{(\langle C_a\rangle + \langle C_{ab}\rangle)(\langle C_b\rangle + \langle C_{ab}\rangle)}\right]^{1/3} \frac{1}{\sqrt{\frac{\pi}{2}}\,w_{0_X}} \left(1 + \frac{4D_{ab}\tau}{w_{0_X}^2}\right)^{-1/2} e^{-\frac{d_x^2}{4D\tau + w_{0_X}^2}}$$

The integration can be performed analogously for the remaining two spatial dimensions, which then leads to an equation for $G_{AB}(\tau)$ identical in form to equation (6.13). Assuming no asymmetric relationship between $F_A(t)$ and $F_B(t)$ fluctuations, the same expression holds also for $G_X(\tau)$.

In our discussion of FCCS, we are interested mainly in the amplitudes of the correlation functions since they carry the information on the abundance of the complexes. From equation (6.13) we get:

$$G_{0_X} = \frac{\langle C_{ab}\rangle}{V_X(\langle C_a\rangle + \langle C_{ab}\rangle)(\langle C_b\rangle + \langle C_{ab}\rangle)} \times \exp\left(-\frac{d_x^2 + d_y^2}{w_{0X}^2} - \frac{d_z^2}{z_{0X}^2}\right) \tag{6.16}$$

Equation (6.16) shows that the displacement of the observation volumes decreases the CCF amplitude. This can be also explained quite intuitively; the double-labelled particles present in the non-overlapping parts of the observation volumes appear effectively as single-labelled particles since they contribute only to the signal in one of the detection channels. Equations (6.9a) and (6.9b) show that an increase in the apparent number of single-labelled particles decreases the ratio of the CCF amplitude to the ACF amplitudes.

Therefore, to be able to use G_{0_X} to quantify the abundance of the double-labelled particles, we need to characterize correctly the influence of the displacement $\vec{d}$. One way to do this relies on finding $\vec{d}$, for example, by scanning fluorescent nanoparticles with fluorescent spectra resembling those of the fluorophores used in the FCCS experiment. In an experimentally simpler approach, we introduce an effective apparent cross-correlation observation volume V_X^{eff}, which satisfies the following equation [9]

$$G_{0_X} = \frac{\langle C_{ab}\rangle}{V_X^{\text{eff}}(\langle C_a\rangle + \langle C_{ab}\rangle)(\langle C_b\rangle + \langle C_{ab}\rangle)} \tag{6.17}$$

Provided a non-zero displacement of the observation volume centres, V_X^{eff} is larger than V_X defined in equation (6.15). V_X^{eff} can be determined by an FCCS measurement in a sample containing only double-labelled particles. If the concentration of the double-labelled particles $\langle C_{ab}\rangle$ is known then V_X^{eff} is directly determined from equation (6.17), assuming the sample contains no single-labelled particles ($\langle C_a\rangle = \langle C_b\rangle = 0$). This condition is not trivially satisfied; samples of molecules labelled each by a fluorophore a and a fluorophore b frequently contain imperfectly labelled molecules bearing only one of the fluorophores or molecules in which one of the fluorophores is not fluorescent (e.g. a misfolded fluorescent protein) as we will discuss later in section 6.2.5. $\langle C_{ab}\rangle$ can be obtained experimentally, for example, from the ACF in either of the detection channels, if the respective observation volume (V_A or V_B) has been calibrated. Observation volume calibration can be performed by FCS using a sample with known concentration of fluorescent particles or a fluorescent tracer with known diffusion coefficient (see section 9.1.2). The latter calibration strategy relies on the dependence of the diffusion time on the observation volume. For example, for a confocal microscope, this dependence is expressed by equation (5.19). Assuming the structure parameter K is constant, the diffusion time scales with the square of the cubic root of the observation volume (see math box 5.2). If the observation volume of one of the detection channels has been calibrated, the remaining two observation volumes can be obtained from the diffusion times of double-labelled particles, using the equality $V_X^{\mathrm{eff}}(\tau_{D_{ab}}^{X})^{-3/2} = V_A(\tau_{D_{ab}}^{A})^{-3/2} = V_B(\tau_{D_{ab}}^{B})^{-3/2}$, where $\tau_{D_{ab}}^{\Lambda}$ is the diffusion time of the double-labelled particles in channel Λ (Λ stands for A, B or X). Using the diffusion times instead of the correlation function amplitudes (equation (6.17)) to find the observation volumes has the advantage of being less sensitive to the presence of imperfect labelling. If the particle is sufficiently large, the absence of one of the fluorophores will cause only a negligible change in the diffusion time. Moreover, if the presence of apparently single-labelled particles is caused by one of the fluorophores being non-fluorescent, the diffusion time will not be affected at all.

Note that as far as the CCF amplitudes are concerned, the complete knowledge of $\vec{d}$ or of V_X^{eff} are equivalent. However, since $\vec{d}$ influences also the decay of the CCF (see equation (6.13)) by changing the shape of the apparent observation volume, the theoretical models derived for the ACF decays are no longer exactly valid in this case. Nevertheless, the displacement $\vec{d}$ is usually small enough and the models derived for ACFs can be applied satisfactorily to fit the CCFs, justifying the use of the V_X^{eff} approach. This approach has an additional advantage of being applicable to all DC-FCCS modalities (not only confocal DC-FCCS, for which equation (6.16) has been derived under the assumption of Gaussian observation volumes) as long as the observation volumes displacement $\vec{d}$ is sufficiently small.

Another fact worth noticing here is the identical functional form of the correlation functions in equations (6.13) and (5.54) (when setting the flow velocity to zero). Equation (5.54) describes the spatial CCF between signals originating from two spatially separated foci in two-focus FCS (see section 5.7.1), which is a situation analogous to our case here. The differences between the two cases are only in the origin and the magnitude of the shifts between the foci; in two-focus FCS, the spatial shift is purposely introduced in a defined manner and is typically much larger than in DC-FCCS where it is an unwanted result of imperfections in the optical setup.

6.2.3 Spectral crosstalk

In the idealized model we have assumed that all photons are unambiguously sorted into their respective detection channels; photons emitted by fluorophores a to channel A and by fluorophores b to channel B. This is, however, rarely satisfied in a real experiment. We will, therefore, need to consider the situation in which some photons are detected in the 'wrong' channel. To describe such a situation, we need to quantify the contribution of a fluorophore to each of the detection channels. We can do this by introducing the brightness matrix $\boldsymbol{\kappa}$; the element κ_{kL} of the matrix is the average number of photons which a fluorophore of the k-th species contributes to the L-th detection channel. In general, the two dimensions of the matrix can be any positive integers, not necessarily equal; in a typical DC-FCCS experiment they are both equal to 2. The diagonal elements give contributions of fluorophores of each population to their 'correct' detection channels, and the off-diagonal elements describe the spectral crosstalk. This means that in the previous sections we have assumed the brightness matrix to be diagonal. In this formalism, we can also introduce a vector of fluorescence intensities in the individual detection channels $\{F_L(t)\}$ and a vector of total concentrations of fluorophores of the individual populations $\{C_k^{tot}(\vec{r}, t)\}$. $\{F_L(t)\}$ can be then calculated as:

$$\{F_L(t)\} = \int W_L(\vec{r})\boldsymbol{\kappa}\{C_k^{tot}(\vec{r}, t)\}d\vec{r} \tag{6.18}$$

The brightness matrix can be obtained experimentally by FCCS measurements in samples containing fluorophores of one species only. If we take a sample containing only fluorophores of the k-th species, then the average fluorescence signal in the L-th detection channel $\langle F_L \rangle$ (measured in photon counts per second) divided by the average particle number N is equal to the element κ_{kL} of the brightness matrix. For correct brightness matrix determination, it is important that the sample does not contain any other fluorophores than those of the k-th population and that there are no particles containing more than a single fluorophore molecule. Particles containing h (where h is an integer) fluorophore molecules are expected to have h-times higher brightness than a single fluorophore molecule; therefore, the presence of such particles in the sample would bias the brightness towards higher values. Note that particles contribute to the ACF with the square of their brightness, making the bias towards higher brightness even larger. In the presence of uncorrelated background (see section 6.2.4), the average fluorescence signal is obtained by subtracting the average background from the average total signal in the respective channel. The brightness matrix element is then calculated as:

$$\kappa_{kL} = \frac{\langle F_L \rangle - \langle B_L \rangle}{N_k} = G_{0_L}\frac{\langle F_L \rangle^2}{\langle F_L \rangle - \langle B_L \rangle} \tag{6.19}$$

where $\langle B_L \rangle$ denotes the average background in the L-th detection channel. Equation (6.19) reflects the fact that, in the presence of uncorrelated background, the particle number N is no longer equal to the inverse of the correlation function amplitude (see section 7.1) and involves corrections derived in math box 7.2.

The model for $G_X(\tau)$ in the presence of spectral crosstalk can be derived by inserting equation (6.18) into the equation for $G_X(\tau)$. Remember that we need to modify the models for the ACFs as well, since, in the presence of spectral crosstalk, particles labelled by fluorophore a only, also contribute to the ACF of channel B (and vice versa). The derivation of the models is shown in math box 6.3, leading to the following expressions for the ACF and CCF amplitudes:

$$G_{0_A} = \frac{(\kappa_{aA} + \kappa_{bA})^2\langle C_{ab} \rangle + \kappa_{aA}{}^2\langle C_a \rangle + \kappa_{bA}{}^2\langle C_b \rangle}{V_A[\kappa_{aA}\langle C_a \rangle + \kappa_{bA}\langle C_b \rangle + (\kappa_{aA} + \kappa_{bA})\langle C_{ab} \rangle]^2} \tag{6.20a}$$

$$G_{0_B} = \frac{(\kappa_{aB} + \kappa_{bB})^2\langle C_{ab} \rangle + \kappa_{aB}{}^2\langle C_a \rangle + \kappa_{bB}{}^2\langle C_b \rangle}{V_B[\kappa_{aB}\langle C_a \rangle + \kappa_{bB}\langle C_b \rangle + (\kappa_{aB} + \kappa_{bB})\langle C_{ab} \rangle]^2} \tag{6.20b}$$

$$G_{0_X} = \frac{(\kappa_{aA} + \kappa_{bA})(\kappa_{aB} + \kappa_{bB})\langle C_{ab} \rangle + \kappa_{aA}\kappa_{aB}\langle C_a \rangle + \kappa_{bA}\kappa_{bB}\langle C_b \rangle}{V_X^{\text{eff}}[\kappa_{aA}\langle C_a \rangle + \kappa_{bA}\langle C_b \rangle + (\kappa_{aA} + \kappa_{bA})\langle C_{ab} \rangle]} \times [\kappa_{aB}\langle C_a \rangle + \kappa_{bB}\langle C_b \rangle + (\kappa_{aB} + \kappa_{bB})\langle C_{ab} \rangle]^{-1} \tag{6.20c}$$

Equations (6.20a)–(6.20c) can be understood in a very straightforward way. The ACF amplitude in channel A is a sum of contributions of all three types of particles to that detection channel: the double-labelled particles ($(\kappa_{aA} + \kappa_{bA})^2\langle C_{ab}\rangle$), the single-labelled particles carrying fluorophore a ($\kappa_{aA}{}^2\langle C_a\rangle$) and single-labelled particles carrying fluorophore b ($\kappa_{bA}{}^2\langle C_b\rangle$). The contribution of each type of particles is weighted by the square of their brightness in channel A (see equation (5.26)). Analogously, we get the amplitude of the ACF of channel B. The contributions of each type of particles to the CCF amplitude are weighted by the products of their brightness in each detection channel; for example, for the single-labelled particles carrying fluorophore a it is $\kappa_{aA}\kappa_{aB}\langle C_a\rangle$.

Equation (6.20c) shows that spectral crosstalk increases the CCF amplitude by adding the terms $\kappa_{aA}\kappa_{aB}\langle C_a\rangle + \kappa_{bA}\kappa_{bB}\langle C_b\rangle$ to the numerator; a positive CCF amplitude can be, thus, observed even in cases when there are no double-labelled particles present in the sample ($\langle C_{ab}\rangle = 0$). Spectral crosstalk not only affects the amplitudes of the correlation functions but also their temporal decays. It can be seen from the equations in math box 6.3 that in the presence of spectral crosstalk the temporal decay of the CCF not only depends on the concentration fluctuations of the double-labelled particles but also on the concentration fluctuations of the single-labelled particles of both species (weighted by their respective brightness in the 'wrong' detection channels). In consequence, the CCF temporal decay is no longer entirely independent of processes which involve only fluorophores of a single species such as fluorophore blinking caused by transitions to triplet or other dark states. This means that a strong spectral crosstalk (high brightness of a fluorophore species in the 'wrong' detection channel) not only affects the analysis of correlation function amplitudes but also the fitting of their temporal decays.

Although the artificial influence of spectral crosstalk on the correlation function amplitudes can be modelled using equation (6.20), provided the brightness matrix $\boldsymbol{\kappa}$ is known, it is a common practice in DC-FCCS studies to measure a negative control, in which nothing other than artificial cross-correlation is expected. A typical negative control consists of a system of non-interacting molecules labelled by the same fluorophores which are used in the actual experiment. The use of a negative control is particularly important in semi-quantitative DC-FCCS studies, in which simple correlation function amplitude ratios (such as Q) are the main output rather than concentrations calculated by solving equation (6.20) (or some of the more complex models presented later in this chapter). The average cross-correlation amount Q_{neg} of the negative control is, then, regarded as the lower detection limit and only Q values significantly higher than Q_{neg} can be interpreted as an indication of complex formation. Again, experimental realisations tend to present more challenges here than the theoretical description. Situations have been reported when molecules presumed to have no direct interaction (and, hence, suitable as a negative control) were found to be recruited together into larger supra-molecular structures (such as plasma membrane nanodomains). This resulted in co-diffusion of those molecules giving rise to elevated cross-correlation amounts [5].

Math box 6.3

The CCF in the presence of spectral crosstalk
We will insert intensities in the individual detection channels calculated according to equation (6.18) to the equation for the CCF (let us for simplicity assume identical detection volume for both channels at first and introduce the difference and displacement later):

$$G_{AB}(\tau) = \frac{\iint W_A(\vec{r})W_B(\vec{r}')\langle(\kappa_{aA}\delta C_a^{tot}(\vec{r},t)+\kappa_{bA}\delta C_b^{tot}(\vec{r},t))(\kappa_{aB}\delta C_b^{tot}(\vec{r}',t+\tau)+\kappa_{bB}\delta C_b^{tot}(\vec{r}',t+\tau))\rangle d\vec{r}d\vec{r}'}{\int W_A(\vec{r})\langle\kappa_{aA}\delta C_a^{tot}(\vec{r},t)+\kappa_{bA}\delta C_b^{tot}(\vec{r},t)\rangle d\vec{r}\int W_B(\vec{r}')\langle\kappa_{aB}C_a^{tot}(\vec{r}',t)+\kappa_{bB}\delta C_b^{tot}(\vec{r}',t)\rangle d\vec{r}'} + G_\infty$$

Denoting the denominator in the equation above as Denominator, we can rewrite it as:

$$\begin{aligned}&(G_{AB}(\tau) - G_\infty) \times \text{Denominator}\\ &= \iint W_A(\vec{r})W_B(\vec{r}')\kappa_{aA}\kappa_{aB}\langle\delta C_a(\vec{r},t)\delta C_a(\vec{r}',t+\tau)\rangle d\vec{r}d\vec{r}'\\ &+ \iint W_A(\vec{r})W_B(\vec{r}')\kappa_{bA}\kappa_{bB}\langle\delta C_b(\vec{r},t)\delta C_b(\vec{r}',t+\tau)\rangle d\vec{r}d\vec{r}'\\ &+ \iint W_A(\vec{r})W_B(\vec{r}')(\kappa_{aA}+\kappa_{bA})(\kappa_{aB}+\kappa_{bB})\langle\delta C_{ab}(\vec{r},t)\delta C_{ab}(\vec{r}',t+\tau)\rangle d\vec{r}d\vec{r}'\end{aligned}$$

Each of the three integrals at the right-hand side of the equation above has the form:

$$\iint W_A(\vec{r})W_B(\vec{r}')\varphi_\varepsilon^2\langle\delta C_\varepsilon(\vec{r},t)\delta C_\varepsilon(\vec{r}',t+\tau)\rangle d\vec{r}d\vec{r}'$$

where the subscript ε stands either for a, b or ab and φ_ε^2 is a quadratic form of the elements of the brightness matrix $\boldsymbol{\kappa}$. Integration gives:

$$\begin{aligned}&\frac{P_AP_Bw_0^2z_0}{2\sqrt{\pi}}\varphi_\varepsilon^2\langle C_\varepsilon\rangle\left(1+\frac{4D_\varepsilon\tau}{w_0^2}\right)^{-1}\left(1+\frac{4D_\varepsilon\tau}{z_0^2}\right)^{-1/2}\\ &= \frac{P_AP_Bw_0^2z_0}{2\sqrt{\pi}}\varphi_\varepsilon^2\langle C_\varepsilon\rangle\tilde{G}_\varepsilon(\tau)\end{aligned}$$

where $\tilde{G}_\varepsilon(\tau)$ is a normalized decay of the correlation function (decaying from 1 at $\tau = 0$ to 0 at $\tau = \infty$). If we insert this expression to the equation for the CCF, we can see that in the presence of spectral crosstalk, the temporal decay of $G_{AB}(\tau)$ is a linear combination of the temporal decays of concentration fluctuations of the double-labelled and both species of the single labelled particles.

However, for the purpose of DC-FCCS data analysis, we are mainly interested in the correlation function amplitudes. If we evaluate the Denominator, we get for the CCF amplitude:

$$\begin{aligned}G_{0_X} &= \frac{(\kappa_{aA}+\kappa_{bA})(\kappa_{aB}+\kappa_{bB})\langle C_{ab}\rangle + \kappa_{aA}\kappa_{aB}\langle C_a\rangle + \kappa_{bA}\kappa_{bB}\langle C_b\rangle}{V_X[\kappa_{aA}\langle C_a\rangle + \kappa_{bA}\langle C_b\rangle + (\kappa_{aA}+\kappa_{bA})\langle C_{ab}\rangle]}\\ &\quad\times[\kappa_{aB}\langle C_a\rangle + \kappa_{bB}\langle C_b\rangle + (\kappa_{aB}+\kappa_{bB})\langle C_{ab}\rangle]^{-1}\end{aligned}$$

Now we reintroduce the displacement and difference of detection volumes by replacing V_X with the effective detection volume V_X^{eff} like we did in equation (6.17), which finally leads us to the model of equation (6.20c).

The spectral crosstalk also affects the autocorrelation functions, for which we can write:

$$G_L(\tau) = \frac{\iint W_L(\vec{r})W_L(\vec{r}')\langle(\kappa_{aL}\delta C_a^{tot}(\vec{r},t) + \kappa_{bL}\delta C_b^{tot}(\vec{r},t))(\kappa_{aL}\delta C_a^{tot}(\vec{r}',t+\tau) + \kappa_{bL}\delta C_b^{tot}(\vec{r}',t+\tau))\rangle d\vec{r}d\vec{r}'}{\left[\int W_L(\vec{r})\langle\kappa_{aL}\delta C_a^{tot}(\vec{r},t) + \kappa_{bL}\delta C_b^{tot}(\vec{r},t)\rangle d\vec{r}\right]^2} + G_\infty$$

where the subscript L stands either for A or B. By evaluating the integral analogously to that for the CCF, we arrive at equations (6.20a) and (6.20b).

The origin of spectral crosstalk is illustrated in figure 6.2, which shows the emission spectra of a pair of fluorescence proteins commonly used in DC-FCCS. It can be seen in the figure that a non-negligible part of fluorophore a spectrum lies in the same spectral region as the maximum of fluorophore b spectrum. On the other hand, very little signal from fluorophore b is expected in the spectral region where the maximum of fluorophore a lies. Therefore, κ_{aB} is expected to be significantly higher than κ_{bA}, which in many cases can be neglected completely. We will return to the problem of spectral crosstalk in FCCS in section 6.3.2 and outline several approaches to obtain CCFs free from its influence.

6.2.4 Non-correlated background

The signal detected in each of the detection channels usually contains some background, besides the fluorescence of molecules of interest. The background can originate, for example, from dark counts of the detector, excitation laser light scattered in the sample or, in the case of biological samples, autofluorescence. The background is typically not correlated in time (on time-scales relevant to FCS) and as such it does not affect the ACF or CCF temporal decays; however, it affects the correlation function amplitudes, in which we are interested here (see section 7.1). We will, therefore, introduce background $B_L(t)$ for each detection channel. The average background values $\langle B_L \rangle$ can be obtained experimentally as the average signal measured in a non-labelled sample. Measurements in water or buffer only are sometimes used to estimate the background; while giving correctly the contribution of the detector noise, such measurements are likely to underestimate the background by neglecting the contributions of scattered light or autofluorescence from the real sample. Note that in inhomogeneous samples such as cells or tissues, the local background levels may vary significantly from the average $\langle B_L \rangle$ (for example, through cell-to-cell variations in autofluorescence). Since we cannot measure the actual background in each point where we perform the DC-FCCS measurements and rely on an average value $\langle B_L \rangle$ from an unlabelled control sample, the local background variations will translate to an additional error in the cross-correlation amount Q or the dissociation constant K_D.

The vector of fluorescence intensities in the individual detection channels $\{F_L(t)\}$ can then be calculated as:

$$\{F_L(t)\} = \int W_L(\vec{r})\boldsymbol{\kappa}\{C_k^{tot}(\vec{r}, t)\}d\vec{r} + \{B_L(t)\} \tag{6.21}$$

Next, we introduce intensities calculated according to equation (6.21) to the derivation of the models for the correlation function amplitudes and use the fact that the background is not correlated in time. This means that for all $\tau \neq 0$ $\langle B_{L_1}(t)B_{L_2}(t+\tau)\rangle = 0$, for $L_1 = L_2$ (ACF) as well as $L_1 \neq L_2$ (CCF). Therefore, the background will only affect the average intensities in the individual detection channels used for normalizing the correlation functions; the average intensity in the L-th detection channel can be written as

$$\langle F_L\rangle = V_L N_A \sum \kappa_{kL}\langle C_k^{tot}\rangle + \langle B_L\rangle \tag{6.22}$$

Equations (6.20a)–(6.20c) then become:

$$G_{0_A} = \frac{(\kappa_{aA} + \kappa_{bA})^2\langle C_{ab}\rangle + \kappa_{aA}{}^2\langle C_a\rangle + \kappa_{bA}{}^2\langle C_b\rangle}{V_A\left[\kappa_{aA}\langle C_a\rangle + \kappa_{bA}\langle C_b\rangle + (\kappa_{aA} + \kappa_{bA})\langle C_{ab}\rangle + \dfrac{\langle B_A\rangle}{V_A}\right]^2} \tag{6.23a}$$

$$G_{0_B} = \frac{(\kappa_{aB} + \kappa_{bB})^2\langle C_{ab}\rangle + \kappa_{aB}{}^2\langle C_a\rangle + \kappa_{bB}{}^2\langle C_b\rangle}{V_B\left[\kappa_{aB}\langle C_a\rangle + \kappa_{bB}\langle C_b\rangle + (\kappa_{aB} + \kappa_{bB})\langle C_{ab}\rangle + \dfrac{\langle B_B\rangle}{V_B}\right]^2} \tag{6.23b}$$

$$\begin{aligned} G_{0_X} = {} & \frac{(\kappa_{aA} + \kappa_{bA})(\kappa_{aB} + \kappa_{bB})\langle C_{ab}\rangle + \kappa_{aA}\kappa_{aB}\langle C_a\rangle + \kappa_{bA}\kappa_{bB}\langle C_b\rangle}{V_X^{\mathrm{eff}}\left[\kappa_{aA}\langle C_a\rangle + \kappa_{bA}\langle C_b\rangle + (\kappa_{aA} + \kappa_{bA})\langle C_{ab}\rangle + \dfrac{\langle B_A\rangle}{V_X^{\mathrm{eff}}}\right]} \\ & \times\left[\kappa_{aB}\langle C_a\rangle + \kappa_{bB}\langle C_b\rangle + (\kappa_{aB} + \kappa_{bB})\langle C_{ab}\rangle + \frac{\langle B_B\rangle}{V_X^{\mathrm{eff}}}\right]^{-1} \end{aligned} \tag{6.23c}$$

Uncorrelated background reduces the amplitudes of all three correlation functions, the ACFs as well as the CCF; its influence on the cross-correlation amount is, therefore, not obvious. The presence of uncorrelated background can cause both increase and decrease of the cross-correlation amount depending on the ratio of the ACF amplitudes and the ratio of background values in the detection channels. The influence of background is the smallest when both channels contain similar background contributions (see figure 6.4); on the other hand, if the background in one of the channels is significantly higher, the corresponding ACF amplitude is strongly reduced resulting in overestimating the cross-correlation amount Q defined by equation (6.11) (see figure 6.5). Note that in extreme cases where the background in one of the channels is much higher than in the other channel, the CCF amplitude can become larger than the ACF amplitude of the channel with higher background. If the background is neglected in the model, this situation will lead to an apparent

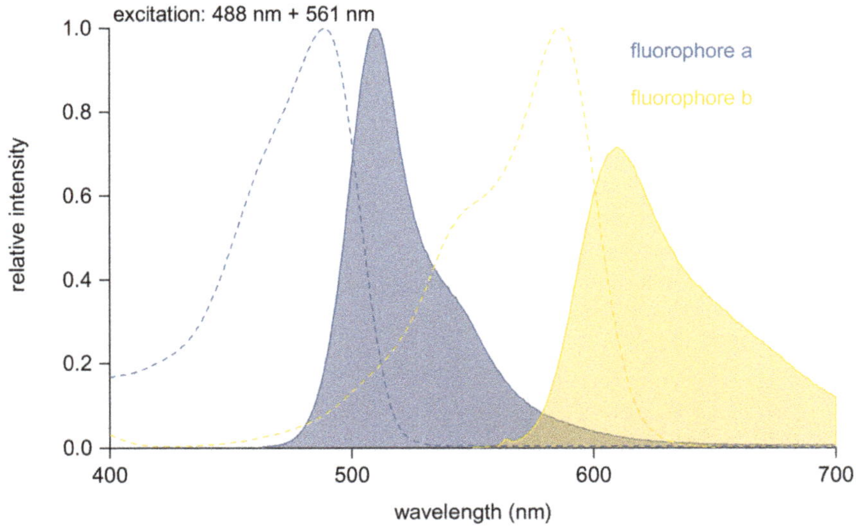

Figure 6.3. An example of fluorescence emission (solid lines) and excitation (dashed lines) spectra of EGFP and mCherry, a pair of fluorophores commonly used in DC-FCCS, illustrating the origin of spectral crosstalk. A non-negligible part of fluorophore a emission spectrum lies in the spectral region where fluorophore b emission has its maximum. At the same time, the emission of fluorophore b is negligible in the region of maximum emission of fluorophore a. The spectra were obtained from an online database [10] and the emission spectra represent a situation of simultaneous excitation by 488 nm and 561 nm laser lines.

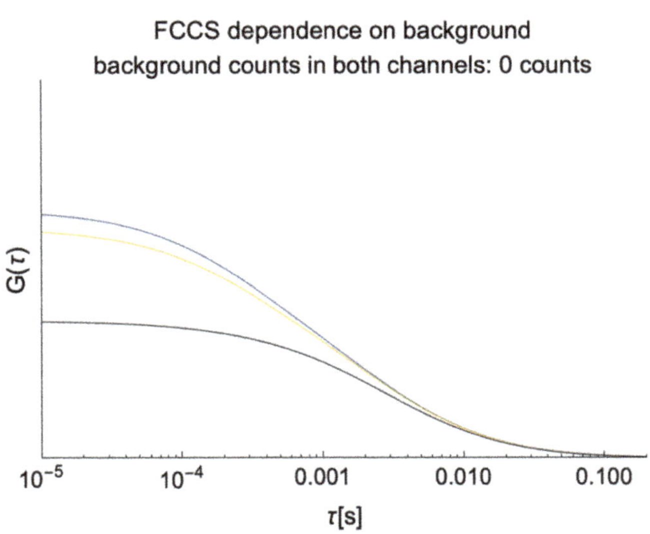

Figure 6.4. Dependence of the ACFs and the CCF (calculated according to equations (6.23a)–(6.23c)) on background counts in both channels, with background ranging from 0–5000 counts/s ($B_A = B_B = 0\ldots5000$). Otherwise, we assumed identical detection volumes for both channels with the following geometrical parameters: $w_{0A} = w_{0B} = w_{0X} = 260$ nm, $K_A = K_B = 5$, $z_{0A} = z_{0B} = z_{0X} = 1.3$ μm, resulting in observation volumes of $V_A = V_B = V_X^{\text{eff}} = 0.5$ fl; the brightness of the two particles in the two detection channels was set to $\kappa_{aA} = 1500$, $\kappa_{aB} = 150$, $\kappa_{bA} = 15$, $\kappa_{bB} = 1500$; the concentration of the particles was $c_a = c_b = c_{ab} = 2.5$ nM; the diffusion times for the two particles were set to $\tau_{Da} = 3$ ms, $\tau_{Db} = 0.3$ ms. Animation available at https://iopscience.iop.org/book/978-0-7503-2080-1.

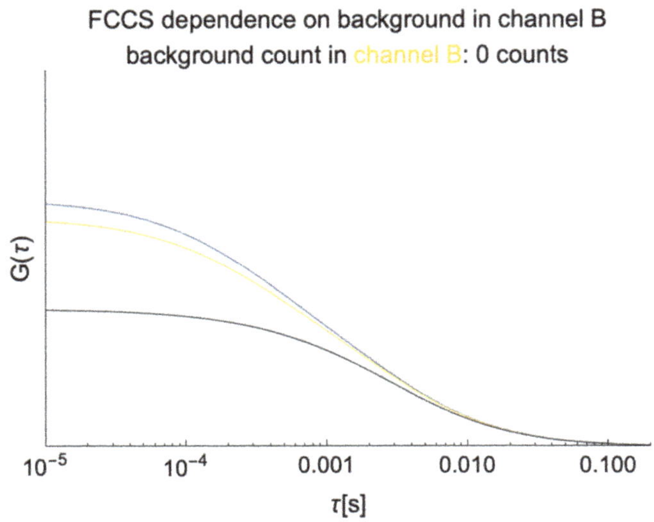

Figure 6.5. Dependence of the ACFs and the CCF (calculated according to equations (6.23a)–(6.23c)) on background counts in one channel only. The background for channel A was 0 ($B_A = 0$) while for channel B it varied from 0–5000 counts/s ($B_B = 0\ldots5000$). Otherwise, we assumed identical detection volumes for both channels with the following geometrical parameters: $w_{0A} = w_{0B} = w_{0X} = 260$ nm, $K_A = K_B = 5$, $z_{0A} = z_{0B} = z_{0X} = 1.3$ µm, resulting in observation volumes of $V_A = V_B = V_X^{\text{eff}} = 0.5$ fl; the brightness of the two particles in the two detection channels was set to $\kappa_{aA} = 1500$, $\kappa_{aB} = 150$, $\kappa_{bA} = 15$, $\kappa_{bB} = 1500$; the concentration of the particles were $c_a = c_b = c_{ab} = 2.5$ nM; the diffusion times for the two particles were set to $\tau_{Da} = 3$ ms, $\tau_{Db} = 0.3$ ms. Animation available at https://iopscience.iop.org/book/978-0-7503-2080-1.

cross-correlation amount Q larger than 1, which has no physical meaning (see equation (6.10)).

6.2.5 Non-fluorescent binding partners and free fluorophores

So far, we have assumed that all molecules of the binding partners M_a and M_b in the reaction scheme of equation (6.7) are labelled by a fluorophore a or b, respectively, and that all individual fluorophores present in the system are fluorescent at all times. In real experiments, these conditions are not always satisfied, for several possible reasons. The presence of unlabelled binding partners results in underestimation of the complex abundance because the complex particles containing an unlabelled binding partner (either M_a or M_b) are in effect single-labelled particles and do not contribute to the cross-correlation (besides the contribution caused by spectral cross-talk). The ratio of the CCF amplitude to the ACF amplitudes is then reduced, giving rise to an artificially lower cross-correlation amount Q or higher apparent complex dissociation constant K_D. We can distinguish two main scenarios leading to the presence of non-fluorescent binding partners. Furthermore, we will discuss a third scenario in which the sample contains free fluorophore molecules that are not bound to any of the binding partners (either M_a nor M_b) and can, therefore, not be involved in the complex formation. We will show that this situation can be considered complementary to the presence of unlabelled binding partners and treated in an analogous way in the theoretical models.

6.2.5.1 A certain ratio of the molecules under study is not fluorescently labelled
This is a very common and generic situation in DC-FCCS, which can result from multiple underlying mechanisms. If the molecules under study are labelled by extrinsic fluorophores such as covalently bound organic dyes, it can be the result of an imperfect labelling procedure that has left a part of the molecules of interest without a fluorophore attached. When studying proteins labelled by genetically encoded fluorescent proteins, imperfect maturation can give rise to a fraction of the fluorescent proteins which are unable to emit fluorescence. A molecule can also appear as non-labelled if the fluorophore attached to it resides in a long-lived dark state, with typical residence times longer than or comparable to the diffusion time τ_D of the studied molecules; for example, some fluorescent proteins, especially those in the red spectral region, are known to have such long-lived dark states. Photobleaching of a fraction of the fluorophores also results in a sub-population of apparently non-labelled molecules of interest.

The common outcome of all these diverse underlying mechanisms is a certain fraction of the molecules that are expected to be labelled by fluorophores of the k-th population either lacking the fluorophore or containing a non-fluorescent fluorophore. In either case, such molecules are 'invisible' in a DC-FCCS experiment. To describe this theoretically we introduce probabilities $\{p_k\}$ with which the target molecule is labelled by a *fluorescent* fluorophore of the k-th population.

A complexed particle will be registered in a DC-FCCS experiment as a complexed particle only if it contains fluorescent fluorophores of both species. It, therefore, does not hold any more that $[M_aM_b] = \langle C_{ab}\rangle/N_A$. Instead, we can write:

$$\langle C_{ab}^{app}\rangle = p_a p_b [M_a M_b] N_A \tag{6.24a}$$

In equation (6.24a) we have denoted the concentration as 'apparent' by adding the superscript *app* to highlight the departure from the assumption of perfect fluorescent labelling. A modified reaction scheme describing the complex formation in the presence of fluorophores in dark states is shown in figure 6.6.

Complexed particles containing only a single fluorescent fluorophore appear as single labelled particles, while particles containing no fluorescent fluorophore are not detected at all. We can write for the remaining concentrations:

$$\langle C_a^{app}\rangle = p_a[M_a]N_A + p_a(1 - p_b)[M_aM_b]N_A \tag{6.24b}$$

$$\langle C_b^{app}\rangle = p_b[M_b]N_A + (1 - p_a)p_b[M_aM_b]N_A \tag{6.24c}$$

The model for the correlation function amplitudes is then obtained by replacing the concentrations $\langle C_\varepsilon\rangle$ in equations (6.23a)–(6.23c) with their apparent counterparts $\langle C_\varepsilon^{app}\rangle$, where the subscript ε stands either for a, b or ab. Figure 6.7 illustrates the influence of a non-fluorescent fraction of one of the fluorophore species on the correlation functions.

It can be shown (see exercise 2) from equations (6.24a)–(6.24c) that the presence of non-fluorescent binding partners (p_a and p_b are smaller than 1) increases the

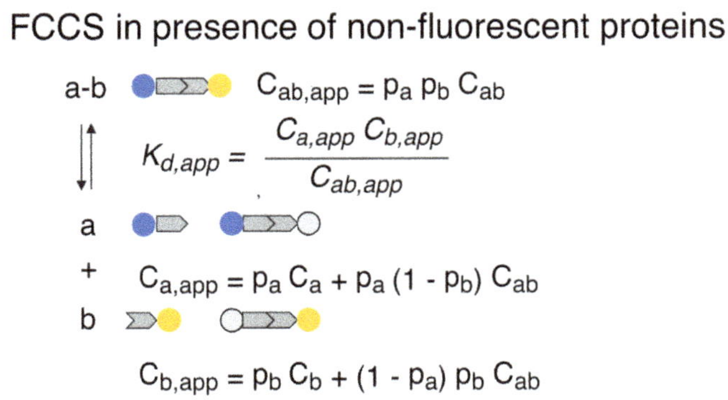

Figure 6.6. A modified reaction scheme of the complex formation reaction in the presence of fluorophores in dark states. C_{ab} denotes the real concentration of the complex, that is the concentration of particles that carry both fluorophores a and b, dark or fluorescent.

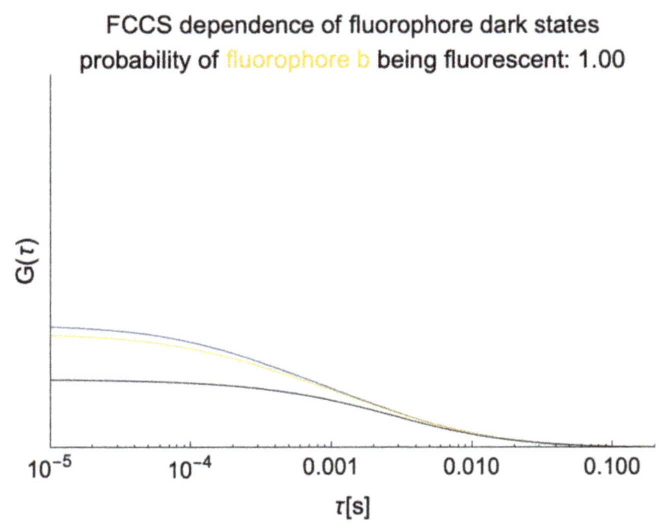

Figure 6.7. Dependence of the CCF on the presence of non-fluorescent binding partners. All fluorophores a are assumed to be fluorescent ($p_a = 1$), while the probability of fluorophore b to be fluorescent ranged from 0 to 1 ($p_b = 0 - 1$). Otherwise, we assumed identical detection volumes for both channels with the following geometrical parameters: $w_{0A} = w_{0B} = w_{0X} = 260$ nm, $K_A = K_B = 5$, $z_{0A} = z_{0B} = z_{0X} = 1.3$ µm, resulting in observation volumes of $V_A = V_B = V_X^{\text{eff}} = 0.5$ fl; the brightness of the two particles in the two detection channels was set to $\kappa_{aA} = 1500$, $\kappa_{aB} = 150$, $\kappa_{bA} = 15$, $\kappa_{bB} = 1500$; the concentration of the particles was $c_a = c_b = c_{ab} = 2.5$ nM; the diffusion times for the two particles were set to $\tau_{Da} = 3$ ms, $\tau_{Db} = 0.3$ ms. Animation available at https://iopscience.iop.org/book/978-0-7503-2080-1.

apparent dissociation constant K_D^{app} of the complex calculated from the apparent concentrations as:

$$K_D^{app} = \frac{\langle C_a^{app} \rangle \langle C_b^{app} \rangle}{N_A \langle C_{ab}^{app} \rangle} \tag{6.25}$$

It is much easier to describe the influence of imperfect labelling theoretically than it is to characterise it experimentally. The probability p_k for the fluorophore of the k-th population can be determined by comparing the brightness κ_k of molecules labelled by a single fluorophore each with the apparent brightness κ_{2k}^{app} of molecules labelled by two fluorophores each (fluorophore tandems); or to be more precise, molecules that would be labelled by two fluorescent fluorophores each if the labelling was perfect (if $p_k = 1$). The brightness can be determined from FCS experiments according to equation (6.19) as the ratio of the average fluorescence signal to the particle number N (see section 6.2.3). Under the assumption of perfect labelling, fluorophore tandems have double the brightness of individual fluorophores ($\kappa_{2k}^{app} = 2\kappa_k$). Imperfect labelling (or fluorophore dark states or photobleaching) results in a part of the tandems being labelled by a single fluorescent fluorophore only (having brightness κ_k), which reduces the apparent average tandem brightness ($\kappa_{2k}^{app} < 2\kappa_k$). Tandems containing at least a single fluorescent fluorophore (those that are entirely dark are not detected by FCS and, therefore, do not influence the average brightness), contain a second fluorescent fluorophore with the probability p_k, which leads to the average tandem brightness equal to:

$$\kappa_{2k}^{app} = (p_k + 1)\kappa_k \tag{6.26}$$

A more detailed derivation of equation (6.26) is presented in math box 6.4. Note that this way of analysing molecular brightness forms the basis of a technique closely related to FCS called number and brightness analysis [11] or, more generally, fluctuations moments analysis [12].

Math box 6.4

The apparent brightness of an imperfectly labelled fluorophore tandem

Denoting the total concentration of the tandem molecules as $\langle C_{tot}^T \rangle$, we can use the analogy with equations (6.24a)–(6.24c) to write the concentrations of tandem molecules with both fluorophores fluorescent ($\langle C_{2k}^T \rangle$) and only one fluorophore fluorescent ($\langle C_k^T \rangle$):

$$\langle C_{2k}^T \rangle = p_k^2 \langle C_{tot}^T \rangle$$

$$\langle C_k^T \rangle = 2\,p_k(1 - p_k)\,\langle C_{tot}^T \rangle$$

The factor 2 in the equation for $\langle C_k^T \rangle$ originates from the fact that either of the two fluorophores in the tandem can be fluorescent while the other is dark ($\langle C_k^T \rangle$ plays the role of both $\langle C_a^{app} \rangle$ and $\langle C_b^{app} \rangle$ in equations (6.24b) and (6.24c)). Alternatively, it follows from more general formulas presented in section 6.2.7.1.

Let us now use equations (6.19) and (5.28a) to derive a formula for the apparent brightness of a mixture of n fluorescent particle populations, the i-th population having brightness κ_i and being present at concentration $\langle C_i \rangle$ in observation volume V. From equation (6.19) we get:

$$\kappa^{app} = G_0 \frac{\langle F\rangle^2}{\langle F\rangle - \langle B\rangle} = \frac{\sum_{i=1}^{n} \kappa_i^2 \langle C_i\rangle V}{\left(\langle B\rangle + \sum_{i=1}^{n} \kappa_i \langle C_i\rangle V\right)^2} \frac{\left(\langle B\rangle + \sum_{i=1}^{n} \kappa_i \langle C_i\rangle V\right)^2}{\sum_{i=1}^{n} \kappa_i \langle C_i\rangle V} = \frac{\sum_{i=1}^{n} \kappa_i^2 \langle C_i\rangle}{\sum_{i=1}^{n} \kappa_i \langle C_i\rangle}$$

We have used here the facts that the average signal $\langle F\rangle$ is the sum of contributions of each population ($\kappa_i\langle C_i\rangle V$) and the background $\langle B\rangle$ and that the denominator of the correlation function amplitude is the square of the average signal.

Next we can apply the formula for κ^{app} in the case of the fluorophore tandem and get equation (6.26):

$$\kappa_{2k}^{app} = \frac{4\,\kappa_k^2\langle C_{2k}^T\rangle + \kappa_k^2\langle C_k^T\rangle}{2\kappa_k\langle C_{2k}^T\rangle + \kappa_k\langle C_k^T\rangle} = (p_k + 1)\kappa_k$$

It is important to make sure that all the experimental conditions match as closely as possible the conditions used in the DC-FCCS experiment. For example, differences in the chemical environment and excitation intensity can alter photobleaching and dark state dynamics; fluorescent protein maturation levels can depend on the tagged protein of interest and covalent bonds between organic fluorophores and the molecules of interest can get hydrolysed over time. Note that the last process mentioned (covalent bond hydrolysis) not only results in increasing the fraction of non-labelled molecules of interest but also in a corresponding population of free organic fluorophore in the sample; see section 6.2.5.3 on a discussion of the presence of free fluorophores on DC-FCCS results.

6.2.5.2 A certain concentration of non-labelled molecules is present in the sample

This is a scenario specific to FCCS studies of protein interactions in living cells or organisms using genetically encoded fluorescent analogues of the proteins of interest. If the plasmids encoding the fluorescent protein analogues are transiently added to the studied cells, the cells may still express the non-fluorescent endogenous proteins of interest. The endogenous proteins can interact with the fluorescently labelled binding partners forming complexes which appear as single-labelled particles as shown in the modified reaction scheme in figure 6.8. If we assume perfect labelling and no dark states of the fluorescent protein analogues ($p_a = p_b = 1$) and denote the endogenous proteins of interest as $M_a^{(e)}$ and $M_b^{(e)}$ to distinguish them from their respective fluorescent analogues $M_a^{(f)}$ and $M_b^{(f)}$, we can write for the apparent concentrations:

$$\langle C_{ab}^{app}\rangle = [M_a^{(f)}M_b^{(f)}]N_A \tag{6.27a}$$

$$\langle C_a^{app}\rangle = [M_a^{(f)}]N_A + [M_a^{(f)}M_b^{(e)}]N_A \tag{6.27b}$$

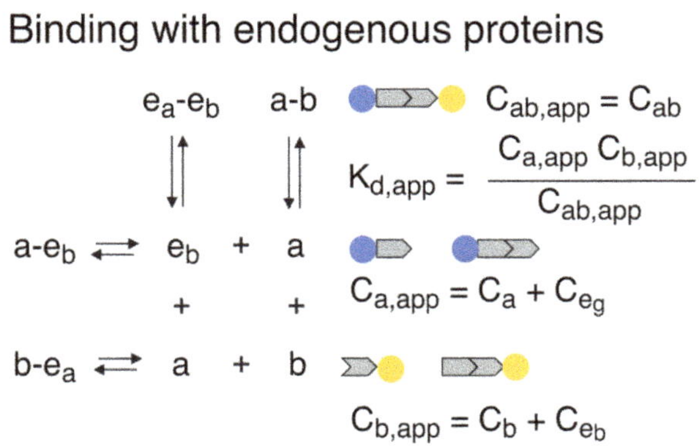

Figure 6.8. A modified reaction scheme of the complex formation reaction in the presence of endogenous proteins. e_a and e_b denote the endogenous forms of the proteins that in their non-endogenous forms are labelled by fluorophores a and b, respectively.

$$\langle C_b^{app} \rangle = [M_b^{(f)}]N_A + [M_a^{(e)} M_b^{(f)}]N_A \tag{6.27c}$$

The influence of the endogenous proteins on the results of a single DC-FCCS measurement is identical to the influence of the phenomena discussed in the previous subsection. By increasing the apparent concentration of the single labelled particles, it decreases the apparent abundance of the complex, or, in other words, increases its apparent dissociation constant K_D^{app}. The two scenarios can be differentiated only by looking at results of multiple measurements taken at different concentrations of the fluorescently labelled protein analogues. Such a set of measurements can be obtained, for example, by taking DC-FCCS measurements in different cells in the same culture and using intrinsic cell-to-cell variations in the expression levels of the fluorescently labelled protein analogues. By plotting the product of $\langle C_a^{app} \rangle$ and $\langle C_b^{app} \rangle$ against $\langle C_{ab}^{app} \rangle$ for each measurement within the set, we obtain the so called K_D^{app} plot. It follows from equation (6.8) that in the ideal case when the apparent concentrations are equal to the actual concentrations of the free binding partners and the complex, respectively, the K_D^{app} plot is linear with a slope equal to the dissociation constant K_D. In the presence of imperfect labelling (or dark states as discussed in section 6.2.5.1) the K_D^{app} plot is convex with K_D^{app} values deviating further from the actual K_D with increasing concentrations of the binding partners (see figure 6.9). On the other hand, in the presence of endogenous proteins, the K_D^{app} plot is concave (see figure 6.9). The deviation of K_D^{app} from K_D decreases with increasing concentrations of the binding partners. This is rather intuitive since the influence of the endogenous proteins is expected to diminish with increasing ratio of fluorescently labelled to endogenous proteins [13].

Note that the assumption of $p_a = p_b = 1$ under which equations (6.27a)–(6.27c) were derived is not very realistic for fluorescent proteins which are frequently used as fluorescent labels in DC-FCCS studies of protein interactions in live cells. In that case the correct description is given by adding p_k values smaller than 1 to equations (6.27a)–(6.27c) in a manner analogous to equations (6.24a)–(6.24c).

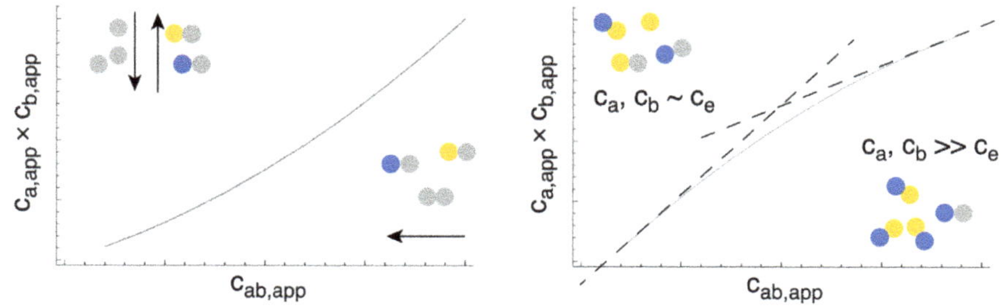

Figure 6.9. An illustration of the shape of K_D plots corresponding to the following scenarios: imperfect labelling or a fraction of the fluorophores residing in a dark state as described by equations (6.24a)–(6.24c) (left-hand side panel); presence of endogenous proteins of interest at concentration $C_e = ([M_a^{(e)}] + [M_b^{(e)}])N_A$, situation described by equations (6.27a)–(6.27c) (right-hand side panel).

$$\langle C_{ab}^{app} \rangle = p_a p_b [M_a^{(f)} M_b^{(f)}] N_A \tag{6.28a}$$

$$\langle C_a^{app} \rangle = p_a [M_a^{(f)}] N_A + p_a (1 - p_b) [M_a^{(f)} M_b^{(f)}] N_A + p_a [M_a^{(f)} M_b^{(e)}] N_A \tag{6.28b}$$

$$\langle C_b^{app} \rangle = p_b [M_b^{(f)}] N_A + (1 - p_a) p_b [M_a^{(f)} M_b^{(f)}] N_A + p_b [M_a^{(e)} M_b^{(f)}] N_A \tag{6.28c}$$

Unlike the probabilities p_k discussed in section 6.2.5.1, the concentrations of the endogenous proteins cannot be determined by control DC-FCCS experiments. Average concentrations of certain proteins in particular cell lines may be known from the results of other methods; however, cell-to-cell variations in endogenous protein concentrations will translate to a large uncertainty in K_D. For this reason, quantitative DC-FCCS experiments are better conducted in the absence of endogenous proteins. Recent advances in genetical manipulations make this requirement easier to satisfy [14, 15].

6.2.5.3 Free fluorophore molecules are present in the sample

The presence of free fluorophores (not bound to molecules M_a or M_b in the case of fluorophores a or b, respectively), will also artificially decrease the apparent dissociation constant by increasing the apparent concentration of single-labelled particles. If we assume that a certain concentration $\langle C_a^{free} \rangle$ of free fluorophore a is present in the sample, equation (6.24b) becomes:

$$\langle C_a^{app} \rangle = \langle C_a^{free} \rangle + p_a [M_a] N_A + p_a (1 - p_b) [M_a M_b] N_A \tag{6.29}$$

Equation (6.24c) could be modified analogously to account for free fluorophore b present at concentration $\langle C_b^{free} \rangle$. By adding either $\langle C_a^{free} \rangle$ or $\langle C_b^{free} \rangle$ as unknown concentrations to the model, the number of unknown variables becomes too high to be unambiguously determined from the three correlation function amplitudes. However, if the molecules of interest, M_a and M_b, are much larger than fluorophores a and b, respectively, the concentrations of the free fluorophores can be determined from the temporal decays of the respective ACFs by fitting them with a model for two diffusion coefficients (see section 5.3). Only the amplitude of the diffusional part of the ACF corresponding to the fluorophore bound to the molecule of interest (either the molecule of interest alone or in the form of a complex) need to be considered in FCCS analysis. For example, in the case of the ACF in channel A we would use as G_{0_A} only the amplitude of the diffusional correlation function corresponding to the diffusion coefficient of molecule M_a and to the diffusion coefficient of the complex M_aM_b.

6.2.5.4 Photobleaching of fluorophores

Fluorophore photobleaching has been mentioned in section 6.2.5.1 as a possible reason for a fraction of the binding partners lacking a fluorescent fluorophore. Equations (6.24a)–(6.24c) are, therefore, applicable in this case to describe the apparent concentrations of the single- and double-labelled particles. We return to the case of photobleaching here to discuss its influence on the temporal decay of the correlation functions. As discussed elsewhere (see section 7.3), fluorophore photobleaching shortens the typical residence time of the fluorophores in the effective observation volume, resulting in an apparently faster diffusion. If the bleaching rate of one of the fluorophore species (fluorophore b, for example) is higher than the bleaching rate of the other fluorophore, the CCFs $G_{AB}(\tau)$ and $G_{BA}(\tau)$ are no longer identical. The origin of this effect can be understood in the following way. The temporal decay of $G_{AB}(\tau)$ is determined (disregarding the influence of spectral crosstalk) by the probability of detecting fluorophore b fluorescence at time τ after detecting fluorescence from fluorophore a. The decay of $G_{BA}(\tau)$ reflects, on the other hand, the probability of detecting fluorescence of a at time τ after detecting fluorophore b fluorescence. If fluorophore b bleaches faster than a, $G_{AB}(\tau)$ decays faster than $G_{BA}(\tau)$. Therefore, the difference in $G_{AB}(\tau)$ and $G_{BA}(\tau)$ decays can be used as an indication of FCCS results being influenced by fluorophore photobleaching. Note that fluorophore photobleaching is not always manifested by a trend in the fluorescence intensity time traces. Especially in confocal FCS, so-called cryptic photobleaching is quite common, leading to artefacts that remain easily unnoticed (see section 7.3 for more details).

6.2.6 Fluorescence quenching and Förster resonance energy transfer (FRET) between fluorophores *a* and *b*

When a complex between molecules labelled by fluorophores a and b is formed, it can lead to changes in the fluorophore brightness. This can be the consequence of collisional fluorescence quenching or of the two fluorophores getting sufficiently close to each other for Förster resonance energy transfer (FRET) to occur between them. By collisional quenching we mean processes in which a fluorophore in the excited state interacts with a molecule in its vicinity (a quencher) and this interaction results in a non-radiative (without emitting a photon) de-excitation of the fluorophore. Quenching, therefore, results in fewer photons being emitted by the fluorophore per unit of time, which is, in other words, a reduction in its brightness (in all detection channels). We will, therefore, introduce quenching factors q_a and q_b, which give the relative change of the brightness of fluorophore a and b, respectively, when those fluorophores are incorporated in a complex. In the case of collisional quenching, both quenching factors are smaller or equal to one. Note that the quencher in this case can be the fluorophore of the other species as well as the molecule of interest; for example, fluorophore a in a complex can be quenched either by fluorphore b or the molecule M_b and vice versa.

When FRET occurs, a fluorophore with shorter-wavelength emission (the one we call here a) transfers non-radiatively its excitation energy to a fluorophore with longer-wavelength emission (denoted here as b), which enters an excited state. The shorter-wavelength fluorophore, denoted in the theory of FRET as donor, is, thus, de-excited by FRET without emitting a photon (similarly to collisional quenching), which results in a reduction in its brightness ($q_a < 1$). The longer-wavelength fluorophore, referred to in FRET as acceptor, is, on the other hand excited at a higher rate and its brightness, therefore, increases ($q_b > 1$). The efficiency of FRET depends among other factors on the inter-fluorophore distance (the efficiency drops sharply at distances on the order of nm), mutual orientation of the two fluorophores as well as on the overlap of the donor emission spectrum with the acceptor excitation spectrum (which condition is satisfied for many fluorophore pairs commonly used in DC-FCCS) [16].

FRET can occur only in those complexes in which fluorophores of both species, the donor and acceptor, are present. In many cases that would be the complexes appearing as double-labelled in FCCS. The concentration of the FRET-active complexes is, then, given by $\langle C_{ab}^{app} \rangle$ and by introducing the factors q_a and q_b to equations (6.23a)–(6.23c), we get the following model for the ACF and the CCF amplitudes:

$$G_{0_A} = \frac{(q_a\kappa_{aA} + q_b\kappa_{bA})^2\langle C_{ab}^{app}\rangle + \kappa_{aA}{}^2\langle C_a^{app}\rangle + \kappa_{bA}{}^2\langle C_b^{app}\rangle}{V_A\left[\kappa_{aA}\langle C_a^{app}\rangle + \kappa_{bA}\langle C_b^{app}\rangle + (q_a\kappa_{aA} + q_b\kappa_{bA})\langle C_{ab}^{app}\rangle + \dfrac{\langle B_A\rangle}{V_A}\right]^2} \tag{6.30a}$$

$$G_{0_B} = \frac{(q_a\kappa_{aB} + q_b\kappa_{bB})^2\langle C_{ab}^{app}\rangle + \kappa_{aB}{}^2\langle C_a^{app}\rangle + \kappa_{bB}{}^2\langle C_b^{app}\rangle}{V_B\left[\kappa_{aB}\langle C_a^{app}\rangle + \kappa_{bB}\langle C_b^{app}\rangle + (q_a\kappa_{aB} + q_b\kappa_{bB})\langle C_{ab}^{app}\rangle + \dfrac{\langle B_B\rangle}{V_B}\right]^2} \tag{6.30b}$$

$$\begin{aligned} G_{0_X} = {} & \frac{(q_a\kappa_{aA} + q_b\kappa_{aB})(q_a\kappa_{aB} + q_b\kappa_{bB})\langle C_{ab}^{app}\rangle + \kappa_{aA}\kappa_{aB}\langle C_a^{app}\rangle + \kappa_{bA}\kappa_{bB}\langle C_b^{app}\rangle}{V_X^{\text{eff}}\left[\kappa_{aA}\langle C_a^{app}\rangle + \kappa_{bA}\langle C_b^{app}\rangle + (q_a\kappa_{aA} + q_b\kappa_{bA})\langle C_{ab}^{app}\rangle + \dfrac{\langle B_A\rangle}{V_X^{\text{eff}}}\right]} \\ & \times\left[\kappa_{aB}\langle C_a^{app}\rangle + \kappa_{bB}\langle C_b^{app}\rangle + (q_a\kappa_{aB} + q_b\kappa_{bB})\langle C_{ab}^{app}\rangle + \frac{\langle B_B\rangle}{V_X^{\text{eff}}}\right]^{-1} \end{aligned} \tag{6.30c}$$

In some situations, however, the assumption may not be true that only apparently double-labelled complexes are FRET-active. A fluorophore in a dark state may still act as a FRET acceptor (depending on the nature of the dark state). In such a case also the apparently single-labelled complexes where only the donor is fluorescent will be FRET-active and the brightness of the donor in those complexes will be reduced. To account for this, we need to modify equations (6.30a)–(6.30c). For simplicity we will show here only the equation for the CCF amplitude; the equations for the ACF amplitude would be obtained by analogous modifications. Equation (6.30c) then becomes:

$$G_{0_X} = \frac{(q_a\kappa_{aA} + q_b\kappa_{aB})(q_a\kappa_{aB} + q_b\kappa_{bB})\langle C_{ab}^{app}\rangle + \kappa_{aA}\kappa_{aB}(\langle C_a^f\rangle + q_a\langle C_a^c\rangle) + \kappa_{bA}\kappa_{bB}\langle C_b^{app}\rangle}{V_X^{\text{eff}}\left[\kappa_{aA}(\langle C_a^f\rangle + q_a\langle C_a^c\rangle) + \kappa_{bA}\langle C_b^{app}\rangle + (q_a\kappa_{aA} + q_b\kappa_{bA})\langle C_{ab}^{app}\rangle + \frac{\langle B_A\rangle}{V_X^{\text{eff}}}\right]} \times\left[\kappa_{aB}(\langle C_a^f\rangle + q_a\langle C_a^c\rangle) + \kappa_{bB}\langle C_b^{app}\rangle + (q_a\kappa_{aB} + q_b\kappa_{bB})\langle C_{ab}^{app}\rangle + \frac{\langle B_B\rangle}{V_X^{\text{eff}}}\right]^{-1} \tag{6.31}$$

where $\langle C_a^f\rangle$ is the concentration of single-labelled particles (molecules M_a) carrying a fluorescent fluorophore a (the donor) and $\langle C_a^c\rangle$ is the concentration of the apparently single-labelled complexes carrying only a fluorescent fluorophore a. From equations (6.24a) and (6.24b) we get:

$$\langle C_a^f\rangle = p_a[A]N_A = \langle C_a^{app}\rangle - \frac{1-p_b}{p_b}\langle C_{ab}^{app}\rangle \tag{6.32a}$$

$$\langle C_a^c\rangle = \langle C_a^{app}\rangle - \langle C_a^f\rangle = \frac{1-p_b}{p_b}\langle C_{ab}^{app}\rangle \tag{6.32b}$$

Quenching can typically happen also in complexes which appear as single-labelled. That is the case if the fluorophores are quenched by the binding partners themselves (fluorophore a quenched by molecule M_b or vice versa). Quenching can also occur in apparently single-labelled particles if the fluorophores quench each other and the reason for the particles appearing as single-labelled is not the absence of one of the fluorophores but rather the presence of the fluorophore in a non-fluorescent state. The function of a fluorophore as quencher is not likely to be affected by the fluorophore being in a dark state or being photobleached. In such cases we need to apply the quenching factors also to the apparently single-labelled complexes. The CCF amplitude is then given by modified equation (6.31) in which $\langle C_b^{app}\rangle$ is replaced with $(\langle C_b^f\rangle + q_b\langle C_b^c\rangle)$. $\langle C_b^f\rangle$ and $\langle C_b^c\rangle$ have meaning analogous to $\langle C_a^f\rangle$ and $\langle C_a^c\rangle$, respectively. Depending on the system under study, the concentrations $\langle C_a^f\rangle$, $\langle C_a^c\rangle$, $\langle C_b^f\rangle$ and $\langle C_b^c\rangle$ can be obtained from equations (6.24a)–(6.24c) (see equations (6.32a) and (6.32b)) or equations (6.27a)–(6.27c) or (6.28a)–(6.28c) (see exercise 3).

The quenching factors q_a and q_b can be obtained by comparing brightness values κ_{aA} and κ_{bB} measured in samples containing only individual fluorophore populations with the corresponding brightness values measured for a sample containing only complexes (a positive control sample). The brightness ratios thus obtained would give the exact values of q_a and q_b only under the assumption of perfect labelling of the positive control. If the probabilities p_a and p_b are smaller than 1, the positive control will also contain a fraction of single-labelled particles, in which the fluorophores do not undergo quenching or FRET. Furthermore, the distance and mutual orientation of the fluorophores in the positive control complex need to match very closely those in the actual complex under study, in order that the fluorophores experience similar FRET or quenching efficiencies, conditions that are not trivial to satisfy. In the view of the considerable challenges of correctly quantifying the influence of FRET and quenching on the DC-FCCS data, it is worthwhile trying to optimise the fluorescent labelling strategy to prevent or at least minimise FRET and quenching. For example, selecting the labelling sites of fluorophore a and b in such a way that the two fluorophores are as far as possible from the interface involved in complex formation not only minimises the chances of FRET between the fluorophores but also reduces the risk that fluorescent labelling would affect the complex stability.

FRET or quenching decrease the CCF amplitude as can be seen in figure 6.10, except for the special case of a perfect positive control (double-labelled particles only) without background, in which case the correlation functions are independent of brightness.

6.2.7 Complex stoichiometry

Until now, we have only considered complexes with 1:1 stoichiometry formed according to the reaction scheme of equation (6.7). In general, however, complexes

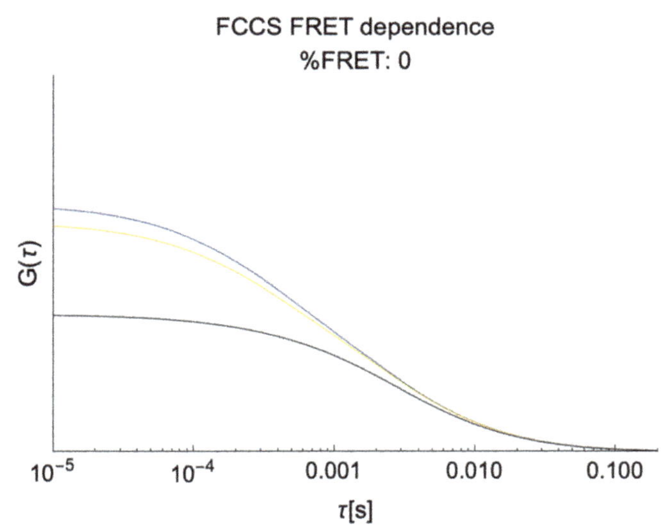

Figure 6.10. Dependence of the CCF on FRET. For the inclusion of FRET, we varied the quenching factors for fluorophore a from 0–1 and for fluorophore b from 1–2 ($q_a = 2 - q_b$, $q_b = 1\ldots2$), mimicking the quenching of the FRET donor and the enhancement of fluorescence of the FRET acceptor. Otherwise, we assumed identical detection volumes for both channels with the following geometrical parameters: $w_{0A} = w_{0B} = w_{0X} = 260$ nm, $K_A = K_B = 5$, $z_{0A} = z_{0B} = z_{0X} = 1.3$ μm, resulting in observation volumes of $V_A = V_B = V_X^{\text{eff}} = 0.5$ fl; the brightness of the two particles in the two detection channels was set to $\kappa_{aA} = 1500$, $\kappa_{aB} = 150$, $\kappa_{bA} = 15$, $\kappa_{bB} = 1500$; the concentration of the particles was $c_a = c_b = c_{ab} = 2.5$ nM; the diffusion times for the two particles were set to $\tau_{Da} = 3$ ms, $\tau_{Db} = 0.3$ ms. Animation available at https://iopscience.iop.org/book/978-0-7503-2080-1.

of molecules M_a and M_b can have any arbitrary stoichiometry $M_{a\alpha}M_{b\beta}$, where α and β are positive integers. If we could assume perfect labelling of all complexes, the complex stoichiometry would be reflected in the model for correlation function amplitudes simply by modifying the brightness of the complex, which would then become $(\alpha\kappa_{gL} + \beta\kappa_{rL})$, where L stands either for A or B. However, since such an assumption is rarely fulfilled, especially when working with fluorescent proteins, we need to consider in the model complexes with all possible combinations of numbers of fluorophores a and b ranging from zero to α and β, respectively. The correlation function amplitudes can be then written as:

$$G_{0_\Lambda} = \frac{\sum_{\rho=0}^{\alpha}\sum_{\sigma=0}^{\beta}(\rho\kappa_{gL_1} + \sigma\kappa_{rL_1})(\rho\kappa_{gL_2} + \sigma\kappa_{rL_2})\left\langle C^{app}_{\rho g\sigma r}\right\rangle}{V^{\mathrm{eff}}_{\Lambda}\left[\sum_{\rho=0}^{\alpha}\sum_{\sigma=0}^{\beta}(\rho\kappa_{gL_1} + \sigma\kappa_{rL_1})\left\langle C^{app}_{\rho g\sigma r}\right\rangle + \frac{\langle B_{L_1}\rangle}{V^{\mathrm{eff}}_{\Lambda}}\right]} \times \left[\sum_{\rho=0}^{\alpha}\sum_{\sigma=0}^{\beta}(\rho\kappa_{gL_2} + \sigma\kappa_{rL_2})\left\langle C^{app}_{\rho g\sigma r}\right\rangle + \frac{\langle B_{L_2}\rangle}{V^{\mathrm{eff}}_{\Lambda}}\right]^{-1} \tag{6.33}$$

where L_1 and L_2 stand either for A or B ($L_1 = L_2$ for the ACFs and $L_1 \neq L_2$ for the CCF), Λ stands either for A, B or X and $\langle C^{app}_{\rho a\sigma b}\rangle$ is the concentration of particles containing ρ fluorophores a and σ red fluorophores b. Note that this definition extends also to non-complexed particles when either ρ is zero and σ is one or vice versa. In equation (6.33), we have not considered quenching or FRET between fluorophores a and b. In general, the quenching and FRET efficiencies can depend on the stoichiometry of the complexes, which means that $\alpha \times \beta$ matrices of quenching factors $q^a_{\rho\sigma}$ and $q^b_{\rho\sigma}$ would need to be introduced for a complete description of the system. While easily done in theory, exact experimental determination of the matrix elements is in most cases impossible because of the unavailability of control samples with defined combinations of ρ and σ. In analogy to equations (6.24a)–(6.24c), we can write for the concentrations:

$$\left\langle C^{app}_{\rho a\sigma b}\right\rangle = \binom{\alpha}{\rho}\binom{\beta}{\sigma}p_a^{\rho}(1 - p_a)^{\alpha-\rho}p_b^{\sigma}(1 - p_b)^{\beta-\sigma}[M_{a\alpha}M_{b\beta}]N_A, \quad \text{for } 1 < \rho + \sigma < \alpha + \beta \tag{6.34a}$$

$$\langle C^{app}_{1a0b}\rangle = p_a[M_a]N_A + \binom{\alpha}{1}p_a(1 - p_a)^{\alpha-1}(1 - p_b)^{\beta}[M_{a\alpha}M_{b\beta}]N_A \tag{6.34b}$$

$$\langle C^{app}_{0a1b}\rangle = p_b[M_b]N_A + \binom{\beta}{1}(1 - p_a)^{\alpha}p_b(1 - p_b)^{\beta-1}[M_{a\alpha}M_{b\beta}]N_A \tag{6.34c}$$

The binomial coefficients $\binom{\alpha}{\rho}$ in equations (6.34a)–(6.34c) account for the number of combinations in which α subunits in a complex can be labelled by ρ fluorophores a; the binomial coefficients $\binom{\beta}{\sigma}$ have an analogous meaning for fluorophores b [9, 17, 18].

Broad distributions of the complex stoichiometries are sometimes encountered in the investigated samples. Distributions of α and β instead of single values are needed to describe the complex stoichiometry. In such cases a rigorous theoretical description of the apparent complex concentrations determined by FCCS becomes increasingly complicated and only semi-quantitative description using, for example, the cross-correlation amount Q would be more likely applied.

6.2.7.1 Oligomerisation studied by FCCS

Oligomerisation can be regarded as a special case of complex formation with stoichiometry $M_{a\alpha}M_{b0}$, where $\alpha > 1$ is the degree of oligomerisation. In order to study such a process by DC-FCCS, we need to have two sub-populations of the monomers M, one labelled with fluorophores a and the other with fluorophores b. The oligomers will, then, carry different combinations of the fluorophores a and b ranging from α fluorophores a to α fluorophores b [9, 18, 19]. The apparent concentrations of the differently labelled oligomers are given analogously to equations (6.34a)–(6.34c) as:

$$\left\langle C^{app}_{\rho a\sigma b}\right\rangle = \binom{\alpha}{\rho}\binom{\alpha-\rho}{\sigma} p_a^{\rho} p_b^{\sigma}(1 - p_a - p_b)^{\alpha-\rho-\sigma}[M_\alpha]N_A, \quad \text{for } 1 < \rho + \sigma < \alpha \tag{6.35a}$$

$$\langle C^{app}_{1a0b}\rangle = p_a[M]N_A + \binom{\alpha}{1} p_a(1 - p_a - p_b)^{\alpha-1}[M_\alpha]N_A \tag{6.35b}$$

$$\langle C^{app}_{0a1b}\rangle = p_b[M]N_A + \binom{\alpha}{1} p_b(1 - p_a - p_b)^{\alpha-1}[M_\alpha]N_A \tag{6.35c}$$

where the probabilities p_a and p_b are the fractions of the total monomer A concentration in the sample (present either as free monomers or in the form of oligomers) that carry fluorophore a or b, respectively:

$$p_a = \frac{\sum_{\rho=1}^{\alpha}\sum_{\sigma=0}^{\alpha-\rho} \rho\left\langle C^{app}_{\rho a\sigma b}\right\rangle}{N_A([M] + \alpha[M_\alpha])} \tag{6.36a}$$

$$p_b = \frac{\sum_{\sigma=1}^{\alpha}\sum_{\rho=0}^{\alpha-\sigma} \sigma\left\langle C^{app}_{\rho a\sigma b}\right\rangle}{N_A([M] + \alpha[M_\alpha])} \tag{6.36b}$$

As said before, the role of the binomial coefficients in equations (6.35a)–(6.35c) is to account for the number of ways in which ρ monomers labelled by fluorophore a and

σ monomers labelled by fluorophore b can be distributed in an oligomer of α subunits. Assuming that the labelling does not affect the ability of the monomers to form an oligomer, we can use the parallel of drawing monomers from a non-transparent bag, commonly used in explaining probability theory. For example, a double-labelled dimer can result from two different sequences of draws: first drawing a monomer labelled by a followed by drawing a monomer labelled by b or vice versa, which is reflected by the product of binomial coefficients $\binom{2}{1}\binom{1}{1} = 2$. On the other hand, there is only one sequence of draws that will give us a dimer with two fluorophores a, $\binom{2}{2}\binom{0}{0} = 1$. Knowing the fractions of monomers in the bag labelled by each fluorophore, p_a and p_b, then allows us to derive the probabilities of drawing ρ monomers labelled by fluorophore a and σ monomers labelled by fluorophore b. Those are the probabilities of finding oligomers with such labelling in the sample, which are given by equations (6.35a)–(6.35c). For higher degrees of oligomerisation α (for longer sequences of draws from the bag of monomers), it is becoming less likely that we will draw only monomers labelled by one of the fluorophores. Assuming for simplicity there are no non-fluorescent monomers in the bag, there is only one way to get an oligomer labelled only by fluorophores a, which is realised with the probability p_a^α. At the same time there are increasingly many sequences of draws leading to a double-labelled oligomer carrying at least one fluorophore of each species. A tetramer with one fluorophore a and 3 fluorophores b can be obtained in four ways, $\binom{4}{1}\binom{3}{3} = 4$, a tetramer with two fluorophores of each species in six ways, $\binom{4}{2}\binom{2}{2} = 6$, and so on.

Let us look now at CCF amplitudes in the case of oligomerisation. For simplicity, we will consider the most idealised assumptions of complete oligomerisation (no free monomers present in the sample), perfect labelling and no dark states ($p_b = 1 - p_a$ and $\sigma = \alpha - \rho$), no background, no crosstalk (diagonal brightness matrix $\boldsymbol{\kappa}$) and identical observation volumes. In the special case of dimerization, when half of the monomers are labelled by fluorophore a and the other half by fluorophore b, a quarter of the particles will consist of two monomers labelled by a, a quarter of two monomers labelled by b and the remaining half of double-labelled particles. Since only half of the dimers are double-labelled, the CCF amplitude is reduced to one half of the CCF amplitude measured in a sample of perfect fluorophore tandems (a sample consisting exclusively of double-labelled particles) of the same concentration. Interestingly, this is not only true for $p_a = p_b = \frac{1}{2}$ but for any value of p_a ($0 < p_a < 1$). This can be easily shown (see exercise 5) by writing explicitly the concentrations in equations (6.35a)–(6.35c) (the equations simplify considerably under the assumption of no non-fluorescent monomers), inserting them into equation (6.6c) for CCF amplitude, applicable under the ideal conditions assumed here, and comparing the resulting CCF amplitude with the CCF amplitude for a perfect fluorophore tandem of the same concentration. According to equation (6.6c) the CCF amplitude for a perfect tandem is simply $G_{0_X}^{(\text{tandem})} = \frac{1}{V\langle C_{ab}\rangle}$, where $\langle C_{ab}\rangle$ is the average tandem concentration. It can be even shown that a similar relationship is true for any degree of oligomerisation α, saying that:

$$G_{0_X}^{(\alpha)} = \left(1 - \frac{1}{\alpha}\right) G_{0_X}^{(\text{tandem})} \tag{6.37}$$

6.3 FCCS modalities derived from DC-FCCS

We have seen in section 6.1 that quantitative analysis of DC-FCCS can be considerably complicated by uncertainties originating from several sources. This motivated the development of FCCS modalities which avoid or at least reduce some of the sources of uncertainties in DC-FCCS.

6.3.1 Single-wavelength FCCS (sw-FCCS)

The development of sw-FCCS was motivated by the goal of reducing the differences in observation volumes for the individual channels by using a single laser line for excitation [20]. The earliest sw-FCCS applications relied on fluorophores or fluorescent proteins with different Stokes shifts to ensure that their emission spectra are spectrally well separated while the single laser line used is close to the maxima of their excitation spectra [17, 21]. Later, sw-FCCS was extended to standard fluorescent protein pairs commonly used in DC-FCCS [5, 13, 19]. By selecting a laser line wavelength located in between the excitation spectra maxima of the two fluorophore species, members of both species can be excited simultaneously (see figure 6.11). The reduction in excitation efficiency compared to using laser lines matching the excitation maxima can be offset by increased laser power.

However, even using a single excitation wavelength does not ensure perfect matching of the observation volumes for individual detection channels. The wavelengths of the photons detected in each channel are still different, which is enough to introduce differences in the volumes as well as a displacement caused by chromatic aberration. For example, Foo *et al* reported $V_A = 0.54$ fL, $V_B = 0.76$ fL and $V_X = 0.93$ fL [13]. This means that the individual volumes still need to be calibrated as discussed in section 6.2.2. Nevertheless, the calibration is greatly simplified by having a single excitation wavelength. All three volumes can be calibrated by a single measurement in a solution of a fluorophore that has similar brightness in both detection channels. This not only reduces the time needed to perform the calibration, but, more importantly, greatly improves the reliability of the calibration. As mentioned in section 6.2.2, the calibration of the cross-correlation effective observation volume using a sample containing purely double-labelled particles is complicated by the difficulty in obtaining such an ideal sample (see section 6.2.5). Note that a single fluorophore calibration can be used in DC-FCCS only under the assumption of a perfect overlap of the two laser foci [7].

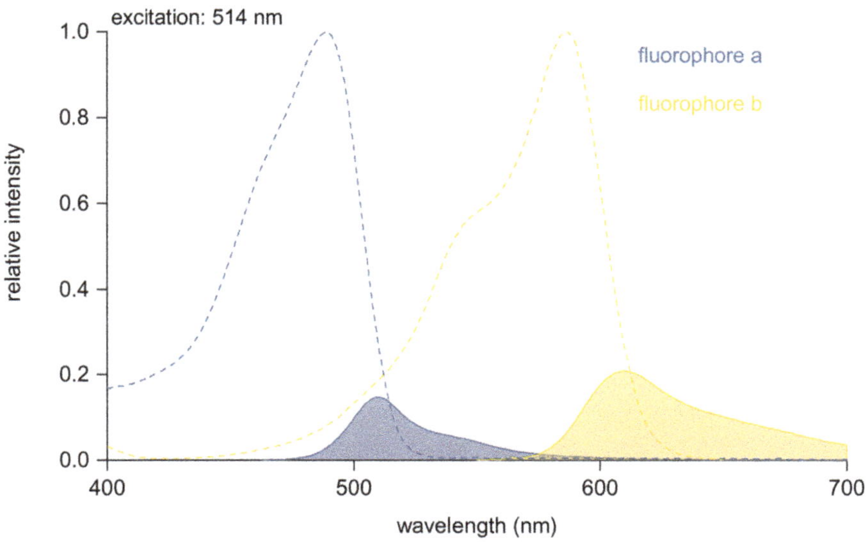

Figure 6.11. An example of fluorescence emission spectra (solid lines) of EGFP and mCherry, a pair of fluorescent proteins successfully used in sw-FCCS [13, 19]. The amplitudes of the individual spectra correspond to excitation by a single laser line of 514 nm. Excitation spectra of both fluorescent proteins are shown for reference (dashed lines). The spectra were obtained from an online database [10].

6.3.2 FCCS with alternating laser excitation

Spectral crosstalk (see section 6.2.3) presents a serious complication in DC-FCCS studies, especially those that aim only at qualitative results. Because of the crosstalk, it is not possible to simply take a positive CCF amplitude as an indication of co-diffusion of the molecules of interest; instead, the CCF amplitude needs to be significantly higher than CCF amplitude measured for a negative control to support such a conclusion. Furthermore, the artificial cross-correlation resulting from spectral crosstalk compromises the sensitivity of DC-FCCS at low complex concentrations. We will explain here how crosstalk free FCCS can be accomplished by temporal separation of excitation of fluorophores *a* and *b*. This approach is known under the acronyms PIE (pulsed interleaved excitation) FCCS [22] and ALEX (alternating laser excitation) FCCS [23, 24]; the term PIE being typically used when the temporal separation happens on the scale of 10 ns, while ALEX preferentially refers to temporal separation on slower timescales (although originally the concept of ALEX extended to the fast timescales as well, denoted as nano-second-ALEX [23]). Another class of crosstalk-free FCCS modalities will be presented in section 6.4.

Let us now look at the principle behind PIE and ALEX. Figure 6.3 showed the emission spectra of a sample containing a pair of fluorophores excited simultaneously by two laser lines selected to be close to their respective excitation spectra maxima (as is the usual situation in DC-FCCS). On the other hand, figure 6.12 shows the spectra of the same sample when it is excited only by either one or the other of the two laser lines. This is exactly what is done in PIE or ALEX. The key

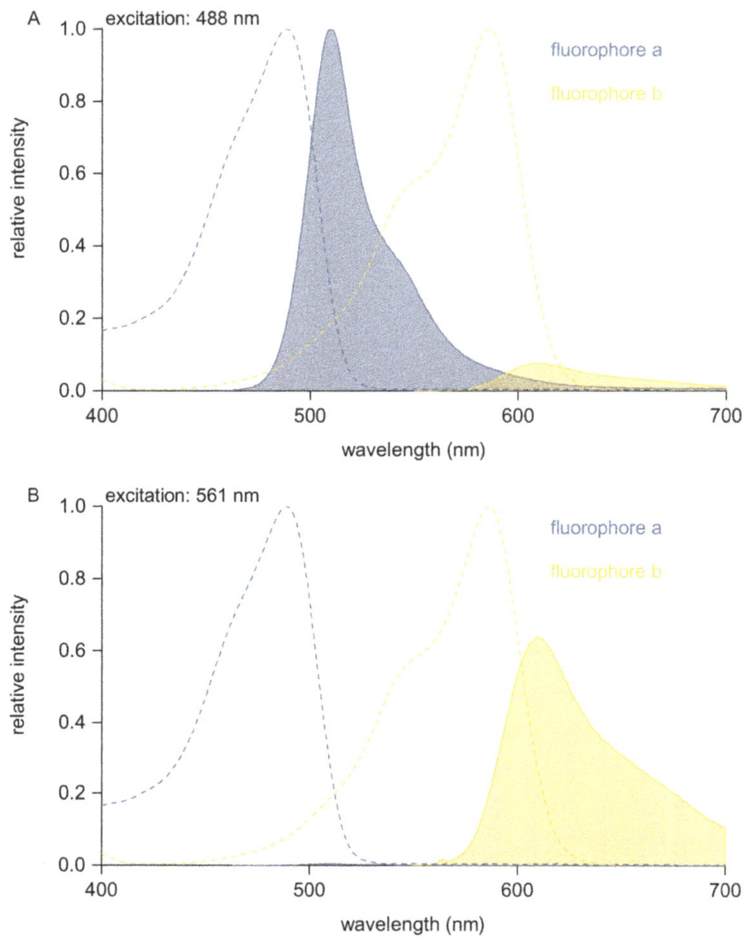

Figure 6.12. An example of fluorescence emission spectra (solid lines) of EGFP and mCherry corresponding to the situation in PIE- or ALEX-FCCS; only one laser line matching the excitation spectrum of one of the fluorescent proteins (either 488 nm or 561 nm) is on at a time. Excitation spectra of both fluorescent proteins are shown for reference (dashed lines). The spectra were obtained from an online database [10].

observation we can make from the figure is the absence of signal from fluorophore *a* under excitation by the longer-wavelength laser line (the line primarily intended for exciting fluorophore *b*). Fluorophore *a* is practically not excited by that laser line, which means that photons collected when only the longer-wavelength laser line is on can be unambiguously assigned to have originated from fluorophore *b*. At the same time, the reduced excitation of fluorophore *b* by the shorter-wavelength line together with the typically intrinsically lower crosstalk of that fluorophore to channel *A* ($\kappa_{bA} < \kappa_{aB}$, see section 6.2.3) ensure that photons detected in channel *A* under the shorter-wavelength excitation can be assumed to originate exclusively from fluorophore *a*. Therefore, by exciting the sample with a single laser line (selected to excite only one of the fluorophore species) at a time, we can unambiguously trace each photon to the fluorophore species by which it has been emitted. As we have discussed in the beginning of this chapter, this is one of the conditions of an ideal FCCS experiment; only by departing from this condition in section 6.2.3, have we introduced spectral crosstalk.

For this principle to be applicable in FCCS, we need to switch repeatedly between the two excitation laser lines to collect intensity time traces for both fluorophore species as illustrated schematically in figure 6.13. The temporal resolution of the correlation functions in PIE- or ALEX-FCCS is equal to the switching period τ_S of the excitation wavelength, i.e. the time interval between two consecutive instances of switching each excitation wavelength on. It has been shown that for a reliable characterisation by FCS, the temporal resolution should be at least an order of magnitude shorter than the typical residence time (diffusion time) τ_D of the molecules of interest in the effective observation volume [25]. Considering a typical confocal setup with approximately 250 nm observation volume radius, the diffusion time of a small organic molecule, such as a free organic fluorophore, would be of the

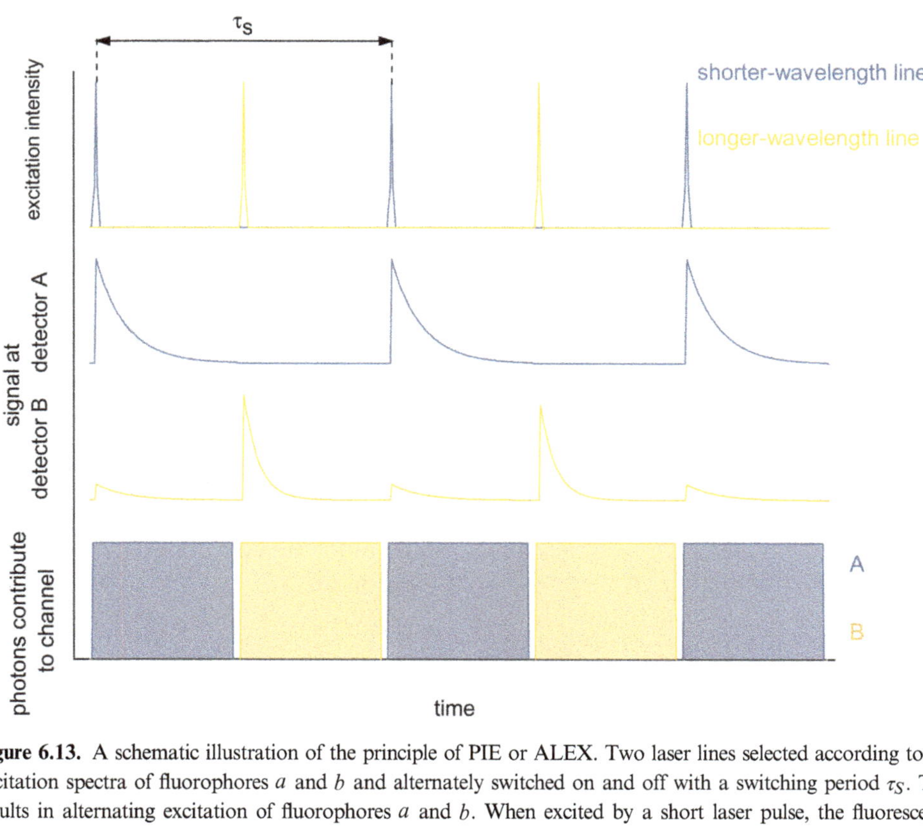

Figure 6.13. A schematic illustration of the principle of PIE or ALEX. Two laser lines selected according to the excitation spectra of fluorophores a and b and alternately switched on and off with a switching period τ_S. This results in alternating excitation of fluorophores a and b. When excited by a short laser pulse, the fluorescence intensity decays approximately exponentially (following a convolution of an exponential decay with the IRF; see section 4.4.1). Photons detected by the detector dedicated for channel B after the longer-wavelength pulses originate only from fluorophores b and generate a crosstalk-free intensity time trace of channel B. Similarly, photons detected by the detector dedicated for channel A after the shorter-wavelength pulses originate only from fluorophores a. When the excitation of fluorophores b by the shorter-wavelength pulses is negligible, it is possible to generate the intensity time trace of channel A from all photons detected after the shorter-wavelength pulses.

order of 10 μs, requiring τ_S of the order of μs or shorter. The requirements on temporal resolution are considerably relaxed when studying molecules with slower diffusion such as large macromolecules or molecules in dense environments such as biological membranes. Such systems are suitable for characterisation by FCS modalities with intrinsically lower temporal resolution. For example, in camera-based FCCS modalities (such as imaging FCCS) the temporal resolution is limited by the camera frame rate to values typically on the order of ms [26, 27]. In such cases, the excitation wavelength would be switched between each consecutive camera frame. If the camera frame rate is 1000 frames per second, the switching time would be 2 ms, reducing the temporal resolution by a factor of two compared to non-ALEX excitation with the same camera frame rate.

PIE excitation is usually performed with pulsed lasers that produce short pulses at repetition rates on the order of 10 MHz. The switching period is then the reciprocal value of the laser repetition frequency, for example, $\tau_S = 25$ ns with a laser pulsing at 40 MHz. While very short switching periods give high temporal resolution of the correlation functions by ensuring nearly simultaneous excitation with both laser lines, consideration of fluorescence lifetimes imposes a lower limit on τ_S practical for PIE-FCCS. As we discussed in section 4.4.1, fluorophores, after being excited, remain for some time in the excited state before emitting a photon and returning to the ground state. This results in an exponential decay of fluorescence intensity (number of emitted photons) following excitation by a short pulse. Ideally, nearly all fluorophores a excited by the shorter-wavelength pulse have returned to ground state before the longer-wavelength pulse is applied to excite fluorophores b (and vice versa). If that is not the case, fluorophores a that remain in the excited state at the time of the longer-wavelength pulse will emit photons during the time window when only fluorophores b are expected to be emitting fluorescence. A fraction (determined by the κ_{aB} element of the brightness matrix) of the photons emitted by fluorophores a during the 'wrong' time window will contribute to the signal in detection channel B, resulting in a residual crosstalk (see exercise 6). Note that an analogy for fluorophores b remaining in the excited state at the time of the shorter-wavelength excitation pulse is also true; however, it is less of a concern since the crosstalk of fluorophores b to detection channel A is typically much weaker than the crosstalk of fluorophores a to channel B ($\kappa_{bA} < \kappa_{aB}$, see section 6.2.3).

6.3.2.1 PIE- or ALEX-FCCS and FRET

So far in discussing how PIE and ALEX eliminate spectral crosstalk we have relied mainly on the fact that fluorophore a is not excited by the longer-wavelength laser line and have not considered whether the shorter-wavelength laser line excites fluorophore b. The case of fluorophore b is not as clear-cut as that of fluorophore a. The typical shape of fluorophore excitation spectra, which decay much more steeply on the long-wavelength side than on the short-wavelength side, results in fluorophore b being much more likely to be excited by the shorter-wavelength laser line than fluorophore a by the longer-wavelength line (see figure 6.12). However, in many cases the excitation probability of fluorophore b by the shorter-wavelength line is negligible and each laser line can be assumed to excite only a single fluorophore species. If that is so, PIE or ALEX can eliminate not only the influence of spectral crosstalk but also the influence of FRET on FCCS results [13, 22]. Since only fluorophore *a* is expected to be directly excited by the shorter-wavelength line, all photons detected in channel *B* after shorter-wavelength excitation result either from spectral crosstalk or from fluorophores *b* that have been excited by FRET, i.e. they received energy from fluorophores *a* in their proximity. Assuming that the decrease in fluorophore *a* brightness caused by FRET results in an equal increase of fluorophore *b* brightness, summing up the signals in both detection channels after shorter-wavelength excitation retrieves a fluorophore *a* signal free from the influence of FRET. Since no fluorophore *a* is excited by the longer-wavelength line, no FRET occurs after the longer-wavelength line excitation and the fluorescence of fluorophore *b* detected in channel *B* after the longer-wavelength excitation is free from the influence of FRET as well.

Note that if the direct excitation of fluorophore *b* by the shorter-wavelength line can be neglected, PIE- or ALEX-FCCS can be performed with a single detector [13, 28]. Spectral separation of fluorescence photons is then replaced by selective excitation of the individual fluorophore species separated in time.

6.3.2.2 Anti-correlation in PIE-FCCS

Negative CCF amplitudes in samples of independently diffusing fluorophores (negative control) were observed in a pioneering work on PIE-FCCS [22]. By finding a linear dependence of the negative amplitude on the average photon count rate, the authors showed that it is caused by the dead time of detection electronics. Photon counting electronics are unable to register any new count within a certain short period after a photon detection event. Any photons reaching the detector within that time are, therefore, overlooked. The higher the photon count rate the more photons tend to be overlooked in this manner. Having detectors for both detection channels connected to the same photon counting card means that, when fluorophores of both species are simultaneously present in the observation volume, detecting a photon from fluorophore *a* prevents for a short time interval registering any photon from fluorophore *b* and vice versa. Thus, the presence of a fluorophore of one species in the observation volume reduces the probability of detecting photons from a fluorophore of the other species, which translates to the signals in the respective detection channels being anti-correlated. Naturally, this effect is not exclusive to negative control samples; CCF amplitudes of samples with positive cross-correlation are also artificially reduced by this mechanism. However, this artefact can be entirely eliminated by modern photon counting cards with independent processing of the signal from each detection channel.

Besides the detection electronics dead time, artificial anti-correlations can be caused also by excluded-volume effects [29]. Each particle occupies a certain volume within the effective observation volume, which is then inaccessible to other particles. The presence of a particle in the observation volume, thus, reduces the probability of another particle being simultaneously observed. In most cases this factor is very small because of the particle volume being typically several orders of magnitude smaller than the effective observation volume. It is likely to be observed only for large particles such as liposomes, large protein complexes or fluorescent microspheres. While this artefact is inherent to FCCS in general and not specific to any FCCS modality, it is typically much smaller than the cross-talk contribution in DC-FCCS and is, therefore, observable only in crosstalk-free FCCS modalities such as PIE- or ALEX-FCCS or FCCS with statistical filtering (see section 6.4.2).

6.4 Statistical filtering in FCCS

Statistical filtering has been introduced in section 4.4 and its goals were explained in analogy to DC-FCCS. In brief, we can regard statistical filtering as an alternative to dispersive elements (such as optical gratings or prisms) or dichroic mirrors and bandpass filters in separating photons into their respective detection channels. It can be used as a stand-alone alternative to spectral separation by optical means [30, 31], or it can supplement it [32].

The main advantage of statistical filtering lies in its ability to handle high levels of ambiguity in photon assignment and overlap of detection channels. To illustrate what we mean by this statement, let us consider the case when we distinguish the two fluorophore populations based on their emission spectra (an approach known as FSCS); statistical filtering, in this case, allows us to distinguish fluorophores with high levels of spectral overlap (see figure 4.10). Using optical means to separate the photons to non-overlapping detection channels (as is the usual practice in DC-FCCS) would not be practical in such a case since each detection channel would receive similar contributions from fluorophores of either population (meaning the brightness matrix would have similar diagonal and off-diagonal elements), which would severely compromise DC-FCCS analysis by causing small relative differences in CCF amplitude between samples with high level of co-diffusion and samples with no co-diffusion at all (see equation (6.20c)).

At the same time, statistical filtering allows us to use other fluorophore characteristics than emission spectra to distinguish between individual fluorophore populations. A notable example is fluorophore lifetime (or more exactly TCSPC histograms) used in a technique called FLCS that has been by a large margin the most frequently used application of statistical filtering. As shown in section 4.4.2, TCSPC histograms of different fluorophore species are, in general, overlapping and statistical filtering is, then, the only way to harness fluorescence lifetime differences for FCCS purposes.

Since the principles of statistical filtering have been outlined in section 4.4, we will not repeat them here, but only discuss the ways in which it deals with some of the artefacts of DC-FCCS introduced in section 6.2 as well as mention specific artefacts of statistical filtering.

6.4.1 Statistical filtering and sources of DC-FCCS artefacts

6.4.1.1 Statistical filtering can eliminate unequal observation volumes

Since statistical filtering does not rely on spectrally separated emission of the individual fluorophore populations to assign the photons to respective detection channels, it can work with pairs of fluorophores that emit both in the same spectral region and both are efficiently excited by the same laser line. The necessary conditions are that the two fluorophore populations have either distinct fluorescence lifetimes (FLCS, see section 4.4.2) or distinct shapes of emission spectra, albeit both located in the same spectral region (FSCS, see section 4.4.3). However, that does not mean that the use of statistical filtering solves the issue with non-identical observation volumes in cases where that issue exists. For example, when statistical filtering is supplementing spectral separation by optical means (such as in the application outlined in section 6.4.3), all the phenomena discussed in section 6.2.2 will have to be considered.

6.4.1.2 Statistical filtering can eliminate spectral crosstalk

As we have already mentioned, the main strength of statistical filtering lies in its ability to retrieve correctly contributions from individual fluorophore species even in situations where the assignment of each individual photon is highly ambiguous. Therefore, statistical filtering can obtain crosstalk free CCFs in situations when the fluorophore species of interest have large spectral overlap. As discussed in section 4.4.2, the quality of crosstalk suppression increases with the number of detected photons (which is in general very high in FCS and FCCS experiments) and decreases with the overlap of area normalised spectra of the two fluorophore species. Therefore, the more similar the spectra of the fluorophore species are, the longer acquisition time is needed to achieve reliable crosstalk suppression by statistical filtering.

6.4.1.3 Statistical filtering can eliminate the influence of non-correlated background

Suppressing background contribution in FCS is probably the most common application of statistical filtering [29]. This relies on the fact that background in most cases has a distinct fingerprint (TCSPC histogram in the case of FLCS or spectrum in the case of FSCS) different from the corresponding fingerprints of the fluorophores of interest. The background fingerprint can be measured in non-labelled samples and then used in statistical filtering to filter out the background contribution. In the case of FLCS, detector contribution to the background is likely to be uniform across all the TCSPC channels [33], scattering will have a narrow TCSPC histogram copying the IRF of the system [34] and autofluorescence has typically a multiexponential TCSPC histogram containing components with shorter lifetime than the lifetime of most fluorophores used for labelling molecules of interest [35].

6.4.2 Statistical filtering and negative CCF amplitudes

In the absence of real cross-correlation (resulting from co-diffusion of fluorophores of different species), artificial negative CCF amplitudes have been reported when using statistical filtering, particularly in FLCS [29, 36]. This artefact is closely related to the negative CCF amplitudes of PIE-FCCS discussed in section 6.3.2. Besides the detector dead-time and excluded volume effects causing the negative CCF amplitudes in PIE-FCCS, there exists an additional FLCS-specific cause for this artefact.

The IRF of some avalanche photodiodes (detectors commonly used in TCSPC experiments) depends on the current photon count-rate. This effect can be quite pronounced in FCS and FCCS experiments where significant fluctuations in photon count-rate are expected to occur within the acquisition time (as a result of concentration fluctuations of the fluorescent molecules). The actual TCSPC histograms are then superpositions of mutually slightly shifted (by as much as 0.5 ns or more) histograms corresponding to parts of the intensity traces with different photon count-rates [37, 38]. This effect also causes artificial negative CCF amplitudes [29]. However, more modern types of photon counting detectors tend to be less prone to this artefact.

6.4.3 Quasi pulsed interleaved excitation FCCS (PIE-FCCS)

In this section, we will present an example of statistical filtering supplementing spectral separation by optical means. Statistical filtering is employed here to eliminate spectral crosstalk, analogously to pulsed interleaved excitation in PIE-FCCS (hence the name). Until recently, pulsed lasers in certain spectral regions (for example, the spectral region coinciding with excitation spectra of red fluorescent proteins) had been less affordable and only continuous-wave lasers had been available for those regions in many setups. This was, for example, the case of the setup used in the original PIE-FCCS paper, where 'pulsed' excitation by that laser was achieved by the means of an acousto-optic modulator [22]. However, such modulation is not available in standard commercial confocal microscopes and for that reason Padilla-Parra *et al* proposed an alternative approach based on statistical filtering [39], which was later named quasi PIE-FCCS [5]. In this approach the fluorophores are excited by two laser lines selected, like in DC-FCCS, to match the excitation spectra of the respective fluorophores. One of the laser lines is pulsed and the other is continuous. This results in the two fluorophore species having distinct TCSPC histograms; the fluorophore species excited by the pulsed laser has an exponentially decaying histogram, while the histogram of the other fluorophore species is constant across all TCSPC channels. Statistical filtering can easily separate photons originating from the individual fluorophore species based on those distinct TCSPC histograms, resulting in crosstalk-free CCFs. Note, however, that since the fluorophores are excited by distinct laser lines and emit in distinct spectral regions, quasi PIE-FCCS suffers from unequal observation volumes (just like standard PIE-FCCS) and only the ACF of the fluorophore species excited by the pulsed laser can be background-free; with continuous-wave excitation, the background and fluorescence TCSPC histograms are both constant across all TCSPC channels and, therefore, indistinguishable.

6.4.4 Reaction kinetics studied by FCCS

So far, we have considered fluorophore species as immutable throughout the FCCS experiments; a fluorophore belongs to a particular species which defines its fluorescence characteristics (excitation and emission spectra and fluorescence lifetime) and those characteristics do not change during the experiment. The CCF, then, reflects on co-diffusion of fluorophores belonging to different species. This is generally the case in DC-FCCS experiments.

Let us now consider that instead of two distinct fluorophore species we have two populations of the same fluorophore (for example, different conformational states of the fluorophore or fluorophore molecules experiencing different local environments) which exhibit different fluorescence characteristics. Assuming that the fluorophore molecules can reversibly switch between the two states on a timescale shorter than the typical residence time in the effective observation volume τ_D, the CCF reflects the equilibrium reaction rates of the switching. If, for example, environmental-sensitive fluorophores are attached to biomolecules undergoing reversible conformational changes which are accompanied by changes in the local environment experienced by the fluorophores, FCCS could be used to study the dynamics of the biomolecule conformational changes [40]. There are not many situations when a fluorophore exhibits large enough reversible changes in its emission spectrum that would allow the two states to be reliably distinguished in DC-FCCS; however, since statistical filtering allows leveraging subtle changes in emission spectra or changes in fluorescence lifetimes, it makes the use of FCCS to study equilibrium reaction kinetics more feasible. For this reason, we mention this type of FCCS experiment only here, in the context of statistical filtering, although, in principle, it is not restricted to any particular FCCS modality. The only requirement is the ability to reliably assign each detected photon to a particular fluorophore state. We will now outline in brief the theoretical models applicable in this type of FCCS experiments. Note that if the switching between the states happens on time scales significantly slower than the typical residence time τ_D, the switching between the states is not observable by FCCS. The theoretical models for distinct fluorophore populations (as presented in the previous sections of this chapter) would apply in such a case and a positive CCF amplitude would be indicative of co-diffusion of fluorophore molecules in different states.

Let us denote the two fluorophore states as a and b. The fluorophore switches reversibly between the two states either as a result of its own internal dynamics or as a result of changes (for example, conformational) of a molecule it is bound to. The model is applicable to both scenarios. Denoting the rate constant of the transition from a to b as k_1 and of the reverse transition as k_2, the system is described by the following reaction scheme:

$$a \xrightarrow{k_1} b \tag{6.38a}$$

$$a \xleftarrow{k_2} b \tag{6.38b}$$

To characterise the temporal evolution of the system, we will introduce the probabilities $p_a(t)$ and $p_b(t)$ of finding a fluorophore in the state a or b, respectively, at time t. The temporal evolution of the probabilities is described by the following system of differential equations:

$$\frac{dp_a(t)}{dt} = -k_1 p_a(t) + k_2 p_b(t) \tag{6.39a}$$

$$\frac{dp_b(t)}{dt} = k_1 p_a(t) - k_2 p_b(t) \tag{6.39b}$$

We will assume that photons from fluorophores in the state a contribute only to detection channel A and photons from fluorophores in the sate b only to channel B (no crosstalk between channels). Further we assume identical observation volumes for both channels and no background. All these assumptions can be expected to be true for an FLCS experiment.

The autocorrelation $G_A(\tau)$ in channel A is given by the diffusional ACF $\tilde{G}_a(\tau)$ (or more exactly the normalised decay of the diffusional ACF, decaying from 1 to 0) of the fluorophore in the state a multiplied by the product of the conditional probability $p_a(t+\tau)|_{p_a(t)=1}$ of finding a fluorophore in the state a at time $t+\tau$ provided that the fluorophore was in the state a at time t and the probability $p_a(t)$ of finding the fluorophore in the state a at time t. Since we assume the system to be in equilibrium, and we only observe fluctuations around equilibrium, the probabilities do not depend on t and we can instead say that $G_A(\tau)$ is proportional to the product of the conditional probability $p_a(\tau)|_{p_a(0)=1}$ of finding a fluorophore in the state a at time τ provided that the fluorophore was in the state a at time 0 and the equilibrium probability p_a^{eq} of the state a. The equilibrium probabilities of the two states can be viewed as the time-averaged probabilities of finding a fluorophore in the respective states. They can be obtained as the static solution of the system of equations (6.39a) and (6.39b), by setting the left-hand sides of the equations to zero (steady-state condition under which the probabilities are not changing in time). This leads to a simple system of linear equations with the following solution:

$$p_a^{eq} = \frac{k_2}{k_1 + k_2} \tag{6.40a}$$

$$p_b^{eq} = \frac{k_1}{k_1 + k_2} \tag{6.40b}$$

The conditional probability $p_a(\tau)|_{p_a(0)=1}$ is obtained from the general solution of the system of equations (6.39a) and (6.39b) (see math box 6.5), leading to:

$$G_A(\tau) = \frac{p_a^{eq} p_a(\tau)|_{p_a(0)=1}}{V\langle C_f^{tot}\rangle\left(p_a^{eq}\right)^2}\tilde{G}_a(\tau) = \frac{1}{V\langle C_f^{tot}\rangle}\left(1 + \frac{k_1}{k_2}e^{-(k_1+k_2)\tau}\right)\tilde{G}_a(\tau) \tag{6.41a}$$

where $\langle C_f^{tot}\rangle$ is the total fluorophore concentration. Following the same reasoning we get for the autocorrelation in channel B:

$$G_B(\tau) = \frac{p_b^{eq} p_b(\tau)|_{p_b(0)=1}}{V\langle C_f^{tot}\rangle\left(p_b^{eq}\right)^2}\tilde{G}_b(\tau) = \frac{1}{V\langle C_f^{tot}\rangle}\left(1 + \frac{k_2}{k_1}e^{-(k_1+k_2)\tau}\right)\tilde{G}_b(\tau) \tag{6.41b}$$

Analogously, the CCF $G_{AB}(\tau)$ is proportional to the product of the conditional probability $p_b(\tau)|_{p_a(0)=1}$ of finding a fluorophore in the state b at time τ provided that the fluorophore was in the state a at time 0 and of the equilibrium probability p_a^{eq} of the state a. The resulting formula is identical for $G_{BA}(\tau)$, allowing us to write for $G_X(\tau)$:

$$\frac{G_X(\tau)}{\tilde{G}_X(\tau)} = \frac{p_a^{eq} p_b(\tau)|_{p_a(0)=1}}{V\left\langle C_f^{tot}\right\rangle p_a^{eq} p_b^{eq}} = \frac{p_b^{eq} p_a(\tau)|_{p_b(0)=1}}{V\left\langle C_f^{tot}\right\rangle p_a^{eq} p_b^{eq}}$$
$$= \frac{1}{V\left\langle C_f^{tot}\right\rangle}\left(1 - e^{-(k_1+k_2)\tau}\right) \tag{6.41c}$$

where the diffusional correlation function $\tilde{G}_X(\tau)$ is a weighted sum of the diffusional correlation functions $\tilde{G}_a(\tau)$ and $\tilde{G}_b(\tau)$. $\tilde{G}_a(\tau)$ can differ significantly from $\tilde{G}_b(\tau)$ since the two fluorophore states can be related to different conformational states of the molecule the fluorophore is bound to; especially in the case of macromolecules, different conformational states can differ largely in diffusion coefficient [40]. According to equation (5.27) the contribution of each state is weighted by the square of its brightness and the fraction of fluorophores in the given state (the corresponding equilibrium probability):

$$\tilde{G}_X(\tau) = \frac{\kappa_a^2 p_a^{eq} \tilde{G}_a(\tau) + \kappa_b^2 p_b^{eq} \tilde{G}_b(\tau)}{\left(\kappa_a p_a^{eq}\right)^2 + \left(\kappa_b p_b^{eq}\right)^2} \tag{6.42}$$

Figure 6.14 shows examples of correlation functions described by equations (6.41a)–(6.41c) for different ratios of the diffusion time τ_D to the characteristic

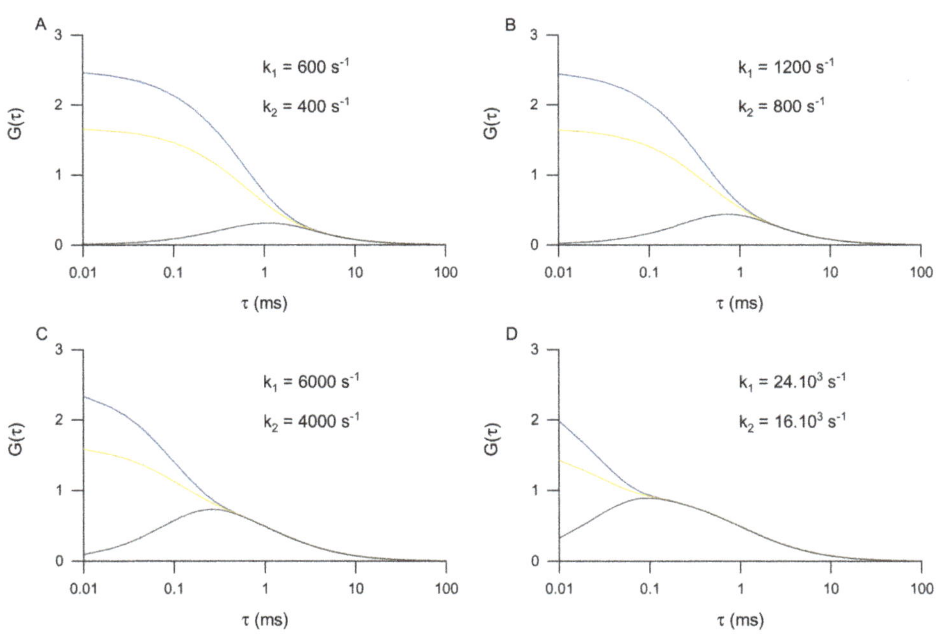

Figure 6.14. Examples of correlation functions of a two-state system exhibiting equilibrium reaction dynamics. The correlation functions were calculated according to equations (6.41a)–(6.41c); the ratio of the rate constant k_1/k_2 was kept constant and equal to 3/2, while the sum of the rate constants varied from $1/\tau_D$ to $40/\tau_D$, where the diffusion time τ_D was kept constant and equal to 1 ms. Otherwise the observation volumes of both channels were assumed identical and the average particle number $V\langle C_f^{tot}\rangle$ was set to 1. The structure parameter $K = 5$.

timescale of the equilibrium reaction $(k_1 + k_2)^{-1}$. The figure shows that the ACFs given by equations (6.41a) and (6.41b) have a maximum at $\tau = 0$, just like usual diffusional ACFs. The CCF (equation (6.41c)), on the other hand, has a maximum shifted to a lag time larger than zero. This shift is related to the typical time needed for a fluorophore to undergo a transition from one of the states to the other. It can be seen as a situation analogous to spatial cross-correlations, where the shift in the CCF maximum to positive lag times reflects the time needed by the fluorophore to travel the distance between the two observation volumes (see section 5.7). Also note that the CCF amplitude diminishes when the reaction kinetics become slower, reflecting the decreasing probability of observing a transition of a fluorophore from one state to the other within the residence time τ_D of the fluorophore in the effective observation volume. As we have mentioned earlier in this section, if the reaction happens on a timescale significantly longer than τ_D, the two forms of the fluorophore will appear as two separate fluorophore populations in FCCS experiments and the cross-correlation between them will be zero as long as all fluorophores diffuse independently as separate particles.

The theoretical model can be extended to apply to more complex reaction schemes than that given by equations (6.38a) and (6.38b). For example, Gregor and Enderlein have derived a four-state model for the case of a fluorophore

Math box 6.5

Temporal evolution of a two-state system described by reaction scheme of equations (6.38a) and (6.38b)

The system of linear differential equations (6.39a) and (6.39b) can be written in matrix form as:

$$\begin{pmatrix} \dfrac{dp_a(t)}{dt} \\ \dfrac{dp_b(t)}{dt} \end{pmatrix} = \begin{pmatrix} -k_1 & k_2 \\ k_1 & -k_2 \end{pmatrix} \begin{pmatrix} p_a(t) \\ p_b(t) \end{pmatrix} = \mathbf{A}\vec{p}(t)$$

and its general solution has the form:

$$\vec{p}(t) = \sum_j \Theta_j \vec{v}^{(j)} e^{\lambda^{(j)} t}$$

where $\vec{v}^{(j)}$ and $\lambda^{(j)}$ are eigenvectors and eigenvalues of matrix $\mathbf{A}$. They can be found by solving equations:

$$|\mathbf{A} - \lambda\mathbf{I}| = 0$$

$$(\mathbf{A} - \lambda\mathbf{I})\vec{v} = \vec{0}$$

where $|\ldots|$ denotes the determinant of a matrix and $\mathbf{I}$ is a unit matrix. Matrix $\mathbf{A}$ has two eigenvalue-eigenvector pairs:

$$\lambda^{(1)} = 0, \ \vec{v}^{(1)} = \begin{pmatrix} k_2 \\ k_1 \end{pmatrix}, \ \lambda^{(2)} = -(k_1 + k_2), \ \vec{v}^{(2)} = \begin{pmatrix} 1 \\ -1 \end{pmatrix}$$

The constants Θ_j are determined by the initial conditions. In the case of the conditional probabilities we are interested in, the initial condition is that one of the probabilities is equal to 1. For example, for the condition $p_a(0) = 1$ we get:

$$p_a(t) = \frac{k_2}{k_1 + k_2} + \frac{k_1}{k_1 + k_2}e^{-(k_1+k_2)t} = p_a^{eq} + p_b^{eq}e^{-(k_1+k_2)t}$$

$$p_a(t) = \frac{k_1}{k_1 + k_2} - \frac{k_1}{k_1 + k_2}e^{-(k_1+k_2)t} = p_b^{eq} - p_b^{eq}e^{-(k_1+k_2)t}$$

Analogously, we could find the solution for the condition $p_b(0) = 1$. Note that for any initial conditions the probabilities converge to their equilibrium values p_a^{eq} and p_b^{eq}, respectively.

switching between two conformations and at each conformation exhibiting different dark state dynamics [37].

6.5 Synopsis

- FCCS measures and analyses cross-correlation between pairs of distinct fluorescence intensity time traces supposed to be signals originating from distinct fluorophore species. The most common form, DC-FCCS, uses differences in emission spectra to distinguish signals from the individual fluorophore species. A positive CCF amplitude is observed only if the fluctuations in the different intensity time traces are correlated with each other, which is in most cases the result of co-diffusion of fluorophores of the two species.
- Typical DC-FCCS experiments study complex formation between molecules of interest labelled by distinct fluorophores. The experiment yields a CCF and ACFs for the individual detection channels. In principle, concentrations of the complex and individual binding partners can be obtained from the amplitudes of the CCF and ACFs, which, then, allow quantification of the complex dissociation constant.
- However, for a fully quantitative interpretation of FCCS data, all factors mentioned further, which affect the FCCS data need to be quantified and included in the theoretical models. If that is not practically feasible, semi-quantitative parameters using ratios of correlation function amplitudes (such as the cross-correlation amount) can be used and their values compared with results for a negative control (a sample containing non-interacting molecules) and a positive control (a sample containing only complexes).
- The observation volumes for different detection channels in DC-FCCS are usually not identical because of wavelength dependence of the volume and a displacement introduced by objective chromatic aberration. This increases the effective observation volume for the CCF and reduces the CCF amplitude.

- The CCF amplitude in a complex-formation FCCS experiment is further reduced if non-labelled binding partners are present (for example, as a result of imperfect labelling or fluorophore photobleaching) to form complexes with fluorescently labelled molecules. A part of the complexes in which only one binding partner is fluorescent, then, does not appear as complexes in FCCS measurements.
- Overlap between spectra of the individual fluorophore species results in ambiguity in assigning photons to the respective detection channels (spectral crosstalk). This is then manifested by an artificial increase in the CCF amplitude. A positive CCF amplitude can be observed even if all the fluorophores diffuse independently of each other.
- Other factors that need to be considered are the non-correlated background in each detection channel, FRET and quenching between fluorophores in complexes and the complex stoichiometry.
- Modifications of DC-FCCS have been developed with the goal of avoiding some of the above-mentioned sources of artefacts and, by doing that, simplifying the quantitative analysis of FCCS data. The differences in effective observation volumes are reduced in sw-FCCS, while PIE-FCCS and ALEX-FCCS data are free from the influence of spectral crosstalk and FRET
- Statistical filtering offers an alternative to hardware optical means of separating the signal from different fluorophore species to their respective detection channels. It allows subtle differences in emission spectra or differences in fluorescence lifetime to be leveraged for signal separation in FCCS.

6.6 Exercises

Note that for exercise 7 you will need to have the computable document format player which is freely available at https://www.wolfram.com/cdf/. With the CDF Player you can open the '.cdf' files available at https://iopscience.iop.org/book/978-0-7503-2080-1.

1. Use equations (6.6a)–(6.6c) and (6.8) to derive an explicit formula for K_D as a function of the CCF and ACF amplitudes measured in an ideal DC-FCCS experiment.

2. Use equations (6.24a)–(6.24c) to show that, when p_a and/or p_b are smaller than 1, the apparent dissociation constant defined by equation (6.25) is larger than the actual dissociation constant calculated from the equilibrium concentrations $[M_a]$, $[M_b]$ and $[M_aM_b]$ using equation (6.8).

3. Consider a DC-FCCS experiment studying complex formation between proteins M_a and M_b in the presence of endogenous forms of the proteins at concentrations $[M_a^{(e)}]$ and $[M_b^{(e)}]$, respectively.
   a. Write a model for the correlation function amplitudes in the case the fluorescent moiety a (which forms a part of $M_a^{(f)}$, the fluorescent

analogue of protein M_a) is quenched by protein M_b. Use equations (6.27a)–(6.27c) to write the concentrations of the different particle types appearing in the model.

b. How does the model change if, instead of being quenched by protein M_b, the fluorescent moiety a forms a FRET pair with the fluorescent moiety b and FRET between them occurs when the two fluorescently labelled protein analogues ($M_a^{(f)}$ and $M_b^{(f)}$) form a complex?

c. How does the model derived in a. change if the fluorescent moieties a and b undergo transitions to dark states? The probabilities of the moieties being fluorescent (residing in a bright state) are p_a and p_b, respectively.

4. Write explicitly the concentrations in equations (6.34a)–(6.34c) for the case of dimer formation ($\alpha = 2$). What is the concentration $\langle C_{0a0b}^{app} \rangle$ of particles that are non-fluorescent? Verify that summing the numbers of monomer units M present in each of the four sub-populations (double-labelled, single-labelled with a, single-labelled with b, non-fluorescent) gives the total number of monomer units in the sample: $N_A([M] + 2\,[M_2])$.

5. Consider an oligomerisation experiment under the assumptions of complete oligomerisation, perfect labelling, no dark states, no background, no cross-talk and identical observation volumes.

   a. Verify the validity of equation (6.37) for the case of dimerisation ($\alpha = 2$). Show that the result is independent on p_a (the fraction of monomers labelled by fluorophore a) for $0 < p_a < 1$.

   b. Derive equation (6.37) for an arbitrary degree oligomerisation α. *Hint*: use the formula for the sum of the binomial distribution:

$$\sum_{i=1}^{n} \binom{n}{i} p^i (1-p)^{n-i} = 1, \text{ for } 0 \leqslant p \leqslant 1$$

6. Consider a PIE-FCCS experiment using two pulsed laser lines with 80 MHz repetition rate (e.g. a Ti-sapphire laser commonly used for two-photon fluorescence excitation) shifted by half of their repetition period with respect to each other. Fluorophore a has fluorescence lifetime of 4 ns. By what factor will be the spectral crosstalk from channel A to channel B reduced in the PIE-FCCS experiment compared to the corresponding DC-FCCS experiment? Consider the excitation pulses to be very short and neglect the IRF. By what factor would the crosstalk be reduced if the laser repetition rate is 40 MHz?

7. In the 'Introduction to FCS—interactive exercises.cdf' (available at https://iopscience.iop.org/book/978-0-7503-2080-1) test the dependences of FCCS

on various parameters. First let us set all parameters to an ideal case so that ACFs and CCF match. For that purpose, set the following parameter values:

Concentration: 50 nM;
Bound fraction: 1
τ_{D1} and τ_{D2}: 400 μs;
Molecular brightness η_{11} and η_{22}: 1500 cts
Crosstalk η_{12} and η_{21}: 0 cts;
Background b_1 and b_2: 0 cts;
Probability for molecules to be in a bright state p_1 and p_2: 1;
FRET factor 1: 1 (corresponding to no FRET);
FRET factor 2: 1 (corresponding to no FRET);
All effective volumes to 5.0×10^{-16}.

Now try varying each parameter individually to test how they affect the ACFs and the CCF. Take some time to systematically change the parameters and explain your observations. Here are some examples:

a. Note how the background in the two channels separates the three CFs. What happens if the background in both channels is the same?
b. Change the three effective volumes. The most common case is that $V_{\text{eff1}} < V_{\text{eff2}} < V_{\text{eff12}}$.
c. Vary p_1 and show that you can cancel the effect by changing V_{eff1}. Why is that so?
d. With the parameters set as above, vary η_{12}, η_{21}, and the FRET factors. Why do you see no change? Now set the bound fraction to 0.5 and try again.

References

[1] Maeder C I, Hink M A, Kinkhabwala A, Mayr R, Bastiaens P I H and Knop M 2007 Spatial regulation of Fus3 MAP kinase activity through a reaction-diffusion mechanism in yeast pheromone signalling *Nat. Cell. Biol.* **9** 1319–26

[2] Shi X, Foo Y H, Sudhaharan T, Chong S-W, Korzh V, Ahmed S and Wohland T 2009 Determination of dissociation constants in living zebrafish embryos with single wavelength fluorescence cross-correlation spectroscopy *Biophys. J.* **97** 678–86

[3] Krieger J W, Singh A P, Garbe C S, Wohland T and Langowski J 2014 Dual-color fluorescence cross-correlation spectroscopy on a single plane illumination microscope (SPIM-FCCS) *Opt. Express* **22** 2358

[4] Muto H 2006 Fluorescence cross-correlation analyses of the molecular interaction between an Aux/IAA protein, MSG2/IAA19, and protein–protein interaction domains of auxin response factors of arabidopsis expressed in HeLa cells *Plant Cell Physiol.* **47** 1095–101

[5] Yavas S, Macháň R and Wohland T 2016 The epidermal growth factor receptor forms location-dependent complexes in resting cells *Biophys. J.* **111** 2241–54

[6] Bumstead J R 2018 Designing a large field-of-view two-photon microscope using optical invariant analysis *Neurophotonics* **5** 1–21

[7] Schwille P, Meyer-Almes F J and Rigler R 1997 Dual-color fluorescence cross-correlation spectroscopy for multicomponent diffusional analysis in solution *Biophys. J.* **72** 1878–86

[8] Kubitscheck U 2017 *Fluorescence Microscopy: From Principles to Biological Applications* (Hoboken, NJ: Wiley)
[9] Weidemann T, Wachsmuth M, Tewes M, Rippe K and Langowski J 2002 Analysis of ligand binding by two-colour fluorescence cross-correlation spectroscopy *Single Mol.* **3** 49–61
[10] Lambert T J 2019 FPbase: a community-editable fluorescent protein database *Nat. Meth.* **16** 277–78
[11] Digman M A, Dalal R, Horwitz A F and Gratton E 2008 Mapping the number of molecules and brightness in the laser scanning microscope *Biophys. J.* **94** 2320–32
[12] Qian H and Elson E L 1990 Distribution of molecular aggregation by analysis of fluctuation moments *Proc. Natl. Acad. Sci. USA* **87** 5479–83
[13] Foo Y H, Naredi-Rainer N, Lamb D C, Ahmed S and Wohland T 2012 Factors affecting the quantification of biomolecular interactions by fluorescence cross-correlation spectroscopy *Biophys. J.* **102** 1174–83
[14] Gaj T, Gersbach C A and Barbas C F III 2013 ZFN, TALEN, and CRISPR/Cas-based methods for genome engineering *Trends Biotechnol.* **31** 397–405
[15] Wang H, La Russa M and Qi L S 2016 CRISPR/Cas9 in genome editing and beyond *Annu. Rev. Biochem.* **85** 227–64
[16] Lakowicz J R 2006 *Principles of Fluorescence Spectroscopy* 3rd edn (Berlin: Springer)
[17] Hwang L C and Wohland T 2005 Single wavelength excitation fluorescence cross-correlation spectroscopy with spectrally similar fluorophores: Resolution for binding studies *J. Chem. Phys.* **122** 114708
[18] Ries J, Petrášek Z, García-Sáez A J and Schwille P 2010 A comprehensive framework for fluorescence cross-correlation spectroscopy *New J. Phys.* **12** 113009
[19] Liu P, Sudhaharan T, Koh R M, Hwang L C, Ahmed S, Maruyama I N and Wohland T 2007 Investigation of the dimerization of proteins from the epidermal growth factor receptor family by single wavelength fluorescence cross-correlation spectroscopy *Biophys. J.* **93** 684–98
[20] Hwang L C and Wohland T 2004 Dual-color fluorescence cross-correlation spectroscopy using single laser wavelength excitation *ChemPhysChem* **5** 549–51
[21] Kogure T, Karasawa S, Araki T, Saito K, Kinjo M and Miyawaki A 2006 A fluorescent variant of a protein from the stony coral Montipora facilitates dual-color single-laser fluorescence cross-correlation spectroscopy *Nat Biotech* **24** 577–81
[22] Müller B K, Zaychikov E, Bräuchle C and Lamb D C 2005 Pulsed interleaved excitation *Biophys. J.* **89** 3508–22
[23] Kapanidis A, Laurence T, Lee N, Margeat E, Kong X and Weiss S 2005 Alternating-laser excitation of single molecules *Acc. Chem. Res.* **38** 523–33
[24] Miller A E, Hollars C W, Lane S M and Laurence T A 2009 Fluorescence cross-correlation spectroscopy as a universal method for protein detection with low false positives *Anal. Chem.* **81** 5614–22
[25] Sankaran J, Bag N, Kraut R S and Wohland T 2013 Accuracy and precision in camera-based fluorescence correlation spectroscopy measurements *Anal. Chem.* **85** 3948–54
[26] Kannan B, Har J, Liu P, Maruyama I, Ding J and Wohland T 2006 Electron multiplying charge-coupled device camera based fluorescence correlation spectroscopy *Anal. Chem.* **78** 3444–51
[27] Burkhardt M and Schwille P 2006 Electron multiplying CCD based detection for spatially resolved fluorescence correlation spectroscopy *Opt. Express* **14** 5013–20

[28] Thews E, Gerken M, Eckert R, Zäpfel J, Tietz C and Wrachtrup J 2005 Cross talk free fluorescence cross correlation spectroscopy in live cells *Biophys. J.* **89** 2069–76

[29] Kapusta P, Macháň R, Benda A and Hof M 2012 Fluorescence lifetime correlation *spectroscopy (FLCS): concepts, applications and outlook IJMS* **13** 12890–910

[30] Benda A, Kapusta P, Hof M and Gaus K 2014 Fluorescence spectral correlation spectroscopy (FSCS) for probes with highly overlapping emission spectra *Opt. Express* **22** 2973

[31] Chen J and Irudayaraj J 2010 Fluorescence lifetime cross correlation spectroscopy resolves EGFR and antagonist interaction in live cells *Anal. Chem.* **82** 6415–21

[32] Paredes J M, Casares S, Ruedas-Rama M J, Fernandez E, Castello F, Varela L and Orte A 2012 Early amyloidogenic oligomerization studied through fluorescence lifetime correlation spectroscopy *IJMS* **13** 9400–18

[33] Enderlein J and Gregor I 2005 Using fluorescence lifetime for discriminating detector afterpulsing in fluorescence-correlation spectroscopy *Rev. Sci. Instrum.* **76** 033102–105

[34] Kapusta P, Wahl M, Benda A, Hof M and Enderlein J 2007 Fluorescence lifetime correlation spectroscopy *J. Fluoresc.* **17** 43–8

[35] Bhatta H, Enderlein J and Rosengarten G 2009 Fluorescence correlation spectroscopy to study diffusion through diatom nanopores *J. Nanosci. Nanotech.* **9** 6760–66

[36] Felekyan S, Kalinin S, Sanabria H, Valeri A and Seidel C A M 2012 Filtered FCS: species auto- and cross-correlation functions highlight binding and dynamics in biomolecules *ChemPhysChem* **13** 1036–53

[37] Gregor I and Enderlein J 2007 Time-resolved methods in biophysics. 3. Fluorescence lifetime correlation spectroscopy *Photochem. Photobiol. Sci.* **6** 13

[38] Otosu T, Ishii K and Tahara T 2013 Note: simple calibration of the counting-rate dependence of the timing shift of single photon avalanche diodes by photon interval analysis *Rev. Sci. Instrum.* **84** 036105–4

[39] Padilla-Parra S, Audugé N, Coppey-Moisan M and Tramier M 2011 Dual-color fluorescence lifetime correlation spectroscopy to quantify protein-protein interactions in live cell *Microsc. Res. Tech.* **74** 788–93

[40] Humpoličková J, Benda A, Sykora J, Macháň R, Kral T, Gasinska B, Enderlein J and Hof M 2008 Equilibrium dynamics of spermine-induced plasmid DNA condensation revealed by fluorescence lifetime correlation spectroscopy *Biophys. J.* **94** L17–9

IOP Publishing

An Introduction to Fluorescence Correlation Spectroscopy

Thorsten Wohland, Sudipta Maiti and Radek Macháň

Chapter 7

Artefacts in FCS

As in any experiment, a good set of FCS data must have a high signal-to-noise ratio, and should be free of artefacts. It is frequently not possible to reach this ideal in real life. Hence it is imperative to understand the source of the noise and the artefacts, so that we can keep them low. It is also necessary to understand how each type of noise and artefact affects our interpretation of the data, so that we know what the acceptable limits may be for each type. This chapter describes the most commonly observed sources of noise and artefacts, their effects on the FCS data, and some ways to minimize them.

7.1 Background

All signals are degraded by background, and FCS is no exception. However, one must understand that a steady 'DC' background does not usually influence the signal-to-noise ratio, as it can be easily subtracted. It has no effect on the correlation curve, because it adds equally to the average and to the signal, and in the definition of the autocorrelation curve, it is the difference between the two that matters. However, a really constant DC background would be a rare phenomenon. Much more likely is background that comes from light sources (which are inherently noisy) which are not linked to the diffusing molecules under investigation. These, for example, may arise from other parts of the sample, or from ambient light leakage to the detector. One source of background is Raman scattering, where the solvent contributes a part of the detected signal (see process 6 in the Jabłoński diagram in figure 3.1). Raman may be expected to be weak, but it comes from all the solvent molecules, which together can be a non-negligible part of the signal. Math box 7.1 provides typical cross sections of Raman and fluorescence processes, and also shows a Raman spectrum together with the fluorescence signal recorded from the dye TPE-TPP in aqueous solution. Temporal fluctuations of such signals have no relationship to the diffusing species in the solution but they can contribute to the ACF. They can

doi:10.1088/978-0-7503-2080-1ch7

not only degrade the signal-to-noise ratio, but they can also introduce artefacts. These 'AC' parts of the background, and not the 'DC' part, are of relevance to FCS.

Math box 7.1.

Raman background in FCS measurements

Raman scattering (process 6 in figure 3.1) can provide considerable background to a fluorescence measurement in some cases. In this process, one part of the energy of the (typically) visible photon creates a vibrationally excited state in a molecule, while the other part of the energy is carried away by a lower energy visible photon. This lower energy photon can frequently overlap with the fluorescence wavelength of a chromophore, which is also at a lower energy (i.e. at a longer wavelength) compared to the excitation photon. The difference of energy between the incoming and the outgoing photon is a constant (given by the vibrational energy levels of the molecule). Since Raman can arise from all the molecules in the solvent, together it can add up to a substantial signal especially if the fluorophore has a weak absorbance at that wavelength, or a low quantum efficiency. Figure 7.1 below shows the emission of a weakly fluorescent dye solution when it is excited at 330 nm in a fluorimeter. While the actual fluorescence peak is at 470 nm, Raman peak is at 375 nm. The Raman band comes from the water, and the difference in energy between the excitation and the Raman frequencies is about 3300–3400 cm^{-1}, which is the frequency of vibration of the water O–H stretch. Fortunately, Raman is a very weak effect compared to fluorescence. Rate of Raman scattering is proportional to the change of polarizability of the molecule as a function of the vibrational (normal) coordinate. This makes it rather small compared to the rate of fluorescence. So the fluorescence signal from most fluorescent dyes, even at a reasonably low concentration, far overrides the Raman signal. However, care should be taken to exclude the sharp Raman peak by using a suitable wavelength of excitation and an appropriate filter.

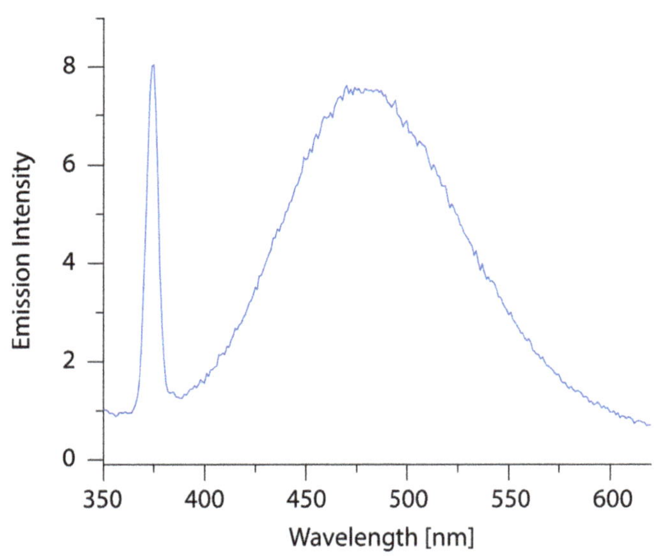

Figure 7.1. Emission spectrum of the dye TPE-TPP in aqueous solution excited at 340 nm. The narrow peak at 375 nm is due to Raman scattering by water, while the wider peak at 470 nm is due to TPE-TPP fluorescence.

For example, let us estimate the background created by water in the fluorescence signal of TPE-TPP. The Raman cross section of water is about 7×10^{-29} cm^2 at 330 nm, while the fluorescence cross section of a typical dye molecule is of the order of 10^{-16} cm^2. However, water is present at a concentration of 55 M. So in a 1 fl probe volume, there would be about 3.3×10^{10} water molecules. If we assume that only a single dye molecule (on an average) is present in this volume, then the signal from the Raman peak of water will be about 0.02 times that of the fluorescence (assuming a quantum efficiency of 1 for the fluorophore). This factor of course may be larger for a dye with a lower cross section or a lower quantum efficiency.

We classify two types of backgrounds: background with noise which is strictly random but whose average is steady, and background with a systematic temporal behaviour (in a time scale which is in a similar order of magnitude to that of the process of interest). The first type of noise directly reduces the amplitude of the signal in FCS. The signal is proportional to $1/N$, where N is the number of fluorescent particles in the probe volume, but this relationship has a background dependent proportionality constant:

$$G_0 = \frac{a^2}{N} \tag{7.1}$$

Here (ignoring the shot noise correlation, see chapter 3) $a = \langle F\rangle/(\langle F\rangle + \langle B\rangle)$, where F is the fluorescence and B is the background, and the angled brackets $\langle\ \rangle$ indicate the time average (math box 7.2). This equation shows that as the background increases, a decreases towards zero, and the signal strength in FCS also decreases to zero. Interestingly, this also explains why the signal in FCS, despite being inversely proportional to the concentration, does not diverge at the limit of low concentrations, i.e. as concentration goes to zero. As the concentration decreases, $\langle F\rangle$ decreases, but $\langle B\rangle$ does not change, and so the factor a converges to zero. However, this type of noise only affects the amplitude, not the temporal characteristics of the autocorrelation signal.

Math box 7.2.

Influence of background on the ACF amplitude
Uncorrelated background will influence the number of particles as recovered from the amplitude, G_0, of the ACF. If the molecular brightness (CPS) is κ, the fluorescence from N molecules is given by $F = \kappa N$. In the absence of background signal we arrive at the following equation for G_0:

$$G_0 = \frac{1}{N} = \frac{\kappa^2 N}{(\kappa N)^2}$$

If background, denoted by B, is present we obtain a background-influenced ACF amplitude, G_0^*, which leads to an apparent number of particles, N_{app}, if this is naively determined by the inverse of G_0^*:

$$G_0^* = \frac{1}{N_{app}} = \frac{\kappa^2 N}{(\kappa N + B)^2}$$

Comparing the actual and apparent number of particles lets us calculate the necessary correction

$$\frac{N}{N_{app}} = \frac{(\kappa N)^2}{\kappa^2 N} \frac{\kappa^2 N}{(\kappa N + B)^2} = \frac{(\kappa N)^2}{(\kappa N + B)^2} = \frac{(F)^2}{(F + B)^2}$$

$$N = N_{app} \frac{(F)^2}{(F + B)^2}$$

Note that with the corrected number of particles N, the correct brightness value κ can also be determined.

The effect of noise that has systematic time dependence in the same time scale as the fluctuations of interest in FCS can cause obvious changes in the temporal decay pattern of the FCS autocorrelation curve. This is more damaging, since in most experiments, it is the temporal decay of the autocorrelation which is of primary interest. Some of the phenomena that cause such fluctuations are discussed below.

7.2 Rare events

Since FCS is a measure of fluctuations, any random fluctuation produces a correlation signal. However, one needs to understand that FCS provides an average timescale for any given type of fluctuation. If the fluctuation occurs only once or a few times during the measurement time, then it will not produce a representative average of that process. This is in principle similar to any measurement which leads to an average quantity, where inadequate sampling can give rise to rather unreliable data. If we measure anything which can have some random variations, we need to average over a sufficient number of measurements to ensure that the average value is a good representation of the actual phenomenon. What is the equivalent of this in FCS? Well, if a phenomenon causes a large fluctuation only once or at best a few times, this will create an extra amplitude in the correlation curve (for example, see the trace in figure 7.2, where a very bright particle comes into the probe volume at about the 8 s time point). This extra amplitude will have a decay time that will be somewhat arbitrary, since only a single (or a few) event determines it. The amplitude contributed by this event will depend on the magnitude of the fluctuation compared to all other fluctuations encoded by the FCS curve. If it happens a considerable number of times, this extra amplitude will decay with the characteristic decay time of the corresponding phenomenon, and this then will truly represent that phenomenon in FCS. For example, if we consider a diffusive process, then a rare event(s) implies a bright particle crossing the probe volume only once (or a few times), taking a particular path. This particular crossing takes an arbitrary amount of time. But the integral in math box 5.1 will result in a proper ACF representing the diffusion time of this bright species only if the particle has unbiasedly sampled all possible paths through the probe volume. So a rare event will present an artefact.

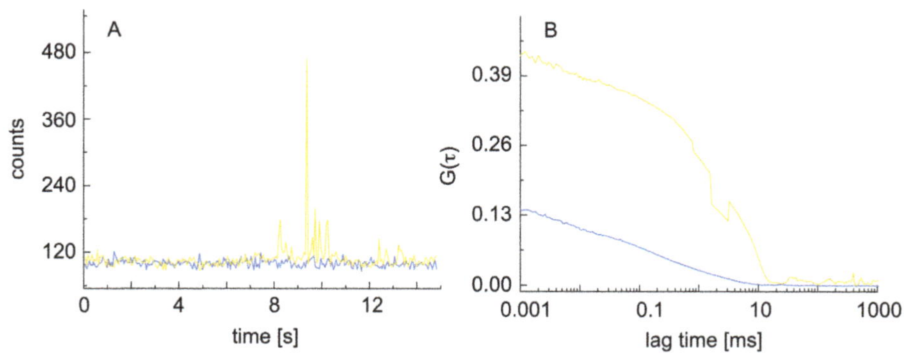

Figure 7.2. The effect of an occasional large, slowly moving, bright particle on the autocorrelation function. (A) The count rate trace of fluorescently labelled amyloid beta attached to small unilamellar lipid vesicles (blue), and the same with a very large multivesicular object traversing the probe volume (yellow). The large particle comes into the volume at about the 8 s time point. (B) The corresponding autocorrelation functions.

We note that this is the essential difference between single molecule measurements and FCS (even if it is performed with a single molecule on average in the probe volume). Both methods can be sensitive to fluctuations caused by single molecules, but the ACF in FCS only provides the characteristic time of particular phenomena, and individual trajectories are averaged out. This is both the strength of FCS, and its weakness. It automatically and robustly provides a good measure of the average timescale of molecular fluctuations even in the solution state, irrespective of whether there is a single molecule or multiple molecules present in the detection volume. Single molecule techniques have to average over a large number of trajectories to measure the same quantity, where there must be only a single molecule at any given time in the probe volume for an extended length of time. On the other hand, single molecule techniques will provide information about rare events and peculiar trajectories, but the ACF will not.

The study of the diffusion of aggregating proteins provides a very relevant example of the 'rare event' artefact. Suppose a protein is aggregating, and most of the particles are small oligomers (i.e. few-mers), while a very small number of large-mers have formed. Since the amplitude of the fluctuation depends on the square of the brightness, a ten-mer will provide 100 times larger fluctuations compared to the monomer (see math box 5.3). If one of these rare particles transits though the volume once, the nature of the correlation curve will completely change. In a repeat of the same experiment, it is possible that this type of particle will not cross the volume during the duration of the experiment, or at the very least, will take a very different path through the probe volume. So two repeats will give completely different results. There are only two ways to tackle such rare events. First, we can ignore those curves which actually contain these large fluctuations. Since no particular amplitude can be predicted, we must analyse the raw data (before correlation) to weed out all large peaks. This will tend to bias the data towards smaller particles, and must be treated with caution. Ries *et al* have described an algorithm to filter out the larger peaks in the time trace to obtain analysable FCS data [3]. A better option is to separate out the larger particles by centrifugation or filtration. This achieves a similar result, but at least the analysis does not have to be deliberately biased. Disappearance of longer time scale fluctuations by controlled centrifugation also provides an independent estimate of the mass of the larger particles [4]. However, this is not possible for samples which cannot be filtered or centrifuged, e.g. for intracellular FCS measurements. The other option is to run the experiment over a long enough time scale, such that the rare events have a chance to recur a large number of times, and then they will produce an analysable signature in the correlation curve. This would be the preferable method. However, given experimental time constraints, this may not be practical in all cases.

7.3 Bleaching

The analysis of the ACF assumes that the system is at least in a quasi-steady state, i.e. the system averages do not change over the time scale of the experiment. This assumption is inherent in the derivation of the autocorrelation function. As given in math box 5.3, the starting point of the derivation for a single species is the diffusion equation:

$$\frac{\partial C(\vec{r}, t)}{\partial t} = D\frac{\partial^2 C(\vec{r}, t)}{\partial x^2} \tag{7.2}$$

This equation contains no population decay term on the right hand side, so it assumes no photobleaching. Even when chemical kinetics is accounted for, such as for triplet dynamics (equation (5.46) and math box 5.5), the equations do not account for any irreversible loss of population. However, there could be physical or chemical processes which can take the system away from this steady state. One of the most common such processes is photobleaching. Photobleaching is difficult to avoid, and every molecular fluorophore will chemically transform into a non-fluorescent molecule/particle after a finite number of photoexcitation cycles. Each fluorophore may have its own pathway towards photobleaching, but it typically involves an oxidation reaction by an available singlet oxygen species in the solution. Since it inexorably reduces the concentration of fluorophores in a solution, it is a non-equilibrium situation, and produces difficulties in performing accurate FCS analysis. If the rate of photobleaching is substantial, then it will be obvious in the ever-decreasing count rate trace. However, even small photobleaching probabilities, sometimes called 'cryptic photobleaching', can change the autocorrelation curve, as described below.

Photobleaching can be described in general as a rate process, whose rate is linearly dependent on the rate of excitation (as photobleaching typically occurs from the excited state). The probability that a particular molecule will photobleach depends on the time it spends in the excitation volume. So the longer a particle stays in the volume, the larger is the probability that it will bleach. The subset of diffusion trajectories which make the particle stay longer than the average time in the probe volume, contributes to the long-time tail (some prefer to call it the 'foot') of the ACF (see arrow in figure 7.3). So photobleaching will be apparent first in the disappearance of these longer than average trajectories, as bleaching makes these particles

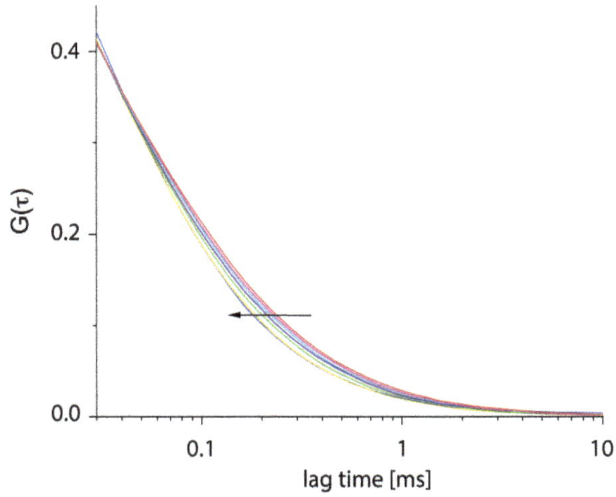

Figure 7.3. Photobleaching: A series of ACFs for fluorescein in aqueous solution at increasing laser powers. The power increases along the direction of the arrow. The ACFs clearly shorten near the foot of the curve (around 0.2 ms) at higher laser powers. This falsely implies the presence of a faster diffusing species, though neither the molecule nor the solution has changed.

effectively 'leave' the probe volume (while they are still physically inside the volume). This effectively shortens the average diffusion time of the particle. So photobleaching begins to be clearly visible at the foot of the correlation curve, whose amplitude starts dipping below the standard fitting curves (figure 7.3).

There are other processes (such as saturation, described later in this chapter) that can *increase* the diffusion time at higher laser powers. So it is always a good practice to check if the measured diffusion time in a system is a function of the excitation laser power. If it is, ideally one should lower the power to a range where the diffusion time is invariant with power.

Is there a simple way to fit the data which can take care of the bleaching, and yield the correct parameters for all the processes? This is not easy. As we have mentioned, the basic equation which describes FCS assumes equilibrium, and that cannot be satisfied if bleaching continuously depletes the population. However, if the bleaching is confined to a small excitation volume, and the rest of the system is very large compared to this volume, then an effective quasi-steady state can be obtained. This is true only for three-dimensional diffusion, as 2D or 1D systems do not attain a steady state (see math box 7.3 and [5, 6]). In such cases of NESS (non-equilibrium steady state), FCS can be carried out in principle with appropriate theoretical considerations [7]. The steady state concentration profile has a dip at the center of the focus, and the profile can be analytically described under some approximations (math box 7.3). This shape must be considered for fitting the ACF. However, no analytical expression for the NESS ACF has been reported for realistic excitation geometries, and so it is hard to arrive at accurate values of the parameters of interest from first principles. If we do obtain an ACF from such a steady state experiment and fit it to the standard equation, we will get a smaller effective diffusion constant,

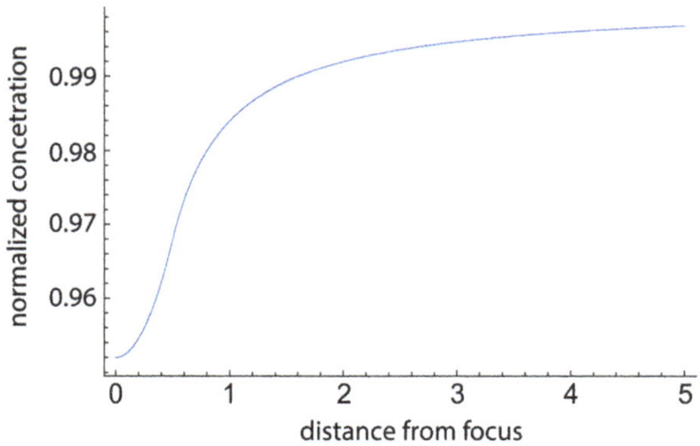

Figure 7.4. Steady state concentration as a function of radial distance r from the centre of focus in presence of photobleaching. The distance from focus is in μm, and the sample is assumed to be uniformly illuminated at the center till a radius of 0.5 μm. The concentration dip at the centre is caused by continuous photobleaching.

as the dwell time of the molecule inside the probe volume is shortened by bleaching to a value, $\tau_D' = \tau_D/(1 + k\tau_D)$, where k is the photobleaching rate (see [5] and math box 7.3).

Math box 7.3.

Effect of photobleaching on the concentration profile

When photobleaching occurs in a confocal geometry, the concentration profile no longer remains flat. This is because the photobleaching rate is dependent on the light intensity, and therefore it is the highest at the center. While photobleaching depletes the concentration at the center, eventually diffusion of molecules from the bulk helps attain a non-equilibrium steady state. The steady state concentration profile has a simple analytic solution under the approximation that illumination is unity inside a radius a, and zero outside. The concentration C as a function of distance r and time t obeys the following equation inside and outside the volume (here D = diffusion constant, k = rate of photobleaching) [6]:

$$D\frac{\partial^2 C_j}{\partial r^2} + \frac{2D}{r}\frac{\partial C_j}{\partial r} - k_j C_j = \frac{\partial C_j}{\partial t}$$

where $j = \text{in} \Rightarrow k_j = k$ and $j = \text{out} \Rightarrow k_j = 0$

The steady state solution to this equation in three dimensions is given by the following (here C_0 is the initial concentration) [6, 11]:

$$C_{\text{in}} = \frac{C_0 \operatorname{Sinh}\left(r\sqrt{k/D}\right)}{r\sqrt{k/D}\operatorname{Cosh}\left(a\sqrt{k/D}\right)}$$

$$C_{\text{out}} = 1 + C_0\left[1 + \frac{\operatorname{Tanh}\left(a\sqrt{k/D}\right)}{r\left(\sqrt{k/D}\right)} - \frac{a}{r}\right]$$

The steady state concentration in three dimensions (as a fraction of the initial concentration) is plotted (figure 7.4) as a function of distance r from the center (in μm), assuming $k = 40\ \mathrm{s}^{-1}$, $a = 0.5\ \mu\mathrm{m}$, and $D = 100\ \mu\mathrm{m}^2\ \mathrm{s}^{-1}$. We note that there is no steady state solution in lower dimensions. The concentration depletion zone at the center keeps expanding and the concentration value continues to go lower for 1 and 2 dimensional cases.

Figure 7.3 shows a dataset of ACFs from a fluorescein solution recorded at different powers. The fitting at the lower powers work well, and produces the right numbers for $G(0)$ and diffusion time (τ_D). However, at higher powers, the foot of the correlation curve cannot be adequately fit by the equation if we assume the same τ_D. If we relax the constraints on τ_D then the data fits reasonably well, but shows a lower diffusion time. As we have already stated, the bleaching tends to increase the value of the apparent diffusion constant, as it adds an additional process that can make a fluorophore disappear from of the volume. It is possible to get the reverse of bleaching, i.e. photoconversion of molecules from a dark state to a brighter state. The probe volume then acts like a 'source' instead of a 'sink', but the equations remain the same, except for the sign of k (the rate of photobleaching), and the associated problems remain the same. Sometimes another alternative process affects the diffusion time in the opposite direction at high excitation powers. This is due to saturation, which will be discussed in a subsequent section.

The standard equation will not fit well if the photobleaching rate is substantial. In such cases, the photon count rate in a finite sample decreases continuously, and the ACF will show pathological features. The shorter timescale fluctuations can still be calculated reasonably accurately if the sloping baseline can be fit well. There have been several approaches by which such data can be corrected to yield reasonable estimates of the diffusion coefficient. For example, wavelet transforms, sliding window approaches or correction by interpolating functions have been used to correct the sloping baseline of the recorded fluorescence count time series [8–10]. Such a fit tries to leave the shorter time scale fluctuations unaffected by taking care of the longer time scale variations. They also help recover the proportionality between $G(0)$ and $1/N$ (see math box 7.4 on a variance conserving bleach correction). It is a powerful but phenomenological approach, so the experimenter should verify whether such data processing affects his/her results in the relevant timescale.

Math box 7.4.

Bleach/trend correction

Assume that we have measured a fluorescence time trace $F(t)$, which contains fast fluctuations of the molecular process under investigation and slow changes due to unwanted artefacts. The artefacts could be due to bleaching, and thus impose a slow decrease of the intensity over time, or they could be other variations due to slow sample movement, not directly linked to the molecular processes. Assume also that the slow changes are described by a function $f(t)$, which varies much slower than the molecular fluctuations we are interested in. In that case the intensity trace $F(t)$ can be corrected in the following way (see also equation (2.21)):

$$F_c(t) = \frac{F(t)}{\sqrt{f(t)/f(0)}} + f(0)\left(1 - \sqrt{f(t)/f(0)}\right)$$

It is important that this correction preserves the variance of the signal of the molecular fluctuations, as otherwise it would change the amplitude of the ACF, G_0 [12]. Assuming a Poisson process for fluorescence, as we have seen earlier, the variance is equal to the mean at equilibrium. Let's divide the intensity time trace into periods which are longer than the diffusive process but shorter than the bleaching process so each period can be assumed to be at equilibrium. Assume we have M segments. Then $f_m(t)$ is approximately constant and

$$Var(F_m(t)) = \langle F_m(t) \rangle = f_m(t)$$

where m represents the time segment number and runs from 1 to M, the total number of segments. This assumes that the molecular fluctuations are distributed around the function $f_m(t)$.

Note that the segment $f_1(t)$ is just the first segment equal to the intensity in the absence of bleaching or changes. It is an estimate of $f(0)$. First let's calculate the mean of the corrected signal

$$\langle F_{c,m}(t) \rangle = \left\langle \frac{F_m(t)}{\sqrt{f_m(t)/f_1(t)}} + f_1(t)\left(1 - \sqrt{f_m(t)/f_1(t)}\right) \right\rangle$$

As all values are constants except $F_m(t)$, and $\langle F_m(t) \rangle = f_m(t)$,

$$\begin{aligned}
\langle F_{c,m}(t) \rangle &= \frac{f_m(t)}{\sqrt{f_m(t)/f_1(t)}} + f_1(t)\left(1 - \sqrt{f_m(t)/f_1(t)}\right) \\
&= \sqrt{f_m(t) f_1(t)} + f_1(t) - f_1(t)\sqrt{f_m(t)/f_1(t)} \\
&= \sqrt{f_m(t) f_1(t)} + f_1(t) - \sqrt{f_m(t) f_1(t)} = f_1(t)
\end{aligned}$$

Now let's calculate the variance of the corrected intensity trace for one of the segments:

$$Var(F_{c,m}(t)) = Var\left(\frac{F_m(t)}{\sqrt{f_m(t)/f_1(t)}} + f_1(t)\left(1 - \sqrt{f_m(t)/f_1(t)}\right) \right)$$

Note that according to our approximation the $f_m(t)$ are constants, therefore

$$Var(F_{c,m}(t)) = Var\left(\frac{F_m(t)}{\sqrt{f_m(t)/f_1(t)}} \right)$$

With $Var(ax) = a^2 Var(x)$

$$\begin{aligned}
Var(F_{c,m}(t)) &= Var\left(\frac{F_m(t)}{\sqrt{f_m(t)/f_1(t)}} \right) \\
&= \frac{f_1(t)}{f_m(t)} Var(F_m(t)) = \frac{f_1(t)}{f_m(t)} f_m(t) = f_1(t)
\end{aligned}$$

But this is exactly what we wanted, namely that the variance after correction, $Var(F_{c,m}(t))$, should be the same as the one for the initial signal before bleaching, with the mean and variance denoted as $f_1(t)$. Therefore

$$Var(F_{c,m}(t)) = \langle F_{c,m}(t) \rangle = f_1(t)$$

As mentioned, the conservation of the variance is important to obtain a proper value for G_0. This allows us to correct the number of particles to its value before bleaching started. But there are several points to be noted. First, we assumed here that bleaching happens on a much slower scale than the molecular fluctuations. Second, there are two effects of photobleaching. There is a slow decrease of the intensity over time as particles are bleached in the sample overall. But there is a faster timescale effect, as particles within the observation volume are bleached during their stay within the volume and thus appear to 'leave' the observation volume earlier than they in fact do by diffusion. Thus the diffusion coefficient will be overestimated. This is sometimes called cryptic photobleaching, and it was dealt with in math box 7.3. This cannot be corrected by this approach. Finally, the correction assumes that the value at the beginning of the intensity trace is the value to normalize to, and the particle number and brightness stays constant. This however does not have to be the case if there is sample movement (see section 7.4), where one could in principle be out of focus at the beginning of the measurement and then sample movement would bring the sample into and out of focus in a random fashion. For example, in measurements on cell membranes by confocal FCS, movement along the focal volume's axial direction (optical axis) puts the membrane in different parts of the laser beam with different intensities and different widths. Thus both the number of particles as well as the brightness change. Nevertheless, despite these caveats, the correction here allows one to at least remove the slow trends from the intensity trace and it yields correlation functions free of artefactual decays, so that one can use standard models for fitting. In cases were photobleaching cannot be avoided, e.g. in measurements in restricted volumes with limited fluorophores, it allows us to extract some information that otherwise would not be accessible. But these results should always be interpreted with caution.

7.4 Sample movement

Another artefact in FCS is sample movement. It is easy to see that sample movement can remove or put in fluorophores inside the probe volume through an external process which is neither linked to diffusion, nor to any of the other processes which have been considered in deriving the equation for fitting the FCS data (math box 5.1). If the movement is monotonic and steady, such as a creeping movement in one direction which can be taken as a steady velocity, then that can be addressed by a simple analytical equation including of a 'flow' component, as done in chapter 5. Indeed, the very first paper on FCS had a description of a 'velocity' of the particles through the probe volume [13]. Of course, if the velocity is large, then one cannot reliably measure diffusion. If it is small compared to the rate with which diffusion can change the population, then one cannot see a substantial effect of this in the correlation curve. A commonly observed sample movement that can affect the autocorrelation curve is the 'sagging' of the objective nosepiece under its own weight. It results in a movement along the z-direction. While an equation can be modified to take care of such movements, it is perhaps best to avoid it if at all possible. Modern microscopes come with stable stages with feedback loops, which keeps the objective-sample distance invariant. However, it is best to enable focus correction in between short FCS measurements, otherwise the active correction mechanism may itself cause fluctuations in the ACF. Keeping the environment around the instrument appropriate (steady temperature, no air flow) usually renders the stage-drift effect negligible.

However, sample movement becomes really bothersome when there is oscillatory movement of the sample, especially with slowly diffusing particles. In this case, one observes an oscillatory pattern in the correlation function. This is the result of slowly moving bright objects repeatedly entering and leaving the probe volume, as the instrument itself vibrates.

We show an actual dataset, recorded with a rhodamine labelled protein in a viscous solution (figure 7.5). The oscillations are obvious in the tail region. Of course, oscillations do not always depend on the sample. Any periodic laser fluctuations, any regular fluctuations of the background, any periodic fluctuations of the optical path, or any periodic electronic noise that can affect the fluorescence counts will give rise to oscillations in the ACF. Incandescent room lights will contribute background that oscillates at twice the frequency of the AC power supply, and this is frequently seen as a background. One easy way to diagnose these systemic fluctuations is to use a dye solution at a relatively high concentration (where the amplitude of the autocorrelation should be nearly zero) as a sample. If the oscillations are still pronounced, then they are likely originating from one of these non-sample sources.

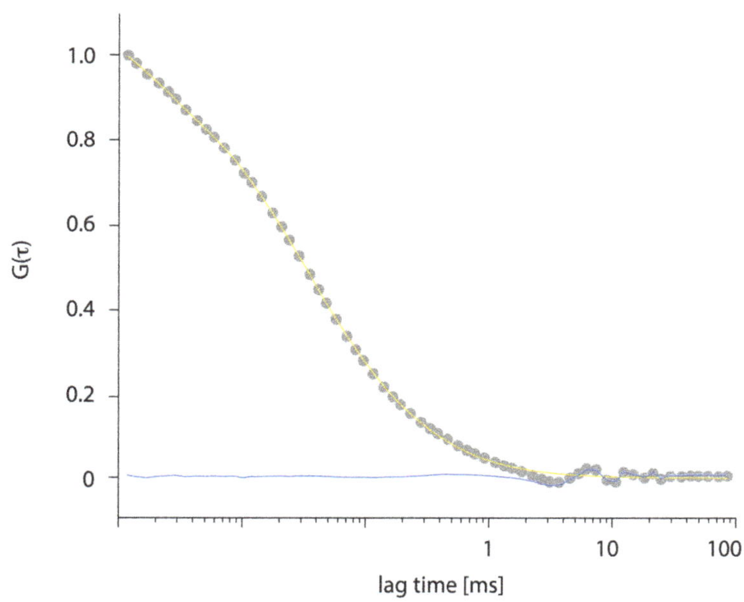

Figure 7.5. Oscillations: ACF (circles) and fit (yellow) for rhodamine B labelled ubiquitin in a viscous 6 M Gdn.HCL solution. The oscillations, due to mechanical vibrations of the setup, are obvious at a few ms time scale. The fit residues (blue) bear the signature of the oscillations.

In a living sample, such as the cell membrane, some movement may be unavoidable. Recording data in short intervals, and averaging the correlation functions later, is one effective way of taking care of such movement related artefacts. In some cases, the periodic traversal may be an inherent feature. Spatz *et al* have monitored fluorescently labelled droplets passing periodically through a capillary using FCS. Here the oscillations faithfully report the transit period of the droplets, even if the droplets are only dimly fluorescent [14].

7.5 Detector-related artefacts: after-pulsing and dead time

The single-photon counting avalanche photo-diode (abbreviated either as SPAD or APD) is the typical detector of choice for point focus confocal FCS. Its low noise and high quantum efficiency of photodetection makes it desirable for most FCS experiments. However, they have two unwanted features, after-pulsing and dead-time, which introduce artefacts in the FCS measurements. Fortunately, these artefacts are in the small time scale, which for most systems is much shorter than the time scale of interest for diffusion. After-pulsing is the finite probability that a spurious second signal pulse will be generated by the detector, after the original pulse that recorded the arrival of a photon. This in effect increases the probability of observing a second photon after the arrival of the first to a value above the average, giving a spurious positive correlation. This usually happens around the time scale of ~1 μs for most APD detectors, and the probability of after-pulsing is less than 1%. However, even that low probability produces a positive correlation that sometimes far outweighs the real signal.

Similarly, there is a spurious negative correlation which can arise due to the dead time of the detector. Dead time is the time interval after the counting of a photon that the detector needs, so that it is ready for the next photon. This arises from the nature of the detector, which needs to be quenched before it is ready for counting the next photon. The result is that the detector is 'dead' for a certain time interval after counting each photon. This is of the order of about 100 ns for APDs. This means the chance of observing a photon within about 100 ns of observing the first photon is actually less than the average value. So the correlation function shows a negative value in this region.

If one is only interested in the process of diffusion, then typically one does not need to worry about most of these artefacts. Even a small dye in aqueous solution will have a diffusion time of >10 μs, assuming a diffraction limited observation volume with sub-μm dimensions. This is more than an order of magnitude longer than the afterpulsing time, and even longer than the dead time. However, if one is interested in the triplet lifetime (which is typically a few μs), or even faster events such as the spontaneous fluctuations of proteins or antibunching effects (~ns timescales, see section 5.6), one has to worry about these artefacts. An example of after-pulsing and dead time artefacts in real FCS data is shown in figure 7.6.

This artefact is fortunately easy to get rid of. One can use two separate detectors in the detection arm, where the photon paths are divided by a 50/50 beam splitter. Then instead of using the ACF, which will contain after-pulsing and dead time artefacts, one uses the CCF between the detectors. The probability that both detectors will see a spurious photon within the short timescale at which this phenomenon occurs (about 1 μs after the real photon appearing) is rather low, given that the photon collection rate is typically of the order of 100 kHz per detector. Therefore, the cross correlation curve does not get affected by after-pulsing, and the spurious positive signal seen in the autocorrelation becomes flat in the cross correlation (figure 7.6). Of course, while the after-pulsing photons are uncorrelated in the two channels, they do contribute to the background, and therefore this can lower the value of *G*() and provide a somewhat erroneous measurement of the concentration (equation (7.1)). Cross correlation also takes care of the dead-time artefact. Although a detector is 'dead' directly after the detection of a photon, the second detector is still active. So there is no minimum time duration between two photons within which a second photon cannot be detected, which eliminates the dead-time induced negative part.

Newer detectors have less after-pulsing and dead time [15]. Use of these may reduce the problem, but using a cross-correlation still remains the most reliable method to get rid of the effects of after-pulsing when using continuous wave excitation. One penalty that one has to pay is that the number of photons in each channel reduces by a factor of two, as one typically uses a 50/50 beam splitter in the emission path for the cross correlation. There is also a way to build an instrument which avoids this penalty. A set up that uses two objectives in a so called '4Pi' configuration actually provides two channels without any loss of counts in a single channel (see figure 7.7). This just brings in another objective from the other side of the sample (usually from the top, since the FCS set-ups are typically implemented in

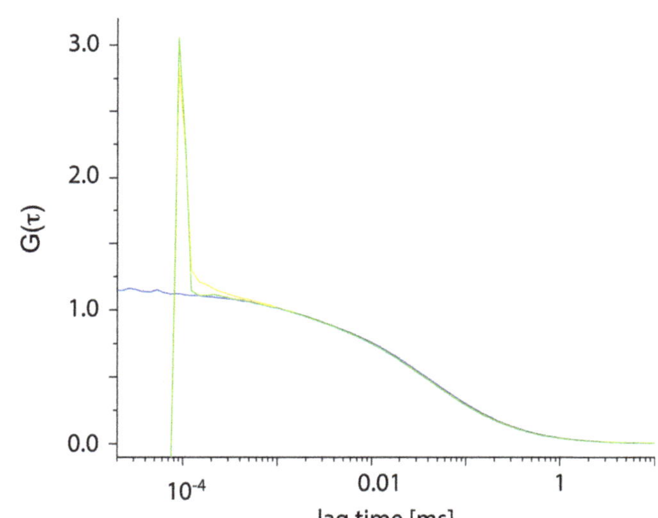

Figure 7.6. Correction of after-pulsing and detector dead time: ACF from individual detection channels (green and yellow) show a strong positive spike at around 200 ns and a negative dip at timescales shorter than that. The positive spike is due to after-pulsing, while the negative spike is due to the dead time of the detector. The cross correlation (blue) is not affected by these two artefacts.

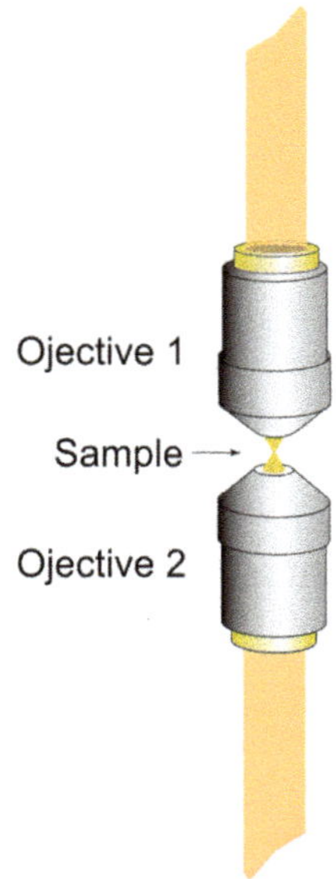

Figure 7.7. 4Pi configuration: A dual objective FCS set-up collects fluorescence from both sides of a sample placed at the common focus, effectively collecting nearly twice as many photons as a single channel set-up. The excitation is typically from a single side only.

'inverted' microscopes, where the first objective is placed vertically below the sample chamber). The focus of the other objective needs to be at the same point as the first one. This effectively doubles (in practice due to imperfections of alignment, it is usually less than double) the count rate, and then a cross-correlation can be performed with the same signal-to-noise as a single channel autocorrelation measurement (see equation (4.9)). The perfect alignment is somewhat tedious to achieve. However, with the help of a precision piezo-controlled three-axis stage to move the second objective, a count rate within 80% of the primary channel can be easily achieved. For details of the construction, please refer to [16] and [17]. While the sample is excited usually from a single side, it is also possible to excite from both sides. Due to interference, this can provide the added advantage of reducing the probe volume [18].

Another method which can get rid of not only after-pulsing, but also the scattering background, is FLCS or fluorescence lifetime correlation spectroscopy. This method combines time correlated single-photon counting (TCSPC) and FCS. This of course requires somewhat more specialized equipment, mainly a pulsed laser source and a high time resolution detector. Since any photon from the actual sample must come within a time period after the excitation pulse that is of the order of the lifetime of the fluorophore, any other photon detected by the detector can easily be neglected as background. After-pulsing usually occurs few 100 ns after the main photon detection event, which is rather long compared to typical fluorescence lifetimes. So the effect of after-pulsing can be eliminated in this mode (see figure 4.9).

7.6 Detector saturation

Another factor that can affect the measurements at a high rate of photon collection is the detector's own counting limits. A typical detector (APD) counts photons at a rate proportional to their flux, but only up to a limit, typically 500 kHz. The count rate then starts to saturate, though it still keeps rising typically up to 10 MHz. This is of course related to the deadtime. With a deadtime of 100 ns we can expect a count rate of up to 10 MHz. However, photons are not equally spaced in time, so there is always a probability that a photon arrives within the deadtime of a previous event. So non-linearities already are noticeable at much lower count rates of ~1 MHz. This sublinear behaviour implies that the probability of detecting each photon at high count rate periods of the experiment is less than the probability of detecting a photon during low count rate periods, which introduces an artefact in the ACF. The basic assumption of FCS, that the count rate fluctuations represent the number fluctuations in the probe volume, no longer holds true. This can have an effect similar to that of fluorophore saturation (see section 7.7), and make the probe volume appear flatter than it actually is, increasing the diffusion time. Note that even if the average count rate is within the limits of linearity, the instantaneous count rate can be high enough to take the detector to saturation. This is rather common if the sample contains particles some of which have low brightness, and some of which have high brightness. When a high brightness particle passes through the probe volume, the detector may get partly saturated, giving a brightness value that is less than the actual. So a user should not only make sure that the average count rate remains in the linear zone, he/she should follow the count rate trace to probe whether there are such transient saturation events. The latter would be important, for example, when one is measuring particle aggregation, where a particular aggregate may be much larger (and much brighter) than the average size. Of course, lowering the excitation laser power should take care of this artefact.

7.7 Fluorophore saturation

How much power should one use for exciting the sample, as long as we do not saturate the detector? The answer lies in the absorption spectrum of the fluorophore. As already mentioned, the figure of merit for the signal-to-noise ratio in an FCS experiment is the number of photons detected per unit time per molecule (*CPS*). So increasing excitation probability should always increase the signal-to-noise (up to a limit, see math box 3.2), all other factors being constant. However, we typically see that *CPS* shows a plateau and then an actual reduction as the power increases. How do we understand this behaviour? To understand this, we need to keep in mind several factors. The first is that the excitation probability cannot scale with power linearly (even for a one-photon excitation) forever. Since the fluorophore has a finite lifetime, a photon that arrives at the fluorophore may find it already in the excited state, and so it would not be able to cause a transition. As the probability of excitation within the fluorophore lifetime becomes a substantial fraction of unity, the probability of excitation starts to become sub-linear with power. This is called 'saturation' of the fluorophore. At what intensity or laser power does a fluorophore saturate? This is given by the molecular cross section (or equivalently the molar absorptivity) of the fluorophore at the excitation wavelength (math box 7.5).

Math box 7.5.

Saturation of fluorophores
If a fluorophore has a molar absorptivity $\varepsilon(\lambda)$ in units of 1 mol^{-1} cm^{-1}, then its molecular cross section (in cm^2) is [19]:

$$\sigma(\lambda) = 3.82 \times 10^{-21}\,\varepsilon(\lambda)$$

The probability that a molecule in the ground state, exposed to intensity $I(\lambda)$, will absorb a photon within a time interval Δt is given by:

$$P(\Delta t) = I(\lambda)\sigma(\lambda)\Delta t$$

Which is simply the probability that a photon will arrive within this cross sectional area within that time interval. Now suppose the fluorophore has a fluorescence lifetime τ. Then the probability that a second photon will arrive at the molecule when it is still in the excited state (and therefore unable to absorb the second photon) is:

$$P(\tau) = I(\lambda)\sigma(\lambda)\tau$$

If $P(\tau)$ is non-negligible, then the rate of excitation will no longer be linear to $I(\lambda)$, and will start to plateau out. This is called saturation. For a tightly focused light beam, molecules at the center of the focus will saturate much before those away from it. So the shape of the probe volume ($\Omega(r)$ in equation (3.4)) becomes flatter at the center, effectively increasing the probe volume. This would be obvious in the ACF, as the diffusion time would become longer with increasing excitation power (as shown in figure 7.8).

The second factor to consider is the profile of the excitation volume. When the fluorophores at the centre are beginning to get saturated, the fluorophores at the edge of the excitation profile are still in the linear regime. So slowly, with increasing power, the probe volume starts to distort and broaden. This gives rise to a lowering of $G(0)$ (since the volume and therefore the number of fluorescent molecules in the probe volume increase), and longer diffusion times (figure 7.8). The volume also no longer remains Gaussian, and therefore the typical equations do not fit it as well. This can give rise to artefacts in determining the true hydrodynamic size of the particles. The third factor is the photophysics and photochemistry of the fluorophore. Fluorophores can reach a triplet state when they are excited. This state is 'dark' and long lived, so they cannot emit detectable photons in the same rate as those in the normally excited S1 state (please see the Jabłoński diagram (figure 3.1)). So the ground state population of the fluorophore begins to get depleted as the population gets transferred to the triplet state. So the *CPS* decreases, and in addition, the triplet state induces microsecond timescale fluctuations in the correlation curve (see chapter 5).

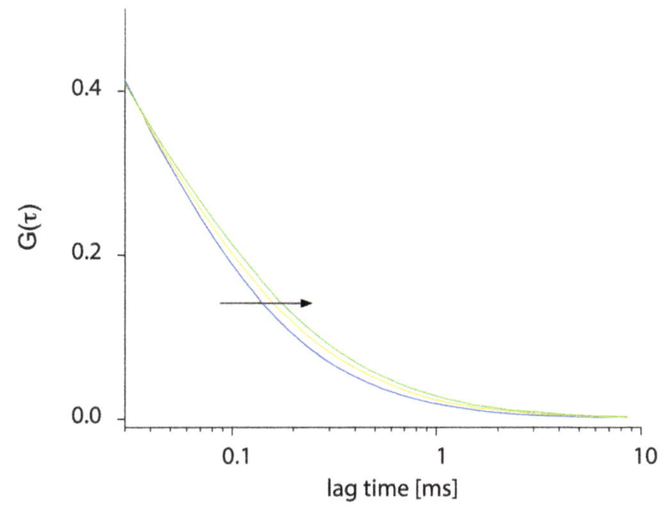

Figure 7.8. Effect of laser-induced fluorophore saturation on the ACF. Sample: rhodamine 110 in aqueous solution. The ACF is recorded as a function of power (direction of the arrow marks increasing powers) at the back aperture of the objective lens. The ACF is normalized at 15 μs, so change of $G(0)$ is not observed.

7.8 Scattering

Scattering from particles with large scattering cross sections can be a problem in FCS, as emission filters may not be able to completely block the scattering. This especially can be a problem for probing samples containing metal nanoparticles, such as for examining protein samples which are attached to gold nanoparticles. The best way to take care of this problem is to time-gate the scattering. This requires a pulsed excitation, and detectors and electronics with high time resolution. For example, if the laser source is a picosecond pulsed laser with 100 ps pulses (obtainable from picosecond diode lasers), then the scattering, being an instantaneous process, lasts also for that 100 ps only. If a detector has a high time resolution, and the processor also can bin the data with high time resolution, then one can compute the autocorrelation by ignoring time bins in the initial (post-pulse) parts of the signal. This at worst decreases the brightness of the fluorophore slightly, but it would have no bearing on the temporal characteristics of the autocorrelation function. If the amplitude was substantially affected by background (see equation (7.1)), then this method of background reduction will increase the autocorrelation amplitude, and provide a better measure of the concentration.

7.9 Autofluorescence

Autofluorescence is fluorescence that is emitted by molecules in the sample other than the fluorescent molecules under investigation. One fundamental problem of fluorescence spectroscopy is the relatively wide fluorescence and excitation spectra of typical fluorophores in the condensed phase. This implies that in a complex mixture, or in a cell, there are many molecules which have a reasonable overlap with the excitation and fluorescence spectra of the molecule(s) of interest. Therefore, any excitation and detection scheme ends up collecting some photons from these unintended targets. Some well-known autofluorescent molecules in the cell are NADH (UV fluorescent), flavins (blue, green) and lipid peroxidation products (wide visible spectrum). All of these will contribute to the background. Fortunately, in most cases, the extent of signal from these molecules will be low compared to what is collected from the actual fluorescent probes. Even then, the background autofluorescence photons will contribute to a lowering of the $G(0)$ in the most common case, and a possible distortion of the autocorrelation function in more severe cases.

How do we tackle autofluorescence? Well it is helpful to have an idea of the fluorescence spectrum of the photons originating from the sample, and to compare this with the spectrum of the intended molecule. This will provide the optimum choices for the emission filters and also in some cases, of the laser. This approach is refined in the so-called filtered FCS techniques (section 4.4). If the autofluorescence is not spectrally separable, lifetime may provide another filter to tackle it. In experiments with pulsed lasers and high time resolution detectors, it is possible to discard photons which arrive too early or too late compared to the known fluorophores. This time filtering is described in the section on fluorescence lifetime correlation spectroscopy (section 4.4).

7.10 Sample topology

FCS assumes a homogeneous solution. Sometimes there are three-dimensional obstacles (i.e. impenetrable structures) occupying some parts of the volume, making the effective volume small. However, their effects on the shape of the curve may largely average out, if the void volume (i.e. volume available for the solution) is connected and reasonably isotropic in the scale of the probe volume. The actual parameters, such as the apparent concentration (as obtained from G_0) or the value of the diffusion constant, may vary with the relative size and nature of the occupied volume, and therefore can change when the probe volume changes. If the particles are confined, i.e. their movement involves crossing of barriers, then their diffusion may become anomalous (see section 5.4). However, useful information about the nature of the barrier may still be obtainable, as has been shown in two-dimensional systems. This is described in another section of this book where we discuss the 'FCS diffusion laws' (section 5.7.8).

If the size of the particle becomes large and comparable to the size of the volume, then another type of artefact occurs [1]. The fluorescence arising from different parts of the particle now do not have uniform brightness, as they are located in different parts of the probe volume. So the assumption made during the derivation of the FCS fitting function (where we assumed that the brightness of a particle is given by the position of its centre of mass in the probe volume, math box 5.3) no longer holds true. This can be theoretically tackled if we know the brightness of the different parts of the particle when the centre of mass is located at a given point. This is easier to do for a solid object (such as a bead), but harder for a flexible chain-like object (such as a long piece of DNA, labelled with multiple dyes along its length). For solid objects of defined shapes, Starchev *et al* have provided some solutions for a 2D Gaussian beam [1], showing that the diffusion time increases with particle radius R (in addition to the change of the diffusion constant with R) as the apparent observation volume is a convolution of the actual observation volume with the particle volume.

$$\tau_D^{app} = \frac{w_0^2 + R^2}{4D} \tag{7.3}$$

Mueller *et al* have provided analytical equations for solid objects of different shapes (ring, disk, sphere and hollow sphere) for a 3D Gaussian case [2]. It is interesting to note that the FCS curve may still fit reasonably well to the standard fitting function (equation (5.3)), but the estimation obtained for the concentration and size can be completely wrong, unless the appropriate correction factors are put in.

Another type of topological problem arises when the diffusion occurs on a 2D surface, which is not flat. This situation arises often when diffusion is probed on a cell membrane. A membrane protein may remain confined to the two-dimensional membrane, but the membrane itself may have folds or tilts in the third dimension. In such cases, the effective area of the membrane that intersects with the excitation light beam can be larger than the area of the focused beam waist. Moreover, the shape of the illuminated cross section of the two-dimensional surface may not remain circular, and the membrane area may be irregularly illuminated. This may require a second diffusing component for fitting the data, falsely implying the presence of a second species [20]. Repeats of the autocorrelation measurement at different places of the sample may yield quite different results. In such situations it is best to perform the FCS measurements at different areas of the membrane and average them. In some areas, the membrane may fold out of the probe volume, which will be apparent from the intensity map of the region. Attempts should be made to record FCS data from those regions of the membrane which are uniformly bright, indicating that the membrane is reasonably flat within the probe volume area.

7.11 Immobile particles

Of course, FCS measures diffusion, and if there is no diffusion, there is no fluctuation from particles moving in and out of the volume. So if a fluorescent particle becomes immobilized within the probe volume, it does not participate in the diffusive fluctuations. One might assume that therefore it will only contribute to the DC background, and may not matter. However, in reality, since it is also a fluorescence source, it does have inherent fluctuations, at least due to the shot noise. This then contributes to the background, and reduces the amplitude of the autocorrelation function. Sometimes such immobilization happens due to aggregation. If this is a large and proportionately bright aggregate, then it may have quite major effects. Slow bleaching of such immobile particles can introduce a sloping baseline in the count rate, thereby affecting the shape and the fit or the correlation curve. If the immovable particle is an aggregate, and can exchange the smaller species with the solution, then another possible complication arises. As has been noted, if there are immobile objects present, and the fluorescent particles can bind and unbind from that object, the ACF can become complicated, and may require very long averaging to report a reasonably accurate $G(0)$ [21].

7.12 Refractive index mismatch

A high numerical aperture microscope objective is a multi-element lens that is exquisitely designed to reduce spherical aberration during imaging. However, the design assumes a certain set of parameters for the light path between the lens and the object. These are typically the immersion medium, the cover glass and the solvent of the sample, respectively. The lens design typically assumes that the refractive index of the immersion medium (which is indicated on the objective) and the sample are the same, and the cover glass has a given refractive index (typically 1.51) and a specific thickness. Under these conditions, an objective can produce near diffraction limited probe volume. However, deviation from these assumptions arises when the refractive index of the solvent and that of the immersion medium are different. The two major types of immersion medium that are used in high NA microscopy are water (refractive index 1.33) and oil (refractive index = 1.51). If an FCS measurement is made with one type of objective, while the object is in a different type of medium (e.g. using oil objective to perform FCS of aqueous solutions), major spherical aberrations will occur, making the probe volume larger and non-Gaussian [22]. Moreover, the degree of this aberration is dependent on the thickness of the 'wrong' medium, and it gets progressively worse as the thickness increases. So if an oil immersion objective lens is used for an aqueous solution, noticeable change will typically happen at depths >10 μm, and it can be a major effect beyond 50 μm. Similar errors can arise from using a cover-glass with non-standard thickness. Most objectives are corrected for a cover glass thickness of 170 μm (which is typically called a cover glass #1.5). However, cover glass of a different thickness may be inadvertently used in an experiment, and even thickness of a cover glass marked #1.5 can actually vary considerably. Also, many samples, such as a living cell, may have an effective refractive index which is in between that of water and oil (the refractive index of the cytoplasm of a living cell is typically 1.36–1.39, which is closer to that of water than of oil [23]). Also, sometimes an FCS experiment is carried out where parameters are compared between samples placed in different solvents with different refractive indices. For example, the concentration of the denaturant can vary in a protein folding experiment (the refractive index of the medium can vary with that), and a size comparison between different denaturant concentrations may be erroneous. A reliable approach in such *in vitro* measurements is to measure the diffusion constant relative to a standard dye, at a position in the sample which is close to the cover glass surface. Fortunately, many objective lenses come with an adjustable collar, which can compensate for these aberrations to a reasonable extent. Whenever possible, one should adjust this collar to obtain the smallest diffusion time for a given sample, which would also help improving the signal-to-noise ratio.

7.13 Exercises

1. Suppose you are performing an FCS experiment with relatively weak fluorophore. The area under the Raman peak is 5% of the area under the fluorescence peak. If you chose emission filters such that the Raman Peak is included in your detection signal, how much error would you have made in the estimation of the real concentration?

2. Suppose you are studying a fluorescently labelled dimeric protein. With time, the protein slowly dissociates. Initially, at the power you were performing the FCS, there were some signs of photobleaching. With time, what changes do you expect to see in the ACFs?

3. If the molar absorptivity of your dye is 100 000 l mol^{-1} cm^{-1}, and its lifetime is 2.62 ns. At what laser power of your 543 nm laser (which is focused through an objective of 1.2 NA) would you expected to start seeing substantial saturation?

References

[1] Starchev K, Zhang J and Buffle J 1998 Applications of fluorescence correlation spectroscopy —particle size effect *J. Colloid Interface Sci.* **203** 189–96

[2] Müller C B, Loman A, Richtering W and Enderlein J 2008 Dual-focus fluorescence correlation spectroscopy of colloidal solutions: influence of particle size *J. Phys. Chem.* B **112** 8236–40

[3] Ries J, Bayer M, Csúcs G, Dirkx R, Solimena M, Ewers H and Schwille P 2010 Automated suppression of sample-related artifacts in fluorescence correlation spectroscopy *Opt. Express* **18** 11073–82

[4] Garai K, Sahoo B, Kaushalya S K, Desai R and Maiti S 2007 Zinc lowers amyloid-beta toxicity by selectively precipitating aggregation intermediates *Biochemistry* **46** 10655–63

[5] Mertz J 1998 Molecular photodynamics involved in multi-photon excitation fluorescence microscopy *Eur. Phys. J.* D **3** 53–66

[6] Balaji J, Sengupta P and Maiti S 2001 *Probing diffusion and photochemical properties through localized photobleaching* ed A Periasamy and P T C So (Bellingham, WA: SPIE) pp 329–8

[7] Qian H and Elson E L 2010 Chemical fluxes in cellular steady states measured by fluorescence-correlation spectroscopy *Single Molecule Spectroscopy in Chemisrty, Physics and Biology* ed A Graeslund, R Rigler and J Widengren (Berlin: Springer) pp 119–37

[8] Lange J J, Wood C J, Marshall W A, Maddera L E, Yu Q E, Bradford W D, Slaughter B D, Unruh J R and Wiegraebe W 2013 Correction of bleaching artifacts in high content fluorescence correlation spectroscopy (HCS-FCS) data ed J Enderlein, I Gregor, Z K Gryczynski, R Erdmann and F Koberling (Bellingham, WA: SPIE) pp 859006–15

[9] Capoulade J, Wachsmuth M, Hufnagel L and Knop M 2011 Quantitative fluorescence imaging of protein diffusion and interaction in living cells *Nat. Biotech.* **29** 835–9

[10] Macháň R, Foo Y H and Wohland T 2016 On the equivalence of FCS and FRAP: simultaneous lipid membrane measurements *Biophys J.* **111** 152–61

[11] Mertz J, Xu C and Webb W W 1995 Single-molecule detection by two-photon-excited fluorescence *Opt. Lett.* **20** 2532

[12] Ries J, Petrášek Z, García-Sáez A J and Schwille P 2010 A comprehensive framework for fluorescence cross-correlation spectroscopy *New J. Phys.* **12** 113009
[13] Magde D, Elson E and Webb W 1972 Thermodynamic fluctuations in a reacting system—measurement by fluorescence correlation spectroscopy *Phys. Rev. Lett.* **29** 705–8
[14] Zamir E, Frey C, Weiss M, Antona S, Frohnmayer J P, Janiesch J-W, Platzman I and Spatz J P 2017 Reconceptualizing fluorescence correlation spectroscopy for monitoring and analyzing periodically passing objects *Anal. Chem.* **89** 11672–8
[15] Ziarkash A W, Joshi S K, Stipčević M and Ursin R 2018 Comparative study of afterpulsing behavior and models in single photon counting avalanche photo diode detectors *Sci. Rep.* **8** 5076
[16] Abhyankar R *et al* 2012 *Amyloid Diagnostics: Probing Protein Aggregation and Conformation with Ultrasensitive Fluorescence Detection* ed S Achilefu and R Raghavachari (Bellingham, WA: SPIE) pp 82330B–38B
[17] Maity B K, Vishvakarma V, Surendran D, Rawat A, Das A, Pramanik S, Arfin N and Maiti S 2018 Spontaneous fluctuations can guide drug design strategies for structurally disordered proteins *Biochemistry* **57** 4206–13
[18] Hell S and Stelzer E 1992 Properties of a 4Pi confocal fluorescence microscope *J. Opt. Soc. Am.* A **9** 2159–66
[19] Parson W W 2015 *Modern Optical Spectroscopy* 2nd edn (Berlin: Springer)
[20] Milon S, Hovius R, Vogel H and Wohland T 2003 Factors influencing fluorescence correlation spectroscopy measurements on membranes: simulations and experiments *Chem. Phys.* **288** 171–86
[21] Ipiña E P and Dawson S P 2014 How long should a system be observed to obtain reliable concentration estimates from the measurement of fluctuations? *Biophys J.* **107** 2674–83
[22] Hell S, Reiner G, Cremer C and Stelzer E H K 1993 Aberrations in confocal fluorescence microscopy induced by mismatches in refractive index *J. Microsc.* **169** 391–405
[23] Choi W, Fang-Yen C, Badizadegan K, Oh S, Lue N, Dasari R R and Feld M S 2007 Tomographic phase microscopy *Nat. Meth.* **4** 717–9

IOP Publishing

An Introduction to Fluorescence Correlation Spectroscopy

Thorsten Wohland, Sudipta Maiti and Radek Macháň

Chapter 8

Data fitting

Data fitting plays a major role in FCS and for this purpose one requires accurate knowledge of the properties of FCS measurements. Therefore, the chapter starts first with a discussion of the FCS data structure, its inherent bias and standard deviation, and how to estimate it. This is followed by a description of non-linear and general least-squares fitting methods, and how they have been implemented in FCS, including the use of the FCS standard deviation. Next comes a discussion of some strategies for improving data fitting, including global fits and the maximum entropy method. With this knowledge the reader is ready to address one of the most difficult issues in FCS data fitting, namely the selection of the right fit model, which can be achieved by statistical tests or by Bayesian model selection. Finally, the chapter ends with some discussion on practical aspects of FCS data analysis.

8.1 Introduction

This chapter is not aiming at an introduction to data fitting in general for which many good textbooks have already been written (e.g. [1, 2]). The goal of this chapter is rather to show which data fitting strategies have successfully been employed for FCS and point out advantages and disadvantages of the different methods.

The extraction of information of a series of data points relies on testing whether the series follows any pattern describable by theoretically understood model functions. The theoretical model will depend on a number of undetermined parameters. By adjusting the parameters one can simultaneously minimize the difference between the model function and all the data points. This fitting of experimental data by theoretical model functions is dependent on various conditions. First, one needs to define how minimizing the difference between data and model function is to be interpreted. We will discuss this very shortly at the beginning of the chapter but will in the end stick to the most common criteria of least squares minimization. Second, the data has inherent errors and bias hindering any attempt at performing the perfect fit to the data. And the errors often are not the same for

doi:10.1088/978-0-7503-2080-1ch8

each data point. So a clear understanding of the errors of the experimental data is important if one wants to be able to perform fits that weigh the influence of data points according to their errors and that determine what is the least complex model that can be used to fit the data without creating artefacts. Third, the model function that is to be fitted is often not known as the measurement is actually conducted to determine what processes are present in a particular system. Therefore, we need a way to decide what the best model is. These topics will be discussed in the sections on quality of fits and on model selection.

8.2 What do we minimize?

In the following we will concentrate on temporal correlations. However, similar treatments would also be valid for spatial correlations. In general any data point (x, y) can have an error associated with both values, x and y. Therefore, a measurement value has in principle two errors associated, one for the x and one for the y coordinate (figure 8.1(a) [3]). However, in temporal correlations, the data exists of a pair of values $(\tau, G(\tau))$, with τ representing the lag time. It so happens that times, and frequencies, are among the parameters we can measure best, with little uncertainty. Therefore, errors in lag time are negligible and are typically not taken into account in FCS. In this latter case we have to account only for the errors in the measured correlation function $G(\tau)$ (figure 8.1(b)).

8.3 The data structure and bias in FCS

In the following we will first introduce the data structure of experimentally determined ACFs and then explain why the autocorrelation functions are biased estimators.

8.3.1 The data structure in FCS

As introduced in chapter 4, FCS data is typically calculated on a semi-logarithmic time scale with symmetric normalization [4, 5]. The bin time for a particular channel in the correlator is given by $\Delta\tau_i$, and the actual measurement time is a multiple of this bin time, $k\Delta\tau_i$. Denoting the number of photons counted in a particular time interval, $\Delta\tau_i$, as $F(k\Delta\tau_i)$, and the mean photon counts as $\langle F \rangle$, we can write the signal fluctuations $\delta F(k\Delta\tau_i)$, as

$$\delta F(k\Delta\tau_i) = F(k\Delta\tau_i) - \langle F \rangle \tag{8.1}$$

Using the fluctuations, we can then write the ACF for the fluctuations as

$$G(m\Delta\tau_i) = \frac{\frac{1}{M-m}\sum_{k=1}^{M-m} \delta F(k\Delta\tau_i)\delta F((k+m)\Delta\tau_i)}{M_{dir,i}M_{del,i}} \tag{8.2}$$

Here M is the total number of measurement points for a particular bin time $\Delta\tau_i$ and a lagtime $m\Delta\tau_i$. Due to the lagtime only $M - m$ points actually contribute to the

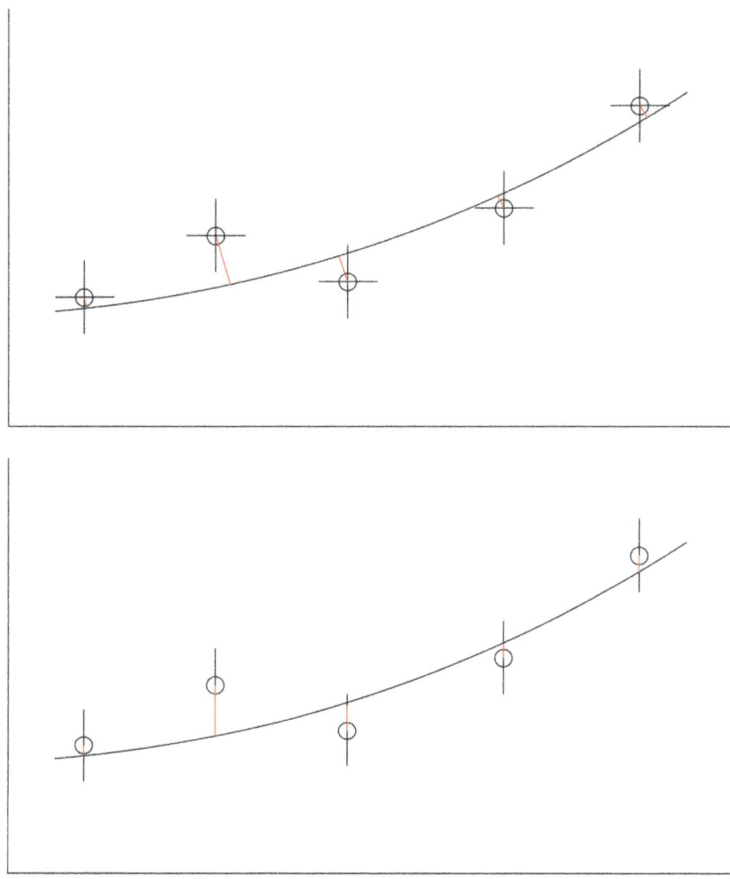

Figure 8.1. (a) If data points (open circles) have errors associated with both parameters along the *x*- and *y*-axes, then for minimization one can search for the length of the shortest line (red line) that connects the data to the fit function (black line). Minimizing these distances will determine the best fit. (b) If the uncertainty in one of the parameters can be neglected, as is typically the case for time in temporal correlations, then the distances to be minimized are just the vertical distances of the data points to the fit function. For temporal correlations this means we need to minimize the distances of the experimentally measured correlation function from the theoretical fit model for a set of parameters.

calculation for $G(m\Delta\tau_i)$. The so-called direct (M_{dir}) and delayed (M_{del}) monitors for each channel i are given by

$$M_{dir,i} = \frac{1}{M - m} \sum_{k=1}^{M-m} F(k\Delta\tau_i) \tag{8.3a}$$

and

$$M_{del,\,i} = \frac{1}{M - m} \sum_{k=m}^{M} F(k\Delta\tau_i) \tag{8.3b}$$

These monitors ensure that the intensity is only averaged over those data points that are actually used in the correlations. Because of the lag time $m\Delta\tau_i$ between the channels that are correlated these are slightly different. Schätzel has introduced this symmetric normalization to avoid bias in the correlation functions [6].

The theoretical model of the correlation function for free 3D diffusion of a single component is given by

$$g(m\Delta\tau_i) = \left(1 + \frac{m\Delta\tau_i}{\tau_D}\right)^{-1}\left(1 + \frac{m\Delta\tau_i}{K^2\tau_D}\right)^{-1/2} \tag{8.4}$$

8.3.2 The bias of correlation functions

Until now we assumed that a measured correlation function is an unbiased estimator. However, correlation functions delivered by hardware or software correlators, as we will show here, are a biased estimator. This means that fits of correlation functions that had different measurement lengths do not provide the same parameters but the fitted parameters converge to the real values with increasing measurement length. Let us write the actual ACF which we will denote here $\overline{G}(\tau)$ by

$$\overline{G}(m\Delta\tau_i) = \frac{\langle F(k\Delta\tau_i)F((k+m)\Delta\tau_i)\rangle}{\langle F\rangle^2} \tag{8.5}$$

The estimate is accurate for infinitely long measurements over infinitely many photons. In an experiment the measurement time is always finite and, therefore, the experimentally obtained ACF, $G(\tau)$, is just an estimator of $\overline{G}(\tau)$, just like a sample average is an estimator of the population average. The values of $G(\tau)$ can be also viewed as random variables; repeated FCS experiments on the same system would yield different values of $G(\tau)$ in the same way as the values of the sample average differ for different samples taken from the population. For further analysis of the experimentally obtained correlation functions (such as fitting with theoretical models as derived in chapter 5) it is, therefore, desirable to characterise the accuracy with which they estimate the actual correlation functions. For that purpose, we will try in this section to determine the variance or standard deviation (which is just the square root of the variance) of $G(\tau)$ for each value of τ.

To study the properties of the estimator $G(\tau)$ it will be useful to write the fluorescence intensity $F(k\Delta\tau_i)$ as the sum of the average fluorescence intensity and the intensity fluctuations $\delta F(ik\Delta\tau_i)$ (see equation (8.1)):

$$F(k\Delta\tau_i) = \langle F\rangle + \delta F(k\Delta\tau_i) \tag{8.6}$$

The formula for the symmetrically normalized ACF (equation (8.2)) then assumes the form:

$$G(m\Delta\tau_i) = \frac{\langle F\rangle^2 + \frac{1}{M-m}\sum_{i=1}^{M-m}(\langle F\rangle(\delta F(k\Delta\tau_i) + \delta F((k+m)\Delta\tau_i)) + \delta F(k\Delta\tau_i)\delta F((k+m)\Delta\tau_i))}{\left(\langle F\rangle + \frac{1}{M-m}\sum_{i=1}^{M-m}\delta F(k\Delta\tau_i)\right)\left(\langle F\rangle + \frac{1}{M-m}\sum_{i=1}^{M-m}\delta F((k+m)\Delta\tau_i)\right)} \quad (8.7)$$

Suppose now we can repeat the measurement of $G(m\Delta\tau_i)$ infinitely many times and, thus, obtain its mean value $\langle G(m\Delta\tau_i)\rangle$. If we expand the denominator in equation (8.7) and perform the averaging (see math box 8.1), we get the following formula:

$$\begin{aligned}\langle G(m\Delta\tau_i)\rangle = 1 &+ \frac{\langle \delta F(k\Delta\tau_i)\delta F((k+m)\Delta\tau_i)\rangle}{\langle F\rangle^2} \\ &- \frac{\langle \sum_{i,j=1}^{M-m}\delta F(k\Delta\tau_i)\delta F((k+m)\Delta\tau_i)\rangle}{(M-m)^2\langle F\rangle^2} + O(M^{-3})\end{aligned} \quad (8.8)$$

We can compare this result with the ideal ACF $\overline{G}(m\Delta\tau_i)$ by inserting equation (8.6) into (8.5):

$$\begin{aligned}\overline{G}(m\Delta\tau_i) &= \frac{\langle\langle F\rangle^2 + \langle F\rangle(\delta F(k\Delta\tau_i) + \delta F((k+m)\Delta\tau_i)) + \delta F(k\Delta\tau_i)\delta F((k+m)\Delta\tau_i)\rangle}{\langle F\rangle^2} \\ &= 1 + \frac{\langle \delta F(k\Delta\tau_i)\delta F((k+m)\Delta\tau_i)\rangle}{\langle F\rangle^2}\end{aligned} \quad (8.9)$$

We have used here the fact that the mean of the intensity fluctuations around δF is zero. Comparing equations (8.8) and (8.9) shows that $G(m\Delta\tau_i)$ is indeed a biased estimator of $\overline{G}(m\Delta\tau_i)$ with a bias given by the third term in equation (8.8) of the order of $1/M$.

Let us now have a look at the variance of the repeated measurements of $G(m\Delta\tau)$, which is defined as:

$$\mathrm{var}(G(m\Delta\tau_i)) = \langle G^2(m\Delta\tau_i)\rangle - \langle G(m\Delta\tau_i)\rangle^2 \quad (8.10)$$

By inserting equations (8.7) and (8.8) into (8.10), we get:

$$\begin{aligned}\mathrm{var}(G(m\Delta\tau_i)) = &\frac{1}{(M-m)^2\langle F\rangle^4} \\ &\Bigg(\langle\sum_{i=1}^{M-m}\delta F(k\Delta\tau_i)\delta F((k+m)\Delta\tau_i) \\ &\sum_{j=1}^{M-m}\delta F(k\Delta\tau_i)\delta F((k+m)\Delta\tau_i)\rangle \\ &- \langle\sum_{i=1}^{M-m}\delta F(k\Delta\tau_i)\delta F((k+m)\Delta\tau_i)\rangle^2\Bigg) + O(M^{-3})\end{aligned} \quad (8.11)$$

To evaluate equation (8.11) we would need to evaluate the ensemble averages of fluorescence fluctuation products. If a theoretical model describing the fluorescence fluctuations is known, the ensemble averages can be calculated using theoretical

Math box 8.1

Bias of the ACF
From equation (8.7) we have

$$G(m\Delta\tau_i) = \frac{\langle F\rangle^2 + \frac{1}{M-m}\sum_{i=1}^{M-m}(\langle F\rangle(\delta F(k\Delta\tau_i) + \delta F((k+m)\Delta\tau_i)) + \delta F(k\Delta\tau_i)\delta F((k+m)\Delta\tau_i))}{\left(\langle F\rangle + \frac{1}{M-m}\sum_{i=1}^{M-m}\delta F(k\Delta\tau_i)\right)\left(\langle F\rangle + \frac{1}{M-m}\sum_{i=1}^{M-m}\delta F((k+m)\Delta\tau_i)\right)}$$

Expanding the numerator and denominator gives

$$G(m\Delta\tau_i) = \frac{\langle F\rangle^2 + \frac{\langle F\rangle}{M-m}\sum_{i=1}^{M-m}(\delta F(k\Delta\tau_i) + \delta F((k+m)\Delta\tau_i)) + \frac{1}{M-m}\sum_{i=1}^{M-m}\delta F(k\Delta\tau_i)\delta F((k+m)\Delta\tau_i)}{\langle F\rangle^2 + \frac{\langle F\rangle}{M-m}\sum_{i=1}^{M-m}(\delta F(k\Delta\tau_i) + \delta F((k+m)\Delta\tau_i)) + \frac{1}{(M-m)^2}\sum_{i=1}^{M-m}\delta F(k\Delta\tau_i)\sum_{i=1}^{M-m}\delta F((k+m)\Delta\tau_i)}$$

We now add a '0' to the numerator:

$$\frac{1}{(M-m)^2}\sum_{i=1}^{M-m}\delta F(k\Delta\tau_i)\sum_{i=1}^{M-m}\delta F((k+m)\Delta\tau_i) - \frac{1}{(M-m)^2}\sum_{i=1}^{M-m}\delta F(k\Delta\tau_i)\sum_{i=1}^{M-m}\delta F((k+m)\Delta\tau_i) = 0$$

Then the equation becomes:

$$G(m\Delta\tau_i) = 1 + \frac{\frac{1}{M-m}\sum_{i=1}^{M-m}\delta F(k\Delta\tau_i)\delta F((k+m)\Delta\tau_i) - \frac{1}{(M-m)^2}\sum_{i=1}^{M-m}\delta F(k\Delta\tau_i)\sum_{i=1}^{M-m}\delta F((k+m)\Delta\tau_i)}{\langle F\rangle^2 + \frac{\langle F\rangle}{M-m}\sum_{i=1}^{M-m}(\delta F(k\Delta\tau_i) + \delta F((k+m)\Delta\tau_i)) + \frac{1}{(M-m)^2}\sum_{i=1}^{M-m}\delta F(k\Delta\tau_i)\sum_{i=1}^{M-m}\delta F((k+m)\Delta\tau_i)} = 1$$

$$+ \frac{\frac{1}{\langle F\rangle^2(M-m)}\sum_{i=1}^{M-m}\delta F(k\Delta\tau_i)\delta F((k+m)\Delta\tau_i) - \frac{1}{\langle F\rangle^2(M-m)^2}\sum_{i=1}^{M-m}\delta F(k\Delta\tau_i)\sum_{i=1}^{M-m}\delta F((k+m)\Delta\tau_i)}{\left(1 + \frac{1}{\langle F\rangle(M-m)}\sum_{i=1}^{M-m}(\delta F(k\Delta\tau_i) + \delta F((k+m)\Delta\tau_i)) + \frac{1}{\langle F\rangle^2(M-m)^2}\sum_{i=1}^{M-m}\delta F(k\Delta\tau_i)\sum_{i=1}^{M-m}\delta F((k+m)\Delta\tau_i)\right)}$$

In the last equation we divided numerator and denominator by $\langle F\rangle^2$. This provides us the denominator in the form $1 + x$, which we now expand and develop it in a Taylor series according to

$$\frac{1}{1+x} = 1 - x + x^2 - x^3 + \ldots$$

Now, we can write the correlation function as

$$G(m\Delta\tau_i) \approx 1 + \left[\frac{\sum_{i=1}^{M-m}\delta F(k\Delta\tau_i)\delta F((k+m)\Delta\tau_i)}{\langle F\rangle^2(M-m)} - \frac{\sum_{i=1}^{M-m}\delta F(k\Delta\tau_i)\sum_{i=1}^{M-m}\delta F((k+m)\Delta\tau_i)}{\langle F\rangle^2(M-m)^2}\right]$$

$$\times\left(1 - \frac{\sum_{i=1}^{M-m}(\delta F(k\Delta\tau_i) + \delta F((k+m)\Delta\tau_i))}{\langle F\rangle(M-m)} - \frac{\sum_{i=1}^{M-m}\delta F(k\Delta\tau_i)\sum_{i=1}^{M-m}\delta F((k+m)\Delta\tau_i)}{\langle F\rangle^2(M-m)^2} + \ldots\right)$$

$$= 1 + \frac{\sum_{i=1}^{M-m}\delta F(k\Delta\tau_i)\delta F((k+m)\Delta\tau_i)}{\langle F\rangle^2(M-m)} - \frac{\sum_{i=1}^{M-m}\delta F(k\Delta\tau_i)\sum_{i=1}^{M-m}\delta F((k+m)\Delta\tau_i)}{\langle F\rangle^2(M-m)^2} + \ldots$$

If we now perform the averaging, we obtain equation (8.8):

$$\langle G(m\Delta\tau_i)\rangle = 1 + \frac{\langle\delta F(k\Delta\tau_i)\delta F((k+m)\Delta\tau_i)\rangle}{\langle F\rangle^2} - \frac{\langle\sum_{i=1}^{M-m}\delta F(k\Delta\tau_i)\sum_{i=1}^{M-m}\delta F((k+m)\Delta\tau_i)\rangle}{\langle F\rangle^2(M-m)^2} + O(M^{-3})$$

higher order correlation functions. Although this approach has been adopted by several authors, we will not follow it here because the theory of higher order correlations is rather complex and, therefore, all the derivations are typically done only for the simplest models of fluorescence fluctuations and often involve over-simplifying approximations, which strongly limit the applicability of this approach. Instead, we will discuss here Koppel's method, which depends on fit model parameters, and several model-independent methods to estimate the standard deviation $\sigma(G(m\Delta\tau_i))$ of the correlation function using only the measured data.

8.4 The standard deviation in FCS

In this section, we explain multiple ways to calculate the standard deviation for an ACF. Which of the methods to use will depend on which data one has access to. Some of the following methods can be applied directly to the measured ACFs, others require that one has access to the full intensity trace from which the ACFs are calculated.

8.4.1 Koppel's standard deviation and its modifications

A first estimate for the noise on correlation functions was derived by Koppel for non-normalized exponential ACFs in the limit of large numbers of particles, negligible background, uniform illumination, and times much faster than the correlation time [7]. This model was then adapted for the typical model correlation functions in FCS (equation (8.4)) and was normalized [8]:

$$\begin{aligned}\sigma^2_{\text{Koppel}}(m\Delta\tau_i) = {} & \frac{1}{M}\frac{1}{N^2}\left[\frac{(1+g^2(\Delta\tau_i))(1+g^2(\mathrm{m}\Delta\tau_i))}{1-g^2(\Delta\tau_i)} + 2mg^2(m\Delta\tau_i)\right] \\ & + \frac{1}{M}\left[\frac{2(1+g^2(m\Delta\tau_i))}{N\langle F\rangle} + \frac{1}{\langle F^2\rangle}(1+\frac{g(m\Delta\tau_i)}{N})\right]\end{aligned} \tag{8.12}$$

Here, $\Delta\tau_i$ is the bin width of the correlator channel, M is the total number of experimental values $M = T/\Delta\tau_i$, with T being the measurement time of the experiment. N is the average number of particles in the observation volume and g is the model function of equation (8.4).

Most of the conditions under which this noise model was derived are actually not fulfilled and so it is no surprise that Koppel's formula does not perfectly describe the standard deviation for FCS. However, it has been proven to be a good approximation for short times (figure 8.2). For long correlation times Koppel's formula overestimates the standard deviation making it not a good approximation when trying to calculate quality of fit measures. Therefore, an approximation of Koppel's formula was devised in which the standard deviation for increasing correlation times was calculated until a minimum is reached. For all larger correlation times, this minimum value is used as the standard deviation. This modified Koppel's formula provides much better estimates for the standard deviation over the full range of correlation times (figure 8.2).

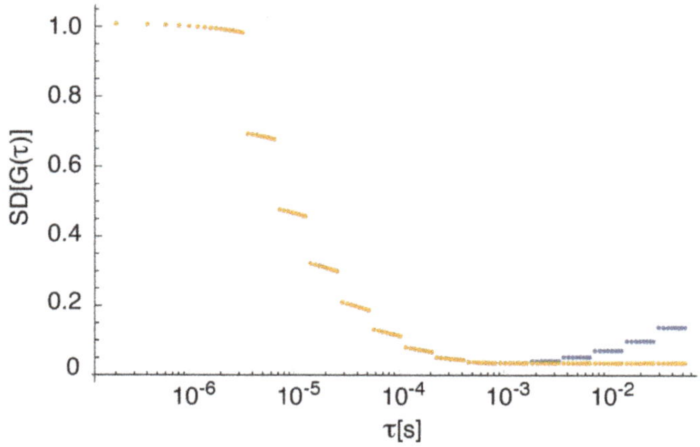

Figure 8.2. The standard deviation according to Koppel (blue) decreases initially with lag time, as would be expected because of the increase in the concomitant increase in bin time. One can verify that the drop of the SD for each set of bin times is just $\sqrt{2}$ as would be expected as the bin time doubles each group. But then it increases again, which is not consistent with measurements, in which a general decrease in the noise on the ACF with lag time is seen. The corrected formula for Koppel (yellow) is kept constant once it reaches a minimum and avoids the increase of the noise at long lag times. Note that at short times the two models do not differ and perfectly overlap.

Note that this approach already needs estimates for the fitting parameters, as they are themselves part of Koppel's formula. A solution to this problem would be a constant update of the parameters and adjustment of the noise estimate during fitting. However, we are not aware that that approach has been used to date and typically only the initial estimate of N over the amplitude and τ_D over the lag time at which the ACF has reached its half-amplitude is used (see chapter 5, equation (5.14)).

8.4.2 Standard deviations from multiple measurements

Another method to estimate the standard deviation of an ACF is to estimate it from multiple measurements. To accomplish this, we either take K repeated measurements of $G(m\Delta\tau_i)$, each of duration T, and calculate the mean and standard deviation of the ensemble of the measured values; or we take a single intensity trace of the duration T and split it into K parts, each of the duration T/K [9]. The second approach differs from the first one only in so far that it has shorter measurement times and thus is inherently less accurate. Both approaches provide adequate estimates of the standard deviation.

Simple as the idea sounds, we need to consider some basic properties of the ACFs to execute it correctly. First of all, the actual correlation value is the difference of $G(m\Delta\tau_i)$ from the accidental correlation background, which is the asymptotic value of the ACF at long correlation times G_∞. For the ACF defined by equation (8.2), the ideal value of G_∞ is 1. However, in real experiments it may differ from 1 because of slow trends in the intensity time traces giving rise to correlation persistent on time scales longer than the maximum correlation lag time (see chapter 7 on further details on such trends in the intensity traces and strategies for their correction).

Another factor to consider is the dependence of the actual correlation value $G(m\Delta\tau) - G_\infty$ on the average number of fluorescent particles in the effective observation volume. The average particle number can change in time, adding an additional source of variations in the correlation values besides the intrinsic error of the ACF calculation, which we aim to estimate. Let us, therefore, denote the normalized ACF of the kth measurement by $\widetilde{G}^k(m\Delta\tau_i)$ and define it in the following way:

$$\widetilde{G}^k(m\Delta\tau_i) = \frac{G^k(m\Delta\tau_i) - G_\infty^k}{G^k(0) - G_\infty^k} \tag{8.13}$$

The amplitudes and asymptotic values can be obtained either by averaging several points at the beginning or the end of the ACF, respectively, or from a non-weighted fit (since the standard deviations of ACFs are not known at this stage) with a suitable theoretical model (chapter 5). The mean amplitude normalized ACF and its standard deviation are then:

$$\langle \widetilde{G}(m\Delta\tau_i) \rangle = \frac{\sum_{k=1}^{K} \widetilde{G}^k(m\Delta\tau_i)}{K} \tag{8.14a}$$

$$\sigma(\langle \widetilde{G}(m\Delta\tau_i) \rangle) = \sqrt{\frac{\sum_{k=1}^{K} (\widetilde{G}^k(m\Delta\tau_i) - \langle \widetilde{G}(m\Delta\tau_i) \rangle)^2}{(K-1)K}} \tag{8.14b}$$

In the formula above, the sum of the squared deviations is normalized by $K-1$ instead of by K, the number of the terms in the sum. This is the so called Bessel's correction for the bias in the standard deviation determined from a sample, which systematically underestimates the population standard deviation. The smaller the size of the sample, the larger the bias. In our case, we do not expect K to exceed the order of 10, which requires a correction for the bias.

Since $\langle \widetilde{G}(m\Delta\tau_i) \rangle$ is the mean of $\widetilde{G}^k(m\Delta\tau_i)$, its standard deviation is calculated as the standard error of the mean of $\widetilde{G}^k(m\Delta\tau_i)$; that is the reason for the $\sqrt{K}$ in the denominator of the formula above. The uncertainty with which $\widetilde{G}^k(m\Delta\tau_i)$ is estimated decreases with the increasing number K of the averaged ACFs. Note also that the standard deviation calculated according to the formula above belongs to the mean amplitude normalized ACF $\langle \widetilde{G}(m\Delta\tau_i) \rangle$; to obtain the standard deviation for the mean ACF without the normalization $\langle G(m\Delta\tau_i) \rangle$, it would have to be rescaled to the corresponding amplitude.

Another important factor to bear in mind when considering estimating the ACF standard deviation by this approach is the fact that $\widetilde{G}^k(m\Delta\tau_i)$ is a biased estimator of the actual ACF with a bias proportional to $1/M$ (see section 8.3.2). This implies that the mean of K ACFs calculated from time traces of T/K duration is not equivalent to a single ACF based on a K-times longer intensity trace recorded for time T. This is quite limiting especially for camera-based FCS modalities, where the numbers of time points in intensity traces are rarely excessive as a result of the limited temporal resolution of the data acquisition. In such cases the whole intensity traces are needed to calculate ACFs with acceptably low bias. At the same time, extending the overall measurement time T is often impossible in biological samples because they undergo photo-bleaching and other changes on time scales comparable to the durations of FCS recordings.

8.4.3 Standard deviation derived from the intensity trace

Originally, most FCS measurements were performed with hardware correlators that saved only the final ACF, reducing the data that needs to be stored considerably. However, with faster computers and cheaper storage in many systems nowadays the full intensity trace is saved and the correlations are calculated by software correlators. With access to the full data not only the ACF but also the standard deviation at each point of the ACF can be directly calculated from the intensity traces. Each point in the ACF is an average value of $M - m$ products (equation (8.2)) and we therefore calculate the standard error of the mean for each ACF point. Following the definition of the standard error of the mean

$$\sigma = \frac{\sqrt{\langle x^2 \rangle - \langle x \rangle^2}}{\sqrt{n}} \tag{8.15}$$

we can write for the standard deviation from the intensity trace

$$\sigma_{it}(m\Delta\tau_i) = \frac{\left[\frac{\sum_{k=1}^{M-m} \delta F^2(k\Delta\tau_i)\delta F^2((k+m)\Delta\tau_i)}{M-m} - \frac{\left(\sum_{k=1}^{M-m} \delta F(k\Delta\tau_i)\delta F((k+m)\Delta\tau_i) \right)^2}{(M-m)^2} \right]^{1/2}}{\sqrt{M-m} \cdot M_{del,\,i} \cdot M_{dir,i}} \tag{8.16}$$

This formula again is an approximation as it is valid only if the products $F(k\Delta\tau_i)F((k+m)\Delta\tau_i)$ are independent, which as we will see is not the case. But it can be used as a first approximation and provides values close to 1 for χ^2_μ, providing an improved standard deviation estimate compared to Koppel's formulas [9].

8.4.4 Standard deviation and bias within the ACF

The experimental ACF converges to the expected ACF but is a biased estimator [6, 10]. Until now we have silently assumed that we always have sufficient statistics so that the experimental ACF is sufficiently close to the theoretical expected one. However, for short measurement times or low molecular brightness the ACF is biased. Saffarian and Elson calculated this bias and also derived an equation as an approximation for the ACF standard deviation. Before we can take a look at those we need to introduce a number of concepts [10]. First we introduce two abbreviations that will simplify the equations. We denote the ratio T of the time bin to the diffusion time as

$$T = \frac{\Delta\tau}{\tau_d} \tag{8.17}$$

Second, we will denote the inverse of the square of the structure factor as

$$\beta = \frac{1}{K^2} \tag{8.18}$$

With these definitions we can now define the number of photons detected from a single particle during the bin time $\Delta\tau$ as the apparent photon yield q_{app}. This photon yield will depend on the diffusion time τ_d of the particle, the excitation intensity I_0, the factor κ describing the absorption and quantum yield of the particle as well as the detection efficiency of the instrument (see chapter 5). Palo *et al* [11] showed that q_{app} can be expressed as

$$q_{app} = \frac{4\kappa I_0 \Delta\tau}{T\beta(1-\beta)^{0.5}}\left[\beta(1+T)\tanh^{-1}\left(\frac{(1-\beta)^{1/2}((1+\beta T)^{1/2}-1)}{\beta+(1+\beta T)^{1/2}-1}\right) - (1-\beta)^{1/2}((1+\beta T)^{1/2}-1)\right] \tag{8.19}$$

Furthermore, we will define the parameter m_{app}, which describes the number of particles that contribute to the signal for a particular bin time, $\Delta\tau$. For $\Delta\tau > \tau_d$, that is when the bin time used in the correlator is larger than the characteristic diffusion time, particles can enter and leave the observation volume during the bin time and thus more particles contribute to the bin than are on average in the observation volume.

$$m_{app} = \frac{cV_{\text{eff}}}{4\Delta\tau}T\beta(1-\beta)^{1/2} \left[\beta(1+T)\tanh^{-1}\left(\frac{(1-\beta)^{1/2}((1+\beta T)^{1/2}-1)}{\beta+(1+\beta T)^{1/2}-1}\right) - (1-\beta)^{1/2}((1+\beta T)^{1/2}-1)\right]^{-1} \tag{8.20}$$

Based on these parameters, Saffarian and Elson derived the standard deviation, σ_{ACF}, for FCS as

$$\sigma_{\text{ACF}} = \left[\frac{1}{(M-m)\langle n\rangle^4}\left((\langle n\rangle(1+\gamma q_{App}))^2 + 2m_{App}^2 q_{App}^4 \gamma^4 \frac{(\beta^{-1}-1)T^{-1} + (1+T^{-1})ln\left[\frac{(1+T^{-1})(M+T^{-1}\beta^{-1})}{(1+T^{-1}\beta^{-1})(M+T^{-1})}\right]}{\beta(T+1)(\beta^{-1}-1)^2}\right)\right]^{1/2} \tag{8.21}$$

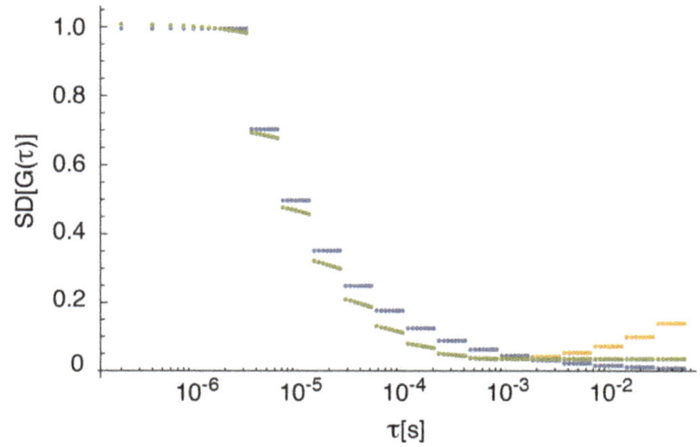

Figure 8.3. Comparison for the standard deviation of FCS for the formula derived by Koppel (yellow; modified Koppel green) and by Saffarian and Elson (blue). It can be seen that in the intermediate range the noise calculated by Saffarian and Elson is higher than Koppel's estimate as they include correlated fluctuation noise. In addition, at long times they continue the trend of a decrease of the noise by a factor $\sqrt{2}$ for each doubling of the bin time as would be expected.

Here, $\gamma = 1/(2\sqrt{2})$ is a correction factor for the intensity distribution in the (confocal) observation volume (math box 5.2). The comparison of Koppel's formula and equation (8.21) is shown in figure 8.3.

The above-mentioned methods provide estimates for the standard deviation of FCS and if available should be included in any fitting procedure as it improves parameter estimation.

8.4.5 The problem of correlations within the ACF

As is evident from equation (8.2), the points in the ACFs are calculated over the same photon count trace. Within the photon count trace it is to be expected that photon counts are correlated when they are within the expected correlation time of the process under investigation. For example, if a particle needs on average 30 μs to cross the confocal FCS observation volume, then one expects that photon counts that were collected within a period of 30 μs are correlated. So the values in the ACF that are calculated over that period are not independent. In this case, the simple definition of the standard variation and variance in equation (8.15) is not valid and covariances need to be taken into account. However, to calculate the covariances one would need to know the exact correlations between the ACF points which can be difficult to evaluate. A solution that circumvents this problem has been given by Peterson *et al* [12] with the so-called blocking transformation. If we have a set of values n_i, then their mean is given by

$$\bar{n} = \frac{1}{M}\sum_{i=1}^{M} n_i \tag{8.22}$$

And their variance is given by

$$\sigma^2 = \left(\frac{1}{M}\sum_{i=1}^{M} n_i^2\right) - \left(\frac{1}{M}\sum_{i=1}^{M} n_i\right)^2 = \left(\frac{1}{M}\sum_{i=1}^{M} n_i^2\right) - \bar{n}^2 \tag{8.23}$$

If we 'block' the values, i.e. we average neighbouring values so that we obtain $M/2$ new values

$$n'_j = \frac{(n_{2j-1} + n_{2j})}{2}; 1 \leqslant j \leqslant M/2 \tag{8.24}$$

then the mean of these new n_j is

$$\begin{aligned}
\frac{1}{M/2}\sum_{j=1}^{M/2} n'_j &= \frac{1}{M/2}\sum_{j=1}^{M/2} \frac{(n_{2j-1} + n_{2j})}{2} \\
&= \frac{1}{M}\sum_{j=1}^{M/2} (n_{2j-1} + n_{2j}) \\
&= \frac{1}{M}\left(\sum_{j=1}^{M/2} n_{2j-1} + \sum_{j=1}^{M/2} n_{2j}\right) \\
&= \frac{1}{M}\sum_{i=1}^{M} n_i = \bar{n}
\end{aligned} \tag{8.25}$$

In the second to last term, the summations range over all odd and all even values, i.e. over all values between 1 and M over all photon counts. So this demonstrates, that the mean of these blocked values is the same as the mean of the original values.

However, the same does not hold for the variance. This becomes immediately apparent when considering the variance of two random variables for which we know that

$$\sigma^2(a, b) = \sigma^2(a) + \sigma^2(b) + 2\sigma_{\mathrm{cov}}(a, b) \tag{8.26}$$

Therefore, the variance of the blocked values is only constant if the covariance, $\sigma_{\mathrm{cov}}(a, b)$, is 0, i.e. if the blocked values are independent. In math box 8.2, we give a longer derivation of the same argument. Although we explained the blocking transformation only for two neighbouring values, the same arguments hold when blocking a larger number of values

$$n'_j = \frac{n_{v(j-1)+1} + n_{v(j-1)+2} + \cdots n_{vj}}{v} = \frac{\sum_{n=1}^{v} n_{v(j-1)+n}}{v}; 1 \leqslant j \leqslant M/v \tag{8.27}$$

where v is the number of values blocked. Here, too, the variance reaches a constant value only once the blocked values become independent. We can use this to our

Math box 8.2

Change of variance in the blocking transformation
The variance for the blocked values is given by

$$\sigma'^2 = \left(\frac{1}{M/2}\sum_{j=1}^{M/2} n_j'^2\right) - \bar{n}^2$$

And the differences between the two variances, for the blocked and non-blocked values, is

$$\sigma'^2 - \sigma^2 = \left(\frac{1}{M/2}\sum_{j=1}^{M/2} n_j'^2\right) - \left(\frac{1}{M}\sum_{i=1}^{M} n_i^2\right)$$

$$= \left(\frac{1}{M/2}\sum_{j=1}^{M/2} \frac{(n_{2j-1} + n_{2j})^2}{4}\right) - \left(\frac{1}{M}\sum_{i=1}^{M} n_i^2\right)$$

$$= \left(\frac{1}{2M}\sum_{j=1}^{M/2} n_{2j-1}^2 + n_{2j}^2 + 2n_{2j-1}n_{2j}\right) - \left(\frac{1}{M}\sum_{i=1}^{M} n_i^2\right)$$

$$= \left(\frac{1}{2M}\sum_{j=1}^{M/2} (n_{2j-1}^2 + n_{2j}^2) + \frac{1}{M}\sum_{j=1}^{M/2} n_{2j-1}n_{2j}\right) - \left(\frac{1}{M}\sum_{i=1}^{M} n_i^2\right)$$

$$= \left(\frac{1}{2M}\sum_{i=1}^{M} n_i^2 + \frac{1}{M}\sum_{j=1}^{M/2} n_{2j-1}n_{2j}\right) - \left(\frac{1}{M}\sum_{i=1}^{M} n_i^2\right)$$

$$= \left(\frac{1}{M}\sum_{j=1}^{M/2} n_{2j-1}n_{2j}\right) + \left(\frac{1}{2M}\sum_{i=1}^{M} n_i^2 - \frac{1}{M}\sum_{i=1}^{M} n_i^2\right)$$

$$= \frac{1}{2}\left(\frac{1}{M/2}\sum_{j=1}^{M/2} n_{2j-1}n_{2j}\right) - \frac{1}{2}\left(\frac{1}{M}\sum_{i=1}^{M} n_i^2\right)$$

$$= \frac{1}{2}\left(\frac{1}{M/2}\sum_{j=1}^{M/2} n_{2j-1}n_{2j}\right) - \frac{1}{2}\overline{n^2}$$

We can now differentiate two cases. In the first case the difference between the variance of the measured values and the blocked values is non-zero and is given by

$$\sigma'^2 - \sigma^2 = \frac{1}{2}\left(\frac{1}{M/2}\sum_{j=1}^{M/2} n_{2j-1}n_{2j}\right) - \frac{1}{2}\overline{n^2}$$

In this case the difference between the variance depends on the first term of the right hand side, which represents the correlation between the values n_{2j} and n_{2j-1}. In the second case, the blocked values are independent from each other. Then there exists no correlation between the values n_{2j} and n_{2j-1}. From our definition of uncorrelated values,

$$\frac{1}{M}\sum_{j=1}^{M}ab = \left(\frac{1}{M}\sum_{j=1}^{M}a\right)\left(\frac{1}{M}\sum_{j=1}^{M}b\right)$$

it follows that we can write for our blocked values

$$\frac{1}{M/2}\sum_{j=1}^{M/2}n_{2j-1}n_{2j} = \left(\frac{1}{M/2}\sum_{j=1}^{M/2}n_{2j-1}\right)\left(\frac{1}{M/2}\sum_{j=1}^{M/2}n_{2j}\right)$$

If we assume, as we typically do in FCS, that the process is stationary then the terms in the two brackets on the right hand side provide the same average value $\bar{n}$.

$$\left(\frac{1}{M/2}\sum_{j=1}^{M/2}n_{2j-1}\right) = \left(\frac{1}{M/2}\sum_{j=1}^{M/2}n_{2j}\right) = \bar{n}$$

Then we obtain,

$$\left(\frac{1}{M/2}\sum_{j=1}^{M/2}n_{2j-1}\right)\left(\frac{1}{M/2}\sum_{j=1}^{M/2}n_{2j}\right) = \bar{n}^2 = \frac{1}{M/2}\sum_{j=1}^{M/2}n_{2j-1}n_{2j} = \overline{n^2}$$

The last step on the right hand side follows from the fact that there are no correlations. Finally, this leads to

$$\sigma'^2 - \sigma^2 = \frac{1}{2}\left(\frac{1}{M/2}\sum_{j=1}^{M/2}n_{2j-1}n_{2j}\right) - \frac{1}{2}\overline{n^2} = \frac{1}{2}\overline{n^2} - \frac{1}{2}\overline{n^2} = 0$$

Therefore, if there are no correlations between the values n_j then the variance is constant, otherwise it will change with blocking. This implies that we can estimate the true variance by continuously blocking our values until we reach a constant value once the blocked values become independent.

advantage by continuing to block the values until their variance does not change anymore. Only then are the values independent and the variance attains a constant value. Let's apply this to the ACF calculation. A point in the ACF is given by

$$G(m\Delta\tau_i) = \frac{1}{M-m}\sum_{k=1}^{M-m}\frac{\delta F(k\Delta\tau_i)\delta F((k+m)\Delta\tau_i)}{M_{dir,i}M_{del,\,i}} \tag{8.28}$$

So, the values to be blocked in this case are the

$$p_k^{i,\,m} = \frac{\delta F(k\Delta\tau_i)\delta F((k+m)\Delta\tau_i)}{M_{dir,i}M_{del,\,i}} \tag{8.29}$$

where the indices i, k, and m represent the binning, the data point, and the lagtime, respectively. The blocking, which happens over the index k, can then be written as

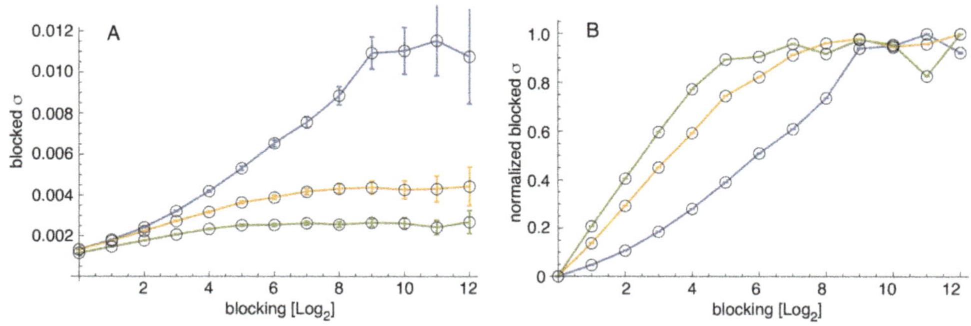

Figure 8.4. Blocking operation for correlations of data of particles with different diffusion coefficients. The data analysed here were imaging FCS simulations of a 2D random walk with a time resolution of 1 ms and 30 000 intensity values and different diffusion coefficients (green: $D = 10\ \mu m^2\ s^{-1}$; yellow: $D = 5\ \mu m^2\ s^{-1}$; blue: $D = 1\ \mu m^2\ s^{-1}$). Due to the semilogarithmic time scheme used in FCS, in which time bins are doubled every eight channels, we use block values which are a power of 2. The x-axis gives the Log_2 of the number of values blocked and the y-axis is the standard deviation after blocking. (A) Absolute values for blocking. (B) Normalized blocking values. It is evident that particles with a slower diffusion coefficient reach the blocking plateau later. This is because they spend more time in the observation volume and thus data is correlated over a longer period. This means more blocking operations have to be conducted before the data points become independent.

$$\mathbb{p}_j^{i,m} = \frac{\sum_{n=1}^{v} p_{v(j-1)+n}^{i,m}}{v}; 1 \leqslant j \leqslant (M - m)/v \tag{8.30}$$

Figure 8.4 shows a typical result of the blocking operation. Once a certain block level has been reached the variance plateaus and stays constant as from then on the blocked values are independent and provide an estimate of the variance of the particular ACF point $G(m\Delta\tau_i)$.

8.5 Non-linear least squares fit

After having introduced a number of ways to calculate the standard deviation for each point in the ACF we can tackle now the issue of fitting the data using the standard deviation as weights.

8.5.1 Least squares and the χ^2 function

The idea behind a non-linear least squares (NLS) fit is to minimize the deviation of the model from the data by optimizing the parameter values. As we have seen earlier in this chapter, we can accept the x-axis in FCS as having negligible errors and thus have to take only the deviation of the experimental values of $G(m\Delta\tau_i)$ from the theoretical points of the model $\gamma(\tau m\Delta\tau_i)$ into account. In NLS methods this is achieved by minimizing the following function:

$$\chi^2 = \sum_i (G(m\Delta\tau_i) - \gamma(m\Delta\tau_i))^2 \tag{8.31}$$

More detailed explanations of fitting procedures can be found in various text books (e.g. [1, 3]). Here, we just want to mention that in the case that the experimental data have a Gaussian distribution, the term in equation (8.31) corresponds to a maximum likelihood estimator (math box 8.3). As FCS data is derived from photon counts that follow Poisson statistics, the number of particles in the focal volume are typically small and follow also Poisson statistics, and the data points are correlated, it has been pointed out that this is only an approximation [9, 13]. Nevertheless, it has been widely used and is probably still the preferred method of data fitting in the field.

Equation (8.21) provides a non-normalized number that does not tell you how well the fit performed, only that for a particular set of parameters the sum of least squares has been minimized. A data set has associated a set of standard deviations, σ_i, for each point. So one would not expect the model to perfectly fit the experimental data, but for a good fit one would expect that the deviation of the model from the data is on the order of the standard deviation. Therefore, we can introduce the so-called reduced χ^2-function:

$$\chi_\mu^2 = \frac{1}{\nu}\sum_i \left(\frac{(G(m\Delta\tau_i) - \gamma(m\Delta\tau_i))}{\sigma_i}\right)^2 \tag{8.32}$$

where ν is the number of degrees of freedom, given by the number of data points minus the number of parameters used in the fit. In the best case we can then expect that χ_μ^2 is 1 for a good fit. If it is much larger than 1, the fit model might not be appropriate. If $\chi_\mu^2 < 1$ then one can strictly speaking make no decision. In this case either the standard deviations for each data point have been overestimated, or possibly more parameters have been included in the model than required.

For FCS data it has been shown that Koppel's estimate of errors as well as the simple calculation of the standard deviation from the intensity trace leads to $\chi_\mu^2 < 1$, even in the case of simulations where in principle the fit parameters are known, and are obtained correctly in the fit. Only the calculation of the standard error from multiple ACFs leads to $\chi_\mu^2 \sim 1$. This leads to a dilemma, as we know that the repeated measurement of the same ACF can be difficult, especially in complex or live systems. But this is also a reflection of the fact that in principle the ACF data is not normally distributed. The least-squared estimator is only a maximum likelihood estimator for normally distributed data. But this is in general not true for FCS and the error of each data point does not follow Gaussian statistics. However, if one records several ACFs then errors between measurements will have a different distribution than the ACF errors of a single measurement and according to the central limit theorem the measured errors will tend towards a normal distribution. This explains why the error measured from the intensity trace of a single ACF has a different distribution than the error estimated from multiple repeated FCS measurements and only the latter leads to χ_μ^2 values close to 1.

Therefore, for practical reasons most experimenters have either just minimized χ_μ^2 irrespective of the fact that it is smaller than one—justified by simulations that

Math box 8.3

Maximum likelihood estimation

Assume we have a model function $f(x|\boldsymbol{\beta})$ depending on a set of parameters $\boldsymbol{\beta}$, and which we would like to fit to a set of N measurement points $\{x_i, y_i\}$. In this case we would like to minimize the difference between the function at the points x_i from the experimental value y_i, i.e. we try to minimize the deviations $|y_i - f(x_i|\boldsymbol{\beta})|$. If the measurements y_i are normally distributed around the values $f(x_i|\boldsymbol{\beta})$, and if we denote the variance for each point by σ_i^2, then the probability for to find a value y_i is given by

$$p(y_i) = \frac{1}{\sqrt{2\pi}\,\sigma_i}\exp\left(-\frac{(y_i - f(x_i \mid \boldsymbol{\beta}))^2}{2\sigma_i^2}\right)$$

If the y_i's can take on continuous values, this is zero. Therefore we need to look for a probability in an interval Δy.

$$p(y_i)\Delta y = \frac{1}{\sqrt{2\pi}\,\sigma_i}\exp\left(-\frac{(y_i - f(x_i \mid \boldsymbol{\beta}))^2}{2\sigma_i^2}\right)\Delta y$$

The combined probability $P(\boldsymbol{y}|\boldsymbol{\beta})$ for the data set $\boldsymbol{y}$ is then the product of all individual probabilities

$$P(\boldsymbol{y} \mid \boldsymbol{\beta}) = \prod_{i=1}^{N}\left[\frac{1}{\sqrt{2\pi}\,\sigma_i}\exp\left(-\frac{(y_i - f(x_i \mid \boldsymbol{\beta}))^2}{2\sigma_i^2}\right)\Delta y\right]$$

Maximizing P is the same as maximizing $\log(P)$ or minimizing $-\log(P)$.

$$-\log(P(\boldsymbol{y} \mid \boldsymbol{\beta})) = \left(\sum_{i=1}^{N}\frac{(y_i - f(x_i \mid \boldsymbol{\beta}))^2}{2\sigma_i^2}\right) + \log((2\pi)^{N/2}\prod_{i=1}^{N}\sigma_i) - N\cdot\log(\Delta y)$$

The second and third term are constant as the σ_i do not change with fitting and both N and Δy are constants. Therefore, this is equivalent to minimizing

$$\chi_\mu^2 = \sum_{i=1}^{N}\frac{(y_i - f(x_i \mid \boldsymbol{\beta}))^2}{\sigma_i^2}$$

Here, χ_μ^2 is the so-called reduced χ^2, in which each squared deviation $(y_i - f(x_i|\boldsymbol{\beta}))^2$ is weighted by the variance σ_i^2. So the least squares method in the case of normally distributed values is a maximum likelihood estimation.

Up to now we assumed that the data points in $\boldsymbol{y}$ are uncorrelated. To generalize this equation to include the full covariance matrix we note that

$$P(\boldsymbol{y} \mid \boldsymbol{\beta}) = \prod_{i=1}^{N}\left[\frac{1}{\sqrt{2\pi}\,\sigma_i}\exp\left(-\frac{(y_i - f(x_i \mid \boldsymbol{\beta}))^2}{2\sigma_i^2}\right)\Delta y\right]$$

$$= \frac{1}{(2\pi)^{N/2}\prod_{i=1}^{n}\sigma_i}\exp\left(-\sum_{i=1}^{N}\frac{[y_i - f(x_i, \boldsymbol{\beta})]^2}{2\sigma_i^2}\right)$$

We can now write the exponent in vector form with the covariance matrix $\boldsymbol{C}$:

$$P(\boldsymbol{y} \mid \boldsymbol{\beta}) = \frac{1}{(2\pi)^{n/2}\sqrt{det(\boldsymbol{C})}}\exp\left(-\frac{1}{2}[\boldsymbol{y} - f(\boldsymbol{x}, \boldsymbol{\beta})]^T\boldsymbol{C}^{-1}[\boldsymbol{y} - f(\boldsymbol{x}, \boldsymbol{\beta})]\right)$$

obtain correct fit parameters also with $\chi_\mu^2 < 1$ – or have started to devise different criteria for the quality of fit.

8.5.2 The Levenberg–Marquardt fitting algorithm

Although there are alternative methods, the Levenberg–Marquardt algorithm is still the most widely used fitting methods for FCS. Alternatives contain the Nelder–Mead or simplex algorithm, or more recently machine learning approaches to FCS data fitting [14].

The Levenberg–Marquardt algorithm minimizes equation (8.32). However, there are a number of points that should be noted. The algorithm can find only local minima and thus is dependent on the starting values given to it. Providing the wrong starting values might lead to fitting artefacts. This has been addressed by using a stochastic global search algorithm that does not require any user input for the start values of the fitting parameters [15]. Although this has not been widely implemented to date, this is a viable strategy for the automation of FCS data evaluation.

The second issue with the Levenberg–Marquardt algorithm is that the user should be aware that the algorithm is typically conditioned by two values that are often not obvious but can in most programs be controlled by the user. The first is the allowed number of iterations the algorithm will perform before stopping. If this number is too low the algorithm might not converge and fail. If it is too large it can increase the time needed for fitting without producing better results as the algorithm will try to converge for data that can actually not be fit. The second value represents the lower limit of the sum of squares that is acceptable. The iteration will stop if the sum of squares falls below this limit. The same reasoning applies here. A low limit can increase the precision of the parameter estimates but it can also increase the time needed for a fit.

8.6 Generalized least squares fit

The NLS fit introduced in the last section works well for uncorrelated data. However, in FCS data points in the ACF are typically correlated. In this case the covariance between the data points cannot be neglected. For correlated data generalized least squares (GLS) fits have been developed [2]. GLS corresponds to transforming the data to remove correlations and provide constant variances first before performing an NLS fit on the transformed data. For a GLS fit one requires the covariance matrix of the data.

Given experimental measurements $\{x_i, y_i\}$, the variance of the measurement σ_i^2, and a theoretical model $f(x_i)$, NLS minimizes the weighted squared deviation:

$$\begin{aligned}\chi^2 &= \sum_i \left(\frac{(y_i - f(x_i))}{\sigma_i}\right)^2 \\ &= (y_1 - f(x_1)\;\; y_2 - f(x_2)\;\; \ldots)\begin{pmatrix}\sigma_1^2 & 0 & \ldots \\ 0 & \sigma_2^2 & \ldots \\ \ldots & \ldots & \ldots\end{pmatrix}^{-1}\begin{pmatrix} y_1 - f(x_1) \\ y_2 - f(x_2) \\ \ldots\end{pmatrix} \\ &= \hat{\beta}^T \boldsymbol{V}^{-1}\hat{\beta}\end{aligned} \tag{8.33a}$$

where

$$\hat{\beta} = \hat{y} - f(\hat{x}) = \begin{pmatrix} y_1 - f(x_1) \\ y_2 - f(x_2) \\ \dots \end{pmatrix} \tag{8.33b}$$

and the covariance matrix is

$$\boldsymbol{V} = \begin{pmatrix} \sigma_1^2 & 0 & \dots \\ 0 & \sigma_2^2 & \dots \\ \dots & \dots & \dots \end{pmatrix} \tag{8.33c}$$

The inverse of the diagonal covariance matrix, in this case of uncorrelated data points, is

$$\boldsymbol{V}^{-1} = \begin{pmatrix} \frac{1}{\sigma_1^2} & 0 & \dots \\ 0 & \frac{1}{\sigma_2^2} & \dots \\ \dots & \dots & \dots \end{pmatrix} \tag{8.34}$$

The extension of this procedure for the case of correlated data points and thus non-zero off-diagonal elements in the covariance matrix is the minimization of

$$\hat{\beta}^T \boldsymbol{V}^{-1} \hat{\beta} \tag{8.35}$$

With the covariance matrix now

$$\boldsymbol{V} = \begin{pmatrix} \sigma_1^2 & \sigma_{12} & \dots \\ \sigma_{21} & \sigma_2^2 & \dots \\ \dots & \dots & \dots \end{pmatrix} \tag{8.36}$$

In a first step, we use the Cholesky decomposition [2] for the covariance matrix

$$\boldsymbol{V} = \boldsymbol{L}\boldsymbol{L}^T \tag{8.37}$$

As long as the covariance matrix is positive-definite and contains only real values a unique Cholesky decomposition exists with $\boldsymbol{L}$ being a lower triangular matrix. This implies that

$$\boldsymbol{V}^{-1} = (\boldsymbol{L}\boldsymbol{L}^T)^{-1} = (\boldsymbol{L}^T)^{-1}\boldsymbol{L}^{-1} = (\boldsymbol{L}^{-1})^T \boldsymbol{L}^{-1} \tag{8.38}$$

and

$$\hat{\beta}^T \boldsymbol{V}^{-1} \hat{\beta} = \hat{\beta}^T (\boldsymbol{L}^{-1})^T \boldsymbol{L}^{-1} \hat{\beta} = (\boldsymbol{L}^{-1}\hat{\beta})^T \boldsymbol{L}^{-1}\hat{\beta} = \hat{\beta}^{*T} \hat{\beta}^* \tag{8.39}$$

The vector $\hat{\beta}^*$ contains the transformed data points $\boldsymbol{L}^{-1}\hat{y}$ and model function $\boldsymbol{L}^{-1}f(\hat{x})$. By using the Cholesky decomposition for the covariance matrix, we transformed the original data into a data set that is independent and thus has no off-diagonal elements and in addition for which all data points have the same variance. For this transformed data we can now apply the standard NLS fit. In the next sections we have to ensure now that our covariance matrix is positive-definite.

8.6.1 The covariance matrix for the ACF

The blocking transformation allowed us to estimate the variance for each point in the ACF by calculating the variance over a set of correlation values, $\delta(k\Delta\tau_i)\delta((k+m)\Delta\tau_i)$ (equations (8.24) and (8.25)), that are independent of each other. These values provide the diagonal elements, i.e. the variances, of the sample covariance matrix. To calculate the off-diagonal elements, i.e. the covariances between points in the ACF at different lag times, we can proceed now as follows:

$$\sigma_{uv} = \frac{1}{M'(M'-1)}\sum_{j=1}^{M'}(\mathbb{p}_j^{i,u} - \overline{\mathbb{p}^{i,u}})(\mathbb{p}_j^{i,v} - \overline{\mathbb{p}^{i,v}}) \tag{8.40}$$

where M' represents the number of (block) transformed samples; $\mathbb{p}_j^{i,u}$ is the blocked sample j at the bin time of channel i and delay u; and $\overline{\mathbb{p}^{i,u}}$ represents the average of $\mathbb{p}_j^{i,u}$ over j.

8.6.2 Regularization

In section 8.6.1, we calculated the sample covariance matrix. However, the sample covariance matrix is only well-conditioned if the number of measurements are much larger than the dimension of the covariance matrix. But here we estimate the covariance matrix from the intensity trace for a single ACF and thus do not fulfil this condition, and the sample covariance matrix is only an insufficient estimator of the true covariance. Therefore, the sample covariance matrix in our case is not guaranteed to be positive definite (i.e. its eigenvalues are non-zero, as expected for the true covariance matrix) or to be invertible. However, the latter case will be required for the GLS fit procedure. We therefore need to find a procedure to arrive at a better estimate of the true covariance matrix that fulfils the above conditions. This is a common problem and so-called regularization techniques have been developed to obtain regularized matrices that are positive definite, and can be inverted [16, 17]. One approach that has been successfully used in FCS [18, 19] is the shrinkage estimation in which the regularized matrix $\boldsymbol{S}^*$ is given by

$$\boldsymbol{S}^* = \lambda\boldsymbol{T} + (1-\lambda)\boldsymbol{S} \tag{8.41}$$

Here, λ is the so-called shrinkage intensity, $\boldsymbol{T}$ is the shrinkage target and $\boldsymbol{S}$ is the sample covariance matrix. This poses the question how to choose the shrinkage target. Guo *et al* tried several variations and finally used the so-called 'two-way' target as it provided smaller uncertainties for model probabilities when used in combination with a Bayes model selection approach [19]. We will therefore use the same here but the reader should be aware that this is not the only possibility. The 'two-way' target is defined as follows:

For the diagonal elements we use

$$v_i^* = \lambda_1 \cdot v_{\text{median}} + (1 - \lambda_1)v_i \tag{8.42a}$$

with

$$\lambda_1 = \min\left(1, \frac{\sum_i \mathrm{Var}(v_i)}{\sum_i (v_i - v_{\text{median}})^2}\right) \tag{8.42b}$$

Here v_i represents the variance, i.e. the diagonal elements of the sample covariance matrix, and v_i^* its regularized values. The value of v_{median} is the median of the variances v_i, i.e. the median of the diagonal elements of the sample covariance matrix.

For the off-diagonal elements we use

$$s_{ij}^* = r_{ij}^* \sqrt{v_i^* v_j^*} \tag{8.43a}$$

$$r_{ij}^* = (1 - \lambda_2) r_{ij} \tag{8.43b}$$

$$\lambda_2 = min\left(1, \frac{\sum_{i \neq j} Var(r_{ij})}{\sum_{i \neq j} r_{ij}^2}\right) \tag{8.43c}$$

where the s_{ij}^* are the off-diagonal elements of the regularized covariance matrix; the r_{ij} are the matrix elements of the sample covariance matrix. This method provides a covariance matrix which is positive-definite, which was not guaranteed for the sample covariance matrix, and allows us to perform fits on correlated data.

8.7 Global fit

In the previous section, we have minimized the squared deviation of the data from a model function by varying the parameters of the model. The same approach can also be used when multiple sets of data are to be fitted which are expected to at least partly depend on the same parameters. For example, assume we conduct FCS measurements in different samples that differ in particle concentration, but exhibit the same diffusion coefficient. In this case, a global fit of all data that minimizes the square deviations with the diffusion coefficient being varied for all samples simultaneously but the concentrations being varied for each data set individually, can provide better consistent fit results. If we assume, without loss of generality, that we minimize

$$\chi_\mu^2 = \frac{1}{\nu} \sum_i \left(\frac{(G(m\Delta\tau_i) - \gamma(m\Delta\tau_i))}{\sigma_i} \right)^2 \tag{8.44}$$

for single FCS measurements, where χ_μ^2 is the so-called reduced χ^2, and ν is the degree of freedom, then, if we have made J total measurements, we can minimize globally

$$\chi_\mu^2 = \frac{1}{J}\sum_j \frac{1}{\nu_j}\sum_i \left(\frac{(G_j(m\Delta\tau_i) - \gamma_j(m\Delta\tau_i \,|\{\beta_1, \beta_2, \ldots, \beta_o\}, \{\beta_{j1}, \beta_{j2}, \ldots, \beta_{jp}\}))}{\sigma_{ij}}\right)^2 \tag{8.45}$$

where the $G_j(m\Delta\tau_i)$ represents the experimental ACF and $\gamma_j(m\Delta\tau_i|\{\beta_1, \beta_2, \ldots, \beta_o\}, \{\beta_{j1}, \beta_{j2}, \ldots, \beta_{jp}\})$ is the theoretical ACF in which the parameters $\{\beta_1, \beta_2, \ldots, \beta_o\}$ are the same for all measurements and $\{\beta_{j1}, \beta_{j2}, \ldots, \beta_{jp}\}$ are minimized for individual measurements. The weighting factor σ_{ij}^2 is the variance for channel i in the jth measurement.

8.8 Maximum entropy method

Until now we have always assumed that we have a discrete number of particles with defined diffusion coefficients. However, there are circumstances where one expects a more or less continuous distribution of particle sizes and thus diffusion coefficients. The challenge is now to find the right distribution of particle sizes that are consistent with the data. For this purpose we can define a set number (N) of diffusion times τ_{Di} with a minimum (τ_{D1}) and maximum (τ_{DN}) value. The correlation function is then given as a sum over the population of particles

$$G(\tau) = \frac{1}{N}\sum_{i=1}^{N}\alpha_i\left(1 + \frac{\tau}{\tau_{Di}}\right)^{-1}\left(1 + \frac{\tau}{K^2\tau_{Di}}\right)^{-1/2} \tag{8.46}$$

with α_i representing the relative amplitudes of the components, and how they contribute to the total ACF. Minimizing χ^2 for model equation (8.46), however, might not lead to a unique result, and multiple distributions might lead to a good fit. To decide between different distributions one can fall back onto the maximum entropy criterion which states that one should choose the distribution with the maximum entropy S as defined by

$$S = -\sum_i p_i \, ln(p_i) \tag{8.47a}$$

with

$$p_i = \frac{\alpha_i}{\sum_i \alpha_i} \tag{8.47b}$$

To obtain a good fit one needs to select appropriate values for the diffusion times τ_{Di}. Sengupta *et al* have shown logarithmically spaced τ_{Di}, which are kept constant, lead to good fits (figure 8.5) [20].

8.9 Pairwise model selection using the *F*-test

We have already discussed χ_μ^2 in the previous section. There then arises the question, how to decide which of two fits, that both have values of $\chi_{\mu1}^2$ and $\chi_{\mu2}^2$, is the better fit. For that purpose, we can use the so-called F-test, with F defined as

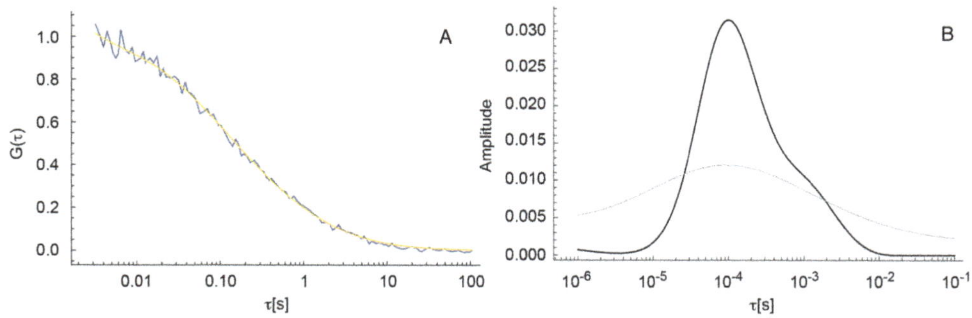

Figure 8.5. Example of an maximum entropy FCS fit. The data has two distributions of diffusion times. The first is centred around 80 μs and represents 70% of the particles. The second is centred at 1.28 ms and represents 30% of the particles. Both distributions follow a Gaussian distribution. (A) ACF (blue) with maximum entropy fit (yellow). (B) Distribution of diffusion times obtained by the fit after 10 (grey) and 30 (black) iterations of the maximum entropy algorithm.

$$F = \frac{\chi_{\mu 2}^2}{\chi_{\mu 1}^2} \tag{8.48}$$

The ratio of two χ_μ^2 values is distributed according to the F distribution. Depending on the value of F and the degrees of freedom for the two fits, ν_1 and ν_2, one can determine the probability to see a particular F statistics (also called critical F value), given a chosen confidence limit. For example, for a confidence level of 95% and for measurement containing 100 data points, we have two models. Model 1 has 3 fit parameters. Model 2 has 4 fit parameters. For this situation we obtain the following F statistics

$$F_{c,\, \alpha=0.95} = 1.4$$

Therefore, if the F-value exceeds this critical value, i.e.

$$F = \frac{\chi_{\mu 2}^2}{\chi_{\mu 1}^2} > 1.4$$

one can justify the extra parameter in model 2. However, if the F-value is smaller than or equal to 1.4, then the extra parameter is not justified and one should stay with model 1.

Values for the integral probability for χ^2 values and the F-test are nowadays easily available in most mathematics software packages and by providing the confidence limit and degrees of freedom one can directly determine the F-statistics.

While the F-test is easy to use it holds two problems for FCS. First, as mentioned before, the χ_μ^2 are often less than one for FCS measurements, as the ACF data points do not fulfil the condition of being normally distributed. Second, the F-test allows only pairwise comparison of fit models. Nevertheless, the F-test has been used as a facile tool to justify fit models. To overcome the limit of only pair-wise model comparison, Bayes model selection was introduced.

8.10 Bayes model selection

Bayes' theorem states that the probability of measurement A given B, written as $P(A \mid B)$, times the probability of observing B, $P(B)$, is equal to the probability of the measurement B given A, $P(B \mid A)$, times the probability of observing A, $P(A)$ [21]:

$$P(A \mid B)P(B) = P(B \mid A)P(A) \tag{8.49}$$

Therefore, one can write

$$P(A \mid B) = \frac{P(B \mid A)P(A)}{P(B)} \tag{8.50}$$

We can apply this to the model selection for FCS measurements, in which we would like to determine which model M_k, out of k different models, has the highest probability given the experimental data $\boldsymbol{y}$.

$$P(M_k \mid \boldsymbol{y}) = \frac{P(\boldsymbol{y} \mid M_k)P(M_k)}{P(\boldsymbol{y})} \tag{8.51}$$

In principle, if we can determine the probability of the data given the model, $P(\boldsymbol{y}|M_k)$, the probability of the model, $P(M_k)$, and the probability of the data $P(\boldsymbol{y})$, we can calculate the probability of the models given the data, $P(M_k|\boldsymbol{y})$. We can simplify our task by noticing two points. First, $P(\boldsymbol{y})$ contributes equally to all model probabilities and is just a normalization factor. Second, $P(M_k)$ describes the prior probability of the model. There are different ways to chooses those priors. But in the absence of any information, we can use a uniform prior, i.e. all model probabilities $P(M_k)$ are equal. Therefore,

$$P(M_k \mid \boldsymbol{y}) \sim P(\boldsymbol{y} \mid M_k) \tag{8.52}$$

and we only have to evaluate $P(\boldsymbol{y}|M_k)$, which can be evaluated by integration over parameter space

$$P(\boldsymbol{y} \mid M_k) = \int P(\boldsymbol{y} \mid \boldsymbol{\beta}, M_k)P(\boldsymbol{\beta} \mid M_k)d\boldsymbol{\beta} \tag{8.53}$$

Here $P(\boldsymbol{y}|\boldsymbol{\beta}, M_k)$ is the probability of the data given the parameters for a particular model. And $P(\boldsymbol{\beta} \mid M_k)$ is the prior probability distribution of the parameters. Using the Laplace approximation (math boxes 8.4 and 8.5) the integration can be performed resulting in

$$P(\boldsymbol{y} \mid M_k) = (2\pi)^{p/2}|\mathbf{C}_{\text{Bayes}}|^{1/2} P(\boldsymbol{y} \mid \boldsymbol{\beta}_{\text{Bayes}}, M_k)P(\boldsymbol{\beta}_{\text{Bayes}} \mid M_k) \tag{8.54}$$

where $\mathbf{C}_{\text{Bayes}}$ is the covariance matrix. To evaluate this further, we need to make assumptions on the prior probability of the parameters. As we do not know anything about the parameters we can assume a uniform prior. In this case, the $\boldsymbol{\beta}_{\text{Bayes}}$ are equal to the maximum likelihood estimates, $\boldsymbol{\beta}_{MLE}$, as, e.g. obtained from a NLS or GLS fit with a particular model M_k and $P(\boldsymbol{y}|\boldsymbol{\beta}_{Bayes}, M_k)$ is a function of the residuals.

Math box 8.4

Laplace approximation
The following integration has to be solved.

$$P(y \mid M_k) = \int P(y \mid \boldsymbol{\beta}, M_k) P(\boldsymbol{\beta} \mid M_k) d\boldsymbol{\beta}$$

One way to achieve this is through the Laplace approximation. The Laplace approximation approximates the integrand, $P(y|\boldsymbol{\beta}, M_k)P(\boldsymbol{\beta} \mid M_k)$, with a Gaussian function around its maxima. For that purpose the Taylor series is applied to the logarithm of the integrand up until the second order around the points $\boldsymbol{\beta}_{\text{Bayes}}$:

$$\log[P(y \mid \boldsymbol{\beta}, M_k)P(\boldsymbol{\beta} \mid M_k)] = \log[P(y \mid \boldsymbol{\beta}_{\text{Bayes}}, M_k)P(\boldsymbol{\beta}_{\text{Bayes}} \mid M_k)] - \frac{(\boldsymbol{\beta} - \boldsymbol{\beta}_{\text{Bayes}})^T \boldsymbol{H}(\boldsymbol{\beta} - \boldsymbol{\beta}_{\text{Bayes}})}{2}$$

Note that the first derivative is 0 as we are developing the series around the maxima. $\boldsymbol{H}$ is the Hessian, the matrix of second-order partial derivatives. It can be shown that the Hessian of the negative of the log-likelihood function of a Gaussian random variable is the inverse of the covariance matrix (see math box 8.5).

$$\boldsymbol{H} = \boldsymbol{C}_{\text{Bayes}}^{-1}$$

Therefore,

$$P(y \mid \boldsymbol{\beta}, M_k)P(\boldsymbol{\beta} \mid M_k) = P(y \mid \boldsymbol{\beta}_{\text{Bayes}}, M_k)P(\boldsymbol{\beta}_{\text{Bayes}} \mid M_k) \times \exp\left[-\frac{(\boldsymbol{\beta} - \boldsymbol{\beta}_{\text{Bayes}})^T \boldsymbol{C}_{\text{Bayes}}^{-1}(\boldsymbol{\beta} - \boldsymbol{\beta}_{\text{Bayes}})}{2}\right]$$

This Gaussian can now be integrated to yield with p parameters

$$P(y \mid M_k) = (2\pi)^{p/2} |\mathbf{C}_{\text{Bayes}}|^{1/2} P(y \mid \boldsymbol{\beta}_{\text{Bayes}}, M_k)P(\boldsymbol{\beta}_{\text{Bayes}} \mid M_k)$$

For an NLS fit this would correspond to $P(y|\boldsymbol{\beta}_{Bayes}, M_k) = e^{-\frac{1}{2}\sum(y_i - f(x_i))^2}$. For $P(\boldsymbol{\beta}_{Bayes}|M_k)$ different priors can be used to determine a parameter range factor (γ_i) together with the standard deviation of the parameters and $P(\boldsymbol{\beta}_{Bayes}|M_k)= [\prod_{i=1}^{nop} \sigma_i \times 2\gamma_i]^{-1}$. If all γ_i are chosen equal, then the prior is uninformative, a method that has been used by He at all for FCS data fitting [18]. However, depending on the system measured, different priors could be chosen. Now $P(y|M_k)$ can be evaluated if the covariance matrix is known. Finally, we can estimate the model probabilities $P(M_k|\mathbf{y})$ with the probability for each model given by

$$P(M_k \mid \mathbf{y}) = \frac{P(y \mid M_k)}{\sum_i P(y \mid M_i)} \tag{8.55}$$

Math box 8.5

Hessian and covariance matrix for a Gaussian random variable
Assume a Gaussian function of a random vector $\boldsymbol{\beta}$ with mean $\boldsymbol{\beta}_*$, and covariance matrix $\mathbf{C}_\beta$.

$$P(\boldsymbol{\beta}) = (2\pi)^{-p/2}|\mathbf{C}_\beta|^{-1/2}\exp\left(-\frac{(\boldsymbol{\beta} - \boldsymbol{\beta}_*)^T \boldsymbol{C}_\beta^{-1}(\boldsymbol{\beta} - \boldsymbol{\beta}_*)}{2}\right)$$

Taking the negative of the logarithm of $P(\boldsymbol{\beta})$ gives

$$-\log[P(\boldsymbol{\beta})] = \frac{p}{2}\ln(2\pi) + \frac{1}{2}\ln(|\mathbf{C}_\beta|) + \frac{(\boldsymbol{\beta} - \boldsymbol{\beta}_*)^T \boldsymbol{C}_\beta^{-1}(\boldsymbol{\beta} - \boldsymbol{\beta}_*)}{2}$$

As this function is a quadratic function of $\boldsymbol{\beta}$, one can calculate the components of the Hessian by taking the partial derivatives in respect to the components β_l and β_m:

$$\boldsymbol{H}_{lm}(\boldsymbol{\beta}_*) = \frac{\partial^2(-\log[P(\boldsymbol{\beta})])}{\partial\beta_l\partial\beta_m}|_{\boldsymbol{\beta}=\boldsymbol{\beta}_*} = (\boldsymbol{C}_\beta^{-1})_{lm}$$

and thus the Hessian is equal to the inverse of the covariance matrix

$$\boldsymbol{H}(\boldsymbol{\beta}_*) = \boldsymbol{C}_\beta^{-1}$$

8.11 Practical aspects

The outcome of a fit should not depend on the person who is doing the fitting. This seems a trivial statement. However, there are a number of decisions a researcher takes while performing a fit that can influence the outcome in unintended ways. The first parameters that influence the outcome of a fit are the number of iterations that the fit algorithm is allowed to take. If there are too few iterations, the fit program might not have converged. If there are too many the fit algorithm just takes longer without improving the results. For this reason there is also a convergence criterion, which states that when the fit parameter changes from iteration to iteration are smaller than this criterion, consider the result adequate and stop fitting. Ideally, the algorithms fulfils the convergence criterion before it reaches the maximum of iterations. Then you have reached a satisfactory result. If it reaches the maximum of iterations first, the result will not fulfil your convergence criterion. Unfortunately, often these parameters are either not accessible in a program or are hidden. Therefore, the user often has no influence on them. In most commercial software these parameters are set to reasonable values and should in most FCS fits play no role. But in complicated fits that have problems in converging, it might be worthwhile to check these parameters.

Other values the user is setting are the start values for the fit parameters. If they are far away from the actual values, the fit might not converge. Typically this is easy to identify as the fit curve does not overlap with the data. But it could also happen that more iterations are needed as one started off far away from the actual values. Then it could happen that the maximum iterations are reached before a satisfactory fit is reached. This can be easily tested, though, by either using different start values and ensuring that the fit parameters converge to the same values, or by using the results of the first fit as starting values for the second fit and see if there is any improvement. It should be noted here, that Rao *et al* have already developed an algorithm for FCS fitting that does not require any selection of starting values [15].

Also note, that one should use weighted fitting for FCS if possible, as this will lead to better parameter optimization. We discussed in this section several approximations to the FCS standard deviation and any of them should provide better fits than an unweighted fit.

Once these preliminaries are dealt with, how does one best go about fitting different models and selecting the correct one? As discussed above, the Bayesian model selection is a good tool and it is included in some available open source software programs [19, 22]. If that is not available the χ^2_μ goodness-of-fit test in combination with the F-test can be used to test competing models, at least in a pairwise fashion. Practically, one advances step by step starting with the most simple model and then increasing model complexity and determining whether the residuals and χ^2_μ improve. Once fits start being satisfactory, one can then use the F-test to determine whether further addition of parameter can be justified. Often this is enough, especially when one tries to decide between two simple cases, e.g. binding or non-binding *in vitro*.

Finally, one can simplify the task by determining fit parameters individually with calibration measurements. For example, it is routine that in confocal FCS measurements a standard calibration dye with known diffusion coefficient is measured at the beginning and the end of every measurement session. The calibration determines the size of the confocal volume and especially the structure factor K. These parameters can then be fixed as long as one works in similar samples. The fixing of these predetermined parameters simplifies the fitting procedure considerably. This is especially the case for the structure factor K as the ACF does not strongly depend on its value and any fit uncertainties typically end up with the fit providing unrealistic K values. This also means that if the fit of the calibration measurements does not converge consistently to a K value in a narrow range (typically 3–8 for most systems) then this is a sign that the system is not properly aligned.

Other parameters that can be determined before measuring a more complex sample could be the diffusion coefficients of the different expected particles in the final sample. For example, the diffusion coefficient of a receptor as well as the ligand could be determined independently. They could then be fixed when the two are mixed to measure binding.

8.12 Synopsis

- To correctly fit FCS data we require the error of the measurements. Errors in time are negligible but we require the error of the experimental ACF.

- There are various models for the error of the ACF from theoretical to experimentally determined ones.
- Typically, FCS data is fit by a non-linear least squares approach using the Levenberg–Marquardt algorithm. This however neglects correlations between data points, as are prone to exist in FCS data.
- A general least square approach can be used for correlated data. It transforms the data if the covariance matrix of the ACF measurement is known. A blocking procedure can be employed to determine the covariance matrix.
- Multi-component data with a distribution of diffusion coefficients can be evaluated by a maximum entropy method (MEMS-FCS).
- If multiple measurements are expected to obtain the same parameters. (e.g. diffusion coefficients) a global fit of the multiple data sets to the same parameter can be used to improve parameter estimation.
- The determination of an appropriate fit model can be performed pair-wise by an *F*-test. For the comparison of multiple models a Bayes model selection methods has been developed.

8.13 Exercises

The accompanying CDF file for fitting (available at https:\\iopscience.iop.org\book\978-0-7503-2080-1) allows testing data fitting in three scenarios: (i) a one-component data set can be fitted to a one-component model; (ii) a two-component data set can be fitted to a one-component model; (iii) a two-component data set can be fitted to a two-component model.

1. Test how the structure parameter K influences the fits.
   a) Note that for high noise or wrong fit model it diverges and needs to be fixed to perform a proper fit.
   b) Investigate how the residuals change when a wrong structure factor is fixed in the fit.

2. Observe the difference in the residuals when a two-component data set is fitted with either a one- or two-component fit.

3. Observe how weighted fits improve on data fitting and how that is influenced by the structure factor and the signal-to-noise ratio (i.e. *cps*).

4. Investigate how well two diffusive components can be solved …
   a) … in dependence of the ratio of D_2/D_1, keeping the fraction $F_2 = 0.5$ and changing the signal-to-noise ratio (i.e. the *cps*).
   b) … in dependence on the fraction F_2, keeping $D_2/D_1 = 5$, and changing the signal-to-noise ratio (i.e. the *cps*).

References

[1] Bevington P R and Robinson D K 1992 *Data Reduction and Error Analysis for the Physical Sciences* (New York: McGraw-Hill)

[2] Seber G and Wild C 1989 *Nonlinear Regression* (New York: Wiley)

[3] Press W H, Teukolsky S A, Vetterling W T and Flannery B P 1992 *Numerical Recipes in C* 2nd edn (Cambridge: Cambridge University Press)

[4] Schätzel K 1990 Noise on photon correlation data: 1. autocorrelation functions *Quantum Opt.* **2** 287–305

[5] Schätzel K and Peters R 1991 Noise on multiple-tau photon corelation data *Proc. SPIE* **1430** 109–15

[6] Schätzel K, Drewel M and Stimac S 1988 Photon correlation measurements at large lag times: improving statistical accuracy *J. Mod. Opt.* **35** 711–18

[7] Koppel D 1974 Statistical accuracy in fluorescence correlation spectroscopy *Phys. Rev.* A **10** 1938

[8] Meseth U, Wohland T, Rigler R and Vogel H 1999 Resolution of fluorescence correlation measurements *Biophys. J.* **76** 1619–31

[9] Wohland T, Rigler R and Vogel H 2001 The standard deviation in fluorescence correlation spectroscopy *Biophys. J.* **80** 2987–99

[10] Saffarian S and Elson E L 2003 Statistical analysis of fluorescence correlation spectroscopy: the standard deviation and bias *Biophys. J.* **84** 2030–42

[11] Palo K, Mets Ü, Jäger S, Kask P and Gall K 2000 Fluorescence intensity multiple distributions analysis: concurrent determination of diffusion times and molecular brightness *Biophys. J.* **79** 2858–66

[12] Flyvbjerg H and Petersen H G 1989 Error estimates on averages of correlated data *J. Chem. Phys.* **91** 461

[13] Rigler R, Mets Ü, Widengren J and Kask P 1993 Fluorescence correlation spectroscopy with high count rate and low background: analysis of translational diffusion *Eur. Biophys. J.* **22** 169–75

[14] Guan X 2018 Machine learning approach to fluorescence correlation *Spectroscopy* 1–62

[15] Rao R, Langoju R, Gösch M, Rigler P, Serov A and Lasser T 2006 Stochastic approach to data analysis in fluorescence correlation spectroscopy *J. Phys. Chem.* A **110** 10674–82

[16] Ledoit O and Wolf M 2004 A well-conditioned estimator for large-dimensional covariance matrices *J. Multivariate Anal.* **88** 365–411

[17] Schäfer J and Strimmer K A shrinkage approach to large-scale covariance matrix estimation and implications for functional genomics *Statis. Appl. Gen. Molec. Biol.* **4** 1175

[18] He J, Guo S-M and Bathe M 2012 Bayesian approach to the analysis of fluorescence correlation spectroscopy data I: theory *Anal. Chem.* **84** 3871–79

[19] Guo S-M, He J, Monnier N, Sun G, Wohland T and Bathe M 2012 Bayesian approach to the analysis of fluorescence correlation spectroscopy data II: application to simulated and *in vitro* data *Anal. Chem.* **84** 3880–88

[20] Sengupta P, Garai K, Balaji J, Periasamy N and Maiti S 2003 Measuring size distribution in highly heterogeneous systems with fluorescence correlation spectroscopy *Bio. Phys. J.* **84** 1977–84

[21] Lee P M 2012 *Bayesian Statistics* 4th edn (New York: Wiley)

[22] Krieger J W, Singh A P, Bag N, Garbe C S, Saunders T E, Langowski J and Wohland T 2015 Imaging fluorescence (cross-) correlation spectroscopy in live cells and organisms *Nat. Protoc.* **10** 1948–74

IOP Publishing

An Introduction to Fluorescence Correlation Spectroscopy

Thorsten Wohland, Sudipta Maiti and Radek Macháň

Chapter 9

FCS and FCCS measurement strategies

The preceding chapters have presented FCS and FCCS theory, typical instrumentation used, data processing to obtain correlation functions and correlation function analysis by fitting with theoretical models. The reader has been, thus, introduced to all aspects of FCS and FCCS needed to design and conduct experiments. This chapter aims at helping to bridge the gap between the principles and theory on one side and their practical application on the other by discussing selected FCS or FCCS measurement strategies. Over the decades, FCS and FCCS have found a broad range of applications some of which lead to highly specialised experimental strategies. In view of that, we do not intent to present an exhaustive overview of all measurement strategies; we will, rather, concentrate on the more widely applicable strategies or on strategies which we find useful in giving the reader a feeling for the possibilities and limitations of FCS. When discussing the individual strategies, we will include notes on the instrumentation needed and whether generally available commercial setups can be used. We will start with measuring concentrations, c (when referring to molar concentration) or C (when we referring to number concentration), and diffusion coefficients, D, by FCS, which will lead us to the topic of effective observation volume calibration needed to obtain absolute values of c and D. This will then bring us to the more specialised topics of calibration-free FCS modalities and the specific issues encountered when studying diffusion confined to two-dimensional structures. In the last sections, we will look at characterising molecular interactions and compare strategies based on FCS with those based on FCCS.

9.1 Measuring concentrations by FCS

According to equation (5.12), the amplitude of the concentration fluctuations ACF, $G_0^{(\mathrm{conc})}$, is inversely proportional to N, the average number of independently diffusing particles in the effective observation volume, V_{eff}:

$$G_0^{(\text{conc})} = \frac{1}{N} = \frac{1}{V_{\text{eff}} \langle C \rangle} \tag{9.1}$$

where we used the fact that the number of particles, N, can be written as the product of the particle concentration, $\langle C \rangle$ and the effective observation volume, V_{eff}. Note that in accordance with the other chapters in this book we use number concentrations here defined as the number of particles in a unit volume; number concentrations, C, are converted to commonly used molar concentrations, c, by dividing them by Avogadro's constant, N_A; $c = C/N_A$. From equation (9.1) we see that, if the effective observation volume is known, absolute concentrations of diffusing fluorescent particles can be directly obtained from the ACF amplitude. In this way, FCS allows us to measure absolute concentrations of molecules of interest in specific locations, e.g. in live cells [1, 2].

Before we proceed to strategies of effective observation volume calibration (section 9.1.2) let us first discuss in more detail what we have briefly outlined above. Firstly, equation (9.1) neglects the influence of background on the ACF amplitude (see section 7.1) and is applicable only if the background contribution to the measured intensity trace is negligible or if the influence of the background on the ACF has been eliminated by the use of statistical filtering (see section 4.4). Otherwise, equation (9.1) needs to be rewritten as (see math box 7.2):

$$G_0^{(\text{conc})} = \frac{1}{N_{app}} = \frac{V_{\text{eff}} \langle C \rangle \kappa^2}{(V_{\text{eff}} \langle C \rangle \kappa + \langle B \rangle)^2} = \frac{1}{V_{\text{eff}} \langle C \rangle \left(1 + \dfrac{\langle B \rangle}{\langle F \rangle - \langle B \rangle}\right)^2} \tag{9.2}$$

and the particle concentration is given by:

$$\langle C \rangle = \frac{1}{V_{\text{eff}} G_0^{(\text{conc})}} \left(\frac{\langle F \rangle - \langle B \rangle}{\langle F \rangle}\right)^2 \tag{9.3}$$

Since in the presence of background, the reciprocal value of the ACF amplitude is no longer equal to the particle number, N, we call it the apparent particle number N_{app}. As before, κ in equation (9.2) is the brightness of the fluorescent particles, $\langle B \rangle$ is the average background and $\langle F \rangle$ is the average of the measured intensity time trace, which is the sum of the average background and the average fluorescence of the particles within the effective observation volume. Using the fact that the average fluorescence of the particles is a product of their number and their brightness, we can write for $\langle F \rangle$:

$$\langle F \rangle = V_{\text{eff}} \langle C \rangle \kappa + \langle B \rangle \tag{9.4}$$

The average background, $\langle B \rangle$, can be determined by a blank measurement in a sample not containing any fluorescent particles, i.e. either pure solvent for FCS measurements in solution or unlabelled cells for live-cell measurement. Note that while the average background value can be determined quite reliably for neat solutions, the background in live cell measurements is likely to be position dependent and possibly varies over time. Differences between the actual background value and its estimate used in equation (9.3) do not compromise the absolute value of concentration only but also the relative differences in $\langle C \rangle$. It is, therefore, advisable to minimise the influence of the background by maximising the $\langle F \rangle / \langle B \rangle$ ratio. This can be achieved by using fluorescent particles with high brightness κ and emission in a spectral region where the background signal from the cells of interest is low.

Secondly, let us look at the ACF amplitude, $G_0^{(\mathrm{conc})}$. In the first sentence of this section, we called it explicitly the concentration fluctuations ACF, by which we mean the ACF of the fluorescence intensity fluctuations caused by fluctuations in the number of particles present in the effective observation volume. The intensity fluctuations are caused by translational movement of particles in and out of the effective observation volume. The movement can be random Brownian diffusion, which is the case in most FCS experiments (see section 5.1), or a combination of random diffusion and oriented flow (see section 5.5). If other processes such as fluorophore blinking (see section 5.6) contribute to the fluorescence intensity fluctuations, equation (9.2) does not hold for the amplitude, G_0, of the overall ACF but only for, $G_0^{(\mathrm{conc})}$, the amplitude of its part corresponding to fluctuations caused by translational motion of particles. $G_0^{(\mathrm{conc})}$ can be obtained by fitting the overall ACF with an appropriate theoretical model, which accounts for all sources of fluorescence intensity fluctuations present in the system. In the most general scenario discussed in section 5.6, the overall ACF is described by the model of equation (5.48). By fitting the experimental ACF with that model, we can obtain the amplitude of the concentration fluctuations ACF, $G_0^{(\mathrm{conc})} = 1/N_{app}$. Note that equation (5.48) does not account for background; therefore, the particle number denoted as N in equation (5.48) is, according to the notation introduced here, the apparent particle number N_{app}. The actual particle concentration would be then obtained according to equation (9.3). What we have just said illustrates that, although the particle concentration determines the ACF amplitude only and not the ACF temporal decay, fitting of the temporal decays is often needed in FCS concentration measurements as this is the only way to reliably disentangle contributions from different processes.

In most FCS experiments, the fluorescence intensity fluctuations originating from different processes are expected to happen on well-separated timescales, allowing the ACF amplitude corresponding to translational motion to be determined reliably by fitting the overall ACF with an appropriate model. While fluctuations caused by photon antibunching and rotational diffusion happen on timescales below the temporal resolution of typical FCS experiments, timescales of photophysical processes leading to fluorophore blinking may in some cases overlap with translational motion timescales. This is likely to introduce high levels of uncertainty in the amplitude of the concentration fluctuations ACF obtained by fitting the overall ACF. In such situations a strategy discussed in section 9.2.4 can be employed to identify the contribution of translational motion to the overall ACF.

9.1.1 Concentration range accessible by FCS

The fluorescence intensity fluctuations which contribute to the concentration fluctuations ACF originate from the movement of individual fluorescent particles in and out of the detection volume. At very low particle concentrations the average particle number in the effective observation volume, $N = V_{\text{eff}} \langle C \rangle$, is much smaller than 1. This means that most of the time there are no particles in the effective observation volume. In the absence of particles in the effective observation volume, no fluorescence fluctuations are detected, only background. Therefore, those portions of the intensity time trace do not contribute any information to the ACF. The larger the fractions of those 'empty' portions in the intensity time trace, the longer is the acquisition time required to record enough intensity fluctuations needed to obtain the ACF with good statistical accuracy.

On the other hand, there are also specific challenges connected with FCS measurements at very high average particle numbers. One of them is the resulting high level of average fluorescence intensity (given by $N\ \kappa$), which is prone to saturate photon counting detectors commonly used in FCS (chapter 3). The obvious solution to this problem, namely reducing particle brightness by decreasing the excitation intensity, is not an ideal one. Particle brightness, κ, is the most critical parameter in determining the ACF signal-to-noise ratio in FCS [3, 4]. A more viable approach is to use detectors with higher saturation thresholds or dividing the fluorescence signal between multiple detectors, each of which then handles only a part of the total fluorescence photon flux [5]. Other potential problems are connected to the low ACF amplitudes at high particle numbers (see equation (9.1)). The contribution of concentration fluctuations to the ACF can become under these conditions comparable to contributions coming from correlated noise in the experimental setup, such as laser intensity fluctuations [5]. Those noise correlations are independent of N and for lower particle numbers their amplitude is negligible compared to $G_0^{(\text{conc})}$. However, with increasing N, $G_0^{(\text{conc})}$ decreases to values comparable to the amplitude of noise correlation thus effectively limiting the concentrations one can measure.

Therefore, particle numbers on the orders of units to hundreds are usually recommended for FCS [6, 7]. For typical confocal setups with femtoliter or sub-femtoliter effective observation volumes this translates to concentrations of nM to 100 nM order (see exercise 1). Setups with larger or smaller effective observation volumes shift this concentration range to lower or higher values, respectively [8]. Removing the sources of correlated noise by stabilising the setup and the laser intensity can also extend the accessible range to higher concentrations and lower ACF amplitudes. Alternatively, ACFs can be corrected for the influence of instable laser intensity following a strategy presented in section 9.2.4.

9.1.2 Effective observation volume calibration

Equations (9.1) or (9.3) allow us to determine the absolute concentrations of independently diffusing particles provided the effective observation volume, V_{eff}, is known. V_{eff} is determined by the instrumental setup used in the FCS experiment. In principle, it can be estimated theoretically based on the parameters of the optical setup used; however, such an approach is not commonly adopted in FCS. Instead, an experimental calibration of V_{eff} is usually performed. The main reason for that is the relatively high sensitivity of FCS to subtle variations in the optical system alignment and sample optical properties [9]. Such effects, which can vary over time, are not easily incorporated in a theoretical estimate but are, at least some of them, intrinsically included in an experimental calibration. Note, however, that for this to be fully true, the optical properties such as the refractive index of the reference sample used for calibration would need to match exactly the optical properties of the sample of interest; we will return to this in section 9.1.2.4. In this way a calibration measurement not only provides the effective observation volume, V_{eff}, but also checks the performance of the setup. Therefore, FCS calibration measurements are routinely performed even if the actual experiments do not require absolute calibration (for example, if the conclusions are based only on relative changes in concentration or diffusion coefficient, rather than their absolute values) and are also suggested as a tool for monitoring confocal microscope performance and alignment [10]. We will present here three strategies of V_{eff} calibration and discuss their advantages and limitations. All three strategies are applicable to confocal FCS and some can be directly or with some modifications extended to other FCS modalities. We will, however, not discuss this here in detail. We will return to the topic of FCS calibration in section 9.2 in the context of absolute diffusion coefficient measurements.

9.1.2.1 Calibration using a fluorophore solution of known concentration

This approach can be regarded as the most straightforward. We perform an FCS measurement in a reference solution of known fluorophore concentration $\langle C^{(\mathrm{ref})}\rangle$. V_{eff} is then obtained from the amplitude of the concentration fluctuations ACF measured in the reference solution, $G_0^{(\mathrm{conc})(\mathrm{ref})}$:

$$V_{\mathrm{eff}} = \frac{1}{\langle C^{(\mathrm{ref})}\rangle G_0^{(\mathrm{conc})(\mathrm{ref})}} \left(\frac{\langle F^{(\mathrm{ref})}\rangle - \langle B^{(\mathrm{ref})}\rangle}{\langle F^{(\mathrm{ref})}\rangle} \right)^2 \tag{9.5}$$

where $\langle F^{(\mathrm{ref})}\rangle$ and $\langle B^{(\mathrm{ref})}\rangle$ are the average intensity and average background of the calibration measurement.

This approach does not depend on any assumptions about the shape of the effective observation volume; it is, therefore, applicable to any FCS modality. Moreover, any deviations of the actual observation volume shape from its theoretical model are intrinsically accounted for as long as they are identical for the measurement in the reference solution and in the actual sample. Nevertheless, any such deviation will still affect the fitting of the ACF temporal decay with a theoretical model.

The key challenge lies in ensuring that the exact concentration of the reference fluorophore sample is known. Beside the errors intrinsically connected with preparing solutions at low concentrations (typically of nM order; see section 9.1.1), many water-soluble fluorophores tend to adsorb to glass surfaces. Cationic fluorophores such as Rhodamine B, are attracted to the surface of glass, which is under neutral pH negatively charged. At small sample volumes and low fluorophore concentrations, commonly used in FCS, adsorption to glass surface leads to significant depletion of the fluorophores from the bulk solution, reducing their concentration in the effective observation volume. Coating the glass surface with positively charged molecules, such as poly-L-lysine or bovine serum albumin, can help to reduce this effect. An additional way to improve the accuracy of the calibration is to prepare a dilution series of the reference solution and perform calibration measurements at each concentration. However, because of the increased number of measurements needed, dilution series calibration is not convenient for routine usage.

9.1.2.2 Calibration using a reference fluorophore with known diffusion coefficient
Besides relating the ACF amplitude to the particle concentration, the effective observation volume also relates the temporal ACF decay to the mobility of the particles. In the case of confocal FCS, this relationship is expressed by equations (5.5)–(5.7), which we will repeat here for convenience:

$$G(\tau) = \frac{1}{N}\left(1 + \frac{\tau}{\tau_D}\right)^{-1}\left(1 + \frac{\tau}{K^2\tau_D}\right)^{-1/2} + G_\infty \tag{5.5}$$

$$\tau_D = \frac{w_0^2}{4D} \tag{5.6}$$

$$K = \frac{z_0}{w_0} \tag{5.7}$$

The effective observation volume is then calculated from w_0 and z_0 as (see math box 5.2):

$$V_{\text{eff}} = \pi^{3/2} w_0^2 z_0 = \pi^{3/2} K w_0^3 \tag{9.6}$$

This allows us to calibrate V_{eff} by an FCS measurement in a solution of fluorescent particles of known diffusion coefficient, D. Reference values of D for fluorophores covering different spectral regions can be found, for example, in [48]. The steps to take are the following: the ACF obtained in the calibration measurement is fitted with an appropriate theoretical model such as equation (5.5) (or equation (5.47) to account for fluorophore blinking) and the values of the diffusion time, τ_D, and the structure parameter, K, are extracted from the fit. w_0 and z_0 are obtained from τ_D and K according to equations (5.6) and (5.7), where D is the known diffusion coefficient of the reference particles, and then used to calculate V_{eff} following equation (9.6).

Note that for other than confocal settings, equations (5.5)–(5.7) and (9.6) need to be revisited. For the case of two-photon FCS, the necessary modification is very small; the expression for the diffusion time in equation (5.6) is for the two-photon case $\tau_D = \frac{w_0^2}{8D}$ [11].

The ACF temporal decay is independent of particle concentration; therefore, we do not need to be concerned about adsorption of the fluorescent particles to glass or about the accuracy of diluting the reference solution. However, it is advisable that the particle concentration lies within the optimal range for FCS (see section 9.1.1). Neither does the ACF temporal decay depend on the uncorrelated background, relieving us from the need to perform a blank measurement. On the other hand, D depends on temperature, T, as expressed by the Stokes–Einstein equation:

$$D = \frac{kT}{6\pi\eta r_h} \tag{9.7}$$

where k is the Boltzmann constant, η the solution viscosity, and r_h the hydrodynamic radius of the particles. Note that besides the explicit dependence on T, the viscosity, η, is also temperature dependent, resulting in a steeper than linear dependence of D on T (see figure 9.1) [48]. Therefore, the temperature at which the FCS measurements are performed needs to be monitored. For this purpose, the temperature within the sample has to be known. When working with heated sample holders, the microscope objective should be kept at the same temperature as the sample. Otherwise it will act as a heatsink and the temperature of the part of the sample closest to the objective, which is where the effective observation volume lies, would be different from the rest of the sample and difficult to monitor.

The main disadvantage of this approach is its dependence on assumptions about the shape of the effective observation volume, which makes it prone to errors if the actual effective observation volume shape deviates from the theoretical model. This is often the case for reasons we mention in section 9.1.2.4.

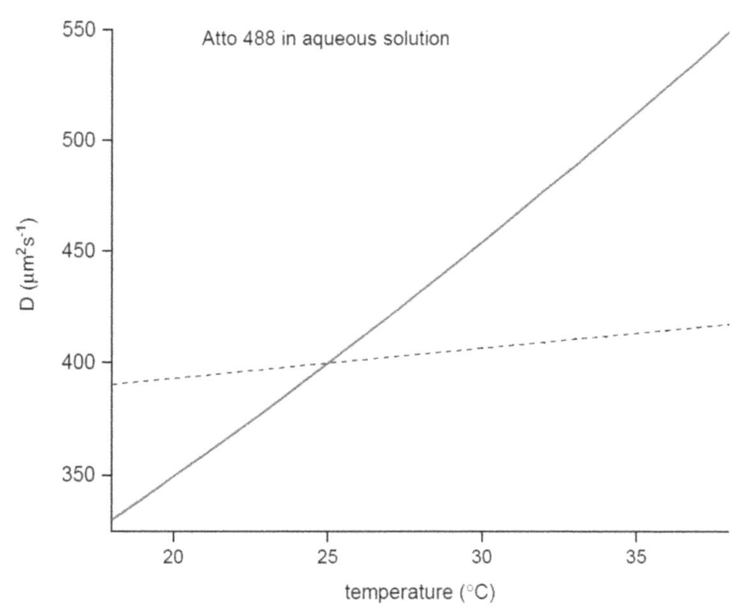

Figure 9.1. The temperature dependence of the diffusion coefficient of Atto 488 carboxylic acid in an aqueous solution. The dashed line shows what the dependence would be if we neglected the temperature dependence of the viscosity of water and used the value of viscosity at 25 °C across the whole temperature range. The plot is based on data from [48].

9.1.2.3 Calibration by scanning a small fluorescent bead

The image in scanning imaging modalities such as confocal or two-photon microscopy can be described as a convolution of the object with the microscope point-spread function (PSF), which is in this case equivalent to the FCS effective observation volume. If the imaged object is significantly smaller than the effective observation volume, the resulting image can be then considered a good approximation of the effective observation volume. V_{eff} can be, therefore, determined from scans of small fluorescent beads. For a confocal setup, where V_{eff} is usually approximated by a three-dimensional Gaussian profile, the dimensions w_0 and z_0 can be obtained as e^{-2} radii of Gaussian fits of the fluorescence intensity distribution in the image and V_{eff} then calculated according to equation (9.6) [12]. Alternatively, the effective observation volume can be extracted from the bead image in a non-parametric way as the volume enclosed by the e^{-2} iso-surface of the fluorescence intensity distribution. Note that the effective observation volume in FCS is defined by the e^{-2} radii or iso-surfaces, while other measures such as $e^{-\frac{1}{2}}$ radii or full width at half maximum (FWHM) values are commonly used to describe microscope PSFs in the context of imaging.

9.1.2.4 General comments about effective observation volume calibration

An important consideration which applies to all three calibration approaches is the spectral dependence of V_{eff}. Therefore, the reference fluorophores used in the first two approaches or fluorescent beads used in the last approach need to be selected to emit fluorescence in the same spectral region as the sample of interest.

We have mentioned repeatedly the potential impact of non-ideal effective observation volume shape on those calibration strategies that rely on theoretical assumption about the effective observation volume. However, deviations in the effective observation volume can compromise the calibration in one more way which applies to all calibration strategies. The effective observation volumes in the reference sample used for calibration and in the sample of interest may not be identical. One of the main factors responsible for that is the dependence of the effective observation volume on refractive index within the sample. The samples of particles of known concentration or known diffusion coefficient used in the first two strategies are typically diluted aqueous solutions of small fluorescent molecules; their refractive index is, therefore, the refractive index of water, approximately 1.333. This is the ideal sample for water immersion objectives and the effective observation volume is expected to be near the theoretical optimum, provided the rest of the system, such as the objective correction collar and the confocal pinhole, are well aligned. The refractive index of the sample of interest may differ from the refractive index of water; for example, in live cells, the refractive index can be approximately 1.360. Although the difference in refractive index between cells and water may seem small, numerical simulations have shown that it is enough to result in an effective volume several times larger than in water, leading to concentrations overestimated by the same factor [9].

In the same paper, the authors also simulated the influence of the depth within the sample (the distance from the coverslip surface) on the distortions of the effective observation volumes. For FCS measurements in solutions it is easy to perform all measurements including the calibration at the same distance from the coverslip surface. This distance is typically selected on the order of 10 μm to ensure the optical interface between the coverslip and the sample does not influence the effective observation volume and at the same time to minimise the optical path in the sample. On the other hand, measurements in living cells or in tissues are performed at the depth where the structure of interest is located, which can be further or closer to the coverslip than the depth at which the calibration measurement in solution was performed. The depth can also vary from sample to sample. It is, therefore, another factor that increases the uncertainty in V_{eff} calibration. However, the simulation has shown that it has smaller influence than the refractive index of the sample [9].

Another factor to consider is the dependence of some optical aberrations on the distance from the optical axis of the objective, which means the distance from the centre of the field of view of the microscope. It was shown that the effective observation volume changes very little with position close to the centre of the field of view but deviates significantly if it is placed near the edges of the field accessible by the confocal scanner [13].

9.1.3 Molecular brightness determined by FCS

Beside measuring the fluorescent particle concentration, the information on average particle number given by the ACF amplitude, can be also used to determine the particle brightness, κ. From equations (9.3) and (9.4) it follows that:

$$\kappa = \frac{\langle F \rangle - \langle B \rangle}{N} = G_0^{(\mathrm{conc})} \frac{\langle F \rangle^2}{\langle F \rangle - \langle B \rangle} \tag{9.8}$$

Applications of the information on particle brightness include quantitative FCCS analysis (see section 6.2.3), benchmarking of fluorescence collection efficiency in confocal microscopes or measuring average degree of oligomerisation of fluorescent molecules by comparing the measured particle brightness with the brightness of the same fluorophore in a monomeric form. However, mutual proximity of multiple fluorophores in an oligomer may result in the fluorophores quenching each other, in which case the brightness of an oligomer is smaller than the brightness of an individual fluorophore multiplied by the degree of oligomerisation.

It can be also used in quantitative fluorescence imaging, in an approach called FCS-calibrated imaging [14, 15]. The approach seeks to characterise quantitatively the distribution of molecules of interest within cells. FCS measures the brightness of the molecules in selected points in the cells; this information is then used to convert fluorescence intensities in all image voxels to concentrations. Similar information is also obtained by a technique called number and brightness analysis [16] which we will introduce in the next section to show how it is related to the analysis of ACF amplitudes in FCS.

9.1.3.1 Number and brightness analysis

This approach uses the mean value and variance of fluorescence intensity time trace to extract information on the brightness and the average number of particles in the corresponding effective observation volume. The principle is illustrated by figure 9.2. The same average fluorescence intensity can be observed with a small number of particles with high brightness or a large number of particles with low brightness. However, the variance in fluorescence intensity would be different in the two cases. If the particle number fluctuates as a result of concentration fluctuations (particle translational motion), this intensity fluctuations are much more pronounced in the case of fewer particles with higher brightness. Quantitatively, the average fluorescence intensity, when neglecting background, is given by (see equation (9.4)):

$$\langle F \rangle = \kappa N \tag{9.9}$$

The fluorescence intensity variance contains contributions from two processes:

(i) The concentration fluctuations resulting in varying number of fluorescent particles present in the effective observation volume. We denote their contribution to the fluorescence intensity variance as σ_N^2.

(ii) Fluctuations in the number of photons detected for a given number of fluorescent particles. This is known as shot noise or photon counting noise (see math box 3.1); we denote its contribution to the intensity fluctuations as σ_P^2.

Both processes are expected to follow Poisson statistics (see math box 3.1), which means that their variance is equal to their mean value. The variance of the measured fluorescence intensity trace is the sum of the contributions of the two processes (see math box 9.1):

$$\sigma^2 = \sigma_N^2 + \sigma_P^2 = \kappa N(\kappa + 1) \tag{9.10}$$

Equations (9.9) and (9.10) lead to the following expressions for the particle number and brightness.

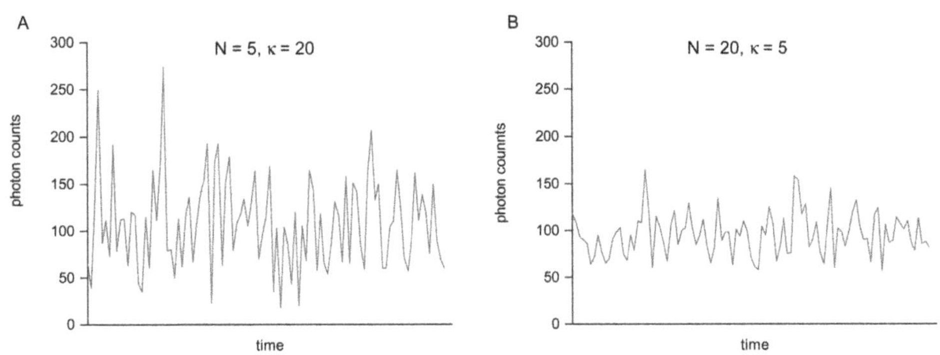

Figure 9.2. Simulated fluorescence intensity traces with the following parameters: $N = 5, \kappa = 20$ (panel A) and $N = 20, \kappa = 5$ (panel B). Both combinations of parameters lead to the same average fluorescence intensity equal to 100; however, the intensity traces differ in the variance of the fluctuations (see exercise 2).

Math box 9.1.

The variance of the fluorescence intensity trace.
Assuming the number of particles in the effective observation volume follows Poisson distribution with the mean value N, its standard deviation is the square root of the mean value, $\sqrt{N}$. This will result in fluctuations in fluorescence intensity characterised by standard deviation σ_N:

$$\sigma_N = \kappa\sqrt{N}$$

The variance is the square of the standard deviation:

$$\sigma_N^2 = \kappa^2 N$$

The mean value of the number of detected photons (the fluorescence intensity) is $\langle F \rangle = \kappa N$. The number of photons also follows Poisson distribution; therefore, the variance of its fluctuations is:

$$\sigma_P^2 = \langle F \rangle = \kappa N$$

The variance of the measured intensity time trace is, thus, given by equation (9.10).

$$\kappa = \frac{\sigma^2}{\langle F \rangle} - 1 \tag{9.11a}$$

$$N = \frac{\langle F \rangle^2}{\sigma^2 - \langle F \rangle} \tag{9.11b}$$

It can be easily shown that this calculation is equivalent to obtaining particle number and brightness from the ACF amplitude in FCS (when neglecting background). If we compare the definition of the intensity trace variance:

$$\sigma^2 = \langle F(i) - \langle F \rangle \rangle^2 = \langle F(i) \rangle^2 - 2\langle F(i) \rangle \langle F \rangle + \langle F \rangle^2 = \langle F(i) \rangle^2 - \langle F \rangle^2 \tag{9.12}$$

with the definition of the ACF (equation (2.20)) and the ACF amplitude (equation (5.13)), we can readily see that the ACF amplitude, $G_0^{(\text{conc})}$, is actually the variance of the fluorescence intensity trace normalised by the square of the average intensity (where $\frac{\langle F \rangle^2}{\langle F \rangle^2} = 1 = G_\infty$). However, since in FCS the amplitude is obtained by extrapolating the correlation function to zero lag time rather than by directly calculating the value for $\tau = 0$ (see section 4.1.4), the value is not affected by shot noise, which is not correlated in time. We can therefore write:

$$G_0^{(\text{conc})} = \frac{\sigma_N^2}{\langle F \rangle^2} = \frac{\kappa^2 N}{(\kappa N)^2} = \frac{1}{N} \tag{9.13}$$

The practical difference between FCS and number and brightness analysis lies in the requirements on temporal resolution. In FCS, the time interval between two consecutive intensity values, $F(i)$ and $F(i + 1)$, needs to be much shorter than the diffusion time of the particles. On the other hand, in number and brightness analysis the time interval can be arbitrarily long. However, the integration time (the time for which photons contributing to $F(i)$ are collected) should still be shorter than the diffusion time to avoid averaging out of concentration fluctuations. This is easily achieved when recording a temporal series of images with a confocal microscope. The pixel dwell time (the integration time) can be arbitrarily short, while the interval between corresponding pixels in the consecutive frames (the time per frame) is typically longer than the diffusion time of most molecules, preventing calculation of temporal correlations in each pixel as it is done in imaging FCS. Let us close with a remark about the detectors and photon counting noise. The assumptions about photon counting noise used in deriving equation (9.10) are satisfied only for photon counting detectors. If the images are acquired with a camera, which is not a true photon counting detector, the camera noise has to be correctly characterised [17].

9.2 Characterising molecular diffusion by FCS

As we have already mentioned here in section 9.1.2.2, information on molecular mobility can be extracted from the temporal decay of the concentration fluctuations ACF. If the particles move by Brownian diffusion, their mobility is characterised by their diffusion coefficient D. D depends on the size and the shape of the particles as well as on the medium they are diffusing in. The information on molecular diffusion obtained by FCS experiments can be used to:

(i) Determine the size of unknown particles by studying their diffusion in a defined medium. FCS is frequently used to measure changes in particle size through changes in their diffusion coefficient; such changes are then related to molecular interactions as we will discuss in section 9.3.

(ii) Characterise the environment in which particles of known size are moving. In this way, FCS is often employed to study complex inhomogeneous environments such as cell nuclei [18] or bacterial biofilms [19, 20]. In such complex environments molecular diffusion is often obstructed, resulting in deviations from Brownian diffusion that are dependent on the particle size as well as the length scale over which the diffusion is probed. For example, when diffusing through a filament network or through porous structures, particles smaller than the size of the typical pore or the typical opening in the net experience little obstruction to their movement. On the other hand, the movement of particles larger than the average pore or opening will be severely restricted. In this way chromatin organisation has been studied by FCS [21]. Biological membranes are another complex environment in which molecular diffusion is often probed by FCS; we will talk more about them in section 9.2.5.

Just like other quantitative readouts from FCS experiments, the diffusion coefficient is typically obtained by fitting the ACF with an appropriate theoretical model. If we take the basic model for confocal FCS as expressed, for example, by equation (5.5), we can notice that D enters the model only in the form of the diffusion time, $\tau_D = \frac{w_0^2}{4D}$.

Diffusion coefficient and the effective observation volume radius, w_0, are completely interdependent in the model and they cannot be determined independently from the fit. To quantify D based on the results of confocal FCS, it is, therefore, necessary to calibrate the effective observation volume radius.

9.2.1 Confocal observation volume radius calibration

The calibration is analogous to the effective observation volume calibration described in section 9.1.2. We can use again the same three approaches: FCS measurement in a solution of known particle concentration or in a solution of particles of known diffusion coefficient or scanning a small fluorescent bead.

In this case, when we are interested primarily in D, the approach using particles of known diffusion coefficient, $D^{(\text{ref})}$, is the most straightforward. By fitting the ACF measured in the reference solution, we obtain the diffusion time of the reference particles $\tau_D^{(\text{ref})}$. The diffusion coefficient, D, of the particles of interest is then calculated from their diffusion time, τ_D, according to equation (5.6) (see also equation (5.20)) as:

$$D = \frac{\tau_D^{(\text{ref})} D^{(\text{ref})}}{\tau_D} \tag{9.14}$$

The structure parameter K can be obtained independently of w_0 by ACF fitting. In principle, it does therefore not require calibration. However, to reduce the number of free parameters involved in fitting the ACFs measured in the samples of interest, K is typically determined by fitting the ACF measured in the reference solution and the value, thus obtained, is used in fitting all other ACFs. This is particularly important when the ACFs measured in the sample of interest require fitting with more complex models such as equation (5.47) for diffusion and blinking or equation (5.27) for multiple particle populations with distinct diffusion coefficients. Reducing the number of free parameters by fixing K can in such cases improve considerably the numerical stability of the fit.

When a solution of fluorescent particles of known concentration is used for the calibration, the structure parameter, K, is determined from the fit of the reference ACF and the effective observation volume from its amplitude according to equation (9.5). The radius of the effective observation volume, w_0, is then calculated according to equation (9.6). It was shown by simulations that the ACF amplitude is more sensitive than the diffusion time to distortions of the effective observation volume by optical aberrations and other imperfections in the setup [9]. This, together with the uncertainties in the reference solution concentration discussed in section 9.1.2.1, make this calibration strategy more prone to errors than the strategy relying on particles of known diffusion coefficient.

The last calibration strategy, imaging a small fluorescent bead, can also be applied. The effective observation volume dimensions, w_0 and z_0 can be obtained as e^{-2} radii of Gaussian fits of the fluorescence intensity distribution; the structure parameter K is then calculated as the z_0/w_0 ratio (equation (5.7)) [12]. In an approach termed numerical FCS, no assumptions are made about the shape of the effective observation volume; instead the actual intensity distribution obtained by scanning a small bead is used and the measured ACF is compared to theoretical ACFs generated by simulating diffusion of particles in the experimentally obtained effective observation volume. Alternatively, the theoretical ACF can be also calculated by numerically evaluating the integrals describing the ACF for diffusion in the measured effective observation volume (a procedure analogous to the ACF derivation described in math box 5.2, where the integrals for a three-dimensional Gaussian observation volume were solved analytically). By varying the diffusion coefficient (and/or any other free parameters of the model) used in the simulations, the value of D is sought for which the best agreement between the measured and the theoretical ACF is reached [22]. This approach is applicable to an arbitrary shape of the effective observation volume or arbitrary deviations of the effective observation volume from its theoretical model, albeit at the price of higher computational complexity of the data analysis, a concern that is alleviated by the increasing computational power of commonly available computers.

9.2.2 Calibration-free FCS modalities

Although, as we have mentioned in the previous section, determination of diffusion coefficient is less affected by the uncertainty in the confocal effective observation volume calibration than concentration measurements [9], the calibration is nevertheless a potential source of errors. Fortunately, there exist several FCS modalities, some of which can be readily implemented at standard commercial setups, able to measure absolute diffusion coefficients without the need of an external calibration, offering improved reproducibility of D measurement compared to standard confocal FCS. Typically, these approaches utilise spatiotemporal correlations, in which the temporal decay of the ACF for particles of a given diffusion coefficient, D, depends not only on the microscope PSF (like in standard confocal FCS), but also on some other distance, which is known with high precision. This second distance then provides an intrinsic length scale against which the PSF and diffusion coefficient are measured. Introducing this distance to the theoretical model for the ACF breaks the interdependence between D and w_0, found in the models for standard confocal FCS (see equations (5.6) and (5.7)), allowing independent fitting of D without an *a priori* knowledge of the PSF. Principles and theory of these techniques have been already explained in chapter 5, we will, therefore, address here briefly just the main practical considerations.

9.2.2.1 Two-focus FCS

Two-focus FCS measures fluorescence intensity fluctuations simultaneously in two observation volumes; the CCF between signals from the two volumes is, for freely diffusing particles, given by equation (5.51), which can be fitted with D, w_0 and z_0 as free parameters, provided the displacement between the two observation volumes is exactly known. In principle the displacement can be arbitrarily large; however, the exponential decrease of the CCF with the displacement imposes practical limitations on upper displacement limit. This reflects the diminishing probability that a

Brownian diffusive trajectory will connect the two observation volumes as their distance increases. The situation is different in the presence of oriented motion, which, for displacements parallel to the preferential motion direction, enhances the probability of particles moving from one observation volume to the other. However, oriented motion is not the case we are interested in here.

To study purely diffusive motion, it is optimal if the two observation volumes are very close to each other, ideally partially overlapping. The proximity of the observation volumes comes with another challenge, crosstalk between the signals from the two observation volumes. An elegant solution was proposed by the group of Jörg Enderlein [23] who used two linearly polarised pulsed lasers emitting alternating pulses of the same wavelength but with polarisation perpendicular to each other. A birefringent prism was placed under the objective to introduce a well-defined small displacement between the foci of the two differently polarised laser beams. Such prisms are commonly available in commercial confocal microscopes, being used for differential interference contrast imaging. The displacement is a characteristic of the prism and once determined accurately, does not need to be repeatedly calibrated. In the setup used in the original work cited above, the displacement was found to be approximately 400 nm, resulting in two slightly overlapping observation volumes. Photons originating from different observation volumes are separated in time and can be, therefore, detected by a single detector. As the two lasers emit alternating pulses, only particles within one of the observation volumes are excited by each pulse. Photons detected following a laser pulse can be, thus, assigned to have originated from the observation volume excited by the given pulse. This concept is analogous to PIE-FCCS explained in section 6.3.2. The same considerations concerning the repetition frequency of the laser pulses which we have discussed there in the context of PIE-FCCS also apply here. Two-focus FCS in this configuration has been used among other applications as a reference technique to determine diffusion coefficient values of common fluorophores used for calibration in standard confocal FCS (see sections 9.1.2.2 and 9.2.1) [48].

9.2.2.2 Scanning FCS and RICS

In these FCS modalities spatiotemporal ACFs are measured by moving a confocal observation volume along a defined trajectory in the sample. The trajectory could be a line or a circle (in scanning FCS) or a whole scanned image (in RICS). As such the techniques can be performed at standard laser scanning confocal microscopes without the need for any specialised modifications. The intrinsic length scale of these technique relies on the calibration of the confocal scanner, which is typically sufficiently accurate and reproducible.

In scanning FCS, the spatiotemporal ACF is a two-dimensional function of two variables; distance along the scanned trajectory and time lag. The profile of the ACF along the spatial distance axis is, for immobile particles, equal to the microscope PSF. For mobile particles, which is the more common situation, it is the PSF broadened by the motion. The faster the motion, the broader the ACF becomes, because the probability that particles will reach a point further along the trajectory at the same time as the scanner increases. For isotropic diffusion the broadening is symmetric; however, it becomes asymmetric in the presence of preferential directions of motion (see section 5.7.2). The profile of the ACF along the time-lag axis is a standard confocal ACF recorded with temporal resolution equal to the scan time (the time needed to complete one repetition of the trajectory) and effective observation volume determined by the microscope PSF. Fitting the whole two-dimensional surface with a suitable theoretical model then yields the diffusion coefficient as well as the dimensions of the effective observation volume [24]. For reliable fitting it is important that the scan time, which determines the resolution along the lag-time axis, is sufficiently shorter than the particle diffusion time. With modern resonant scanners achieving scan times of the order of 10 μs this does not present a serious limitation.

Furthermore, cross-correlation between fluorescence intensity time traces recorded at different positions along the scan trajectory provides information on the presence of barriers to particle movement. Asymmetry in the cross correlation is an indication of semi-permeable barriers that can be crossed by the particles only in one direction (see figure 9.3) [25].

The correlation function in RICS is obtained by spatially correlating pixel values in the scanned image. The temporal information enters into the spatial correlation function through the process of scanning, which means that the intensity values in different pixels are recorded at different times. The spatial correlation function is again a two-dimensional function which depends on spatial distances along two

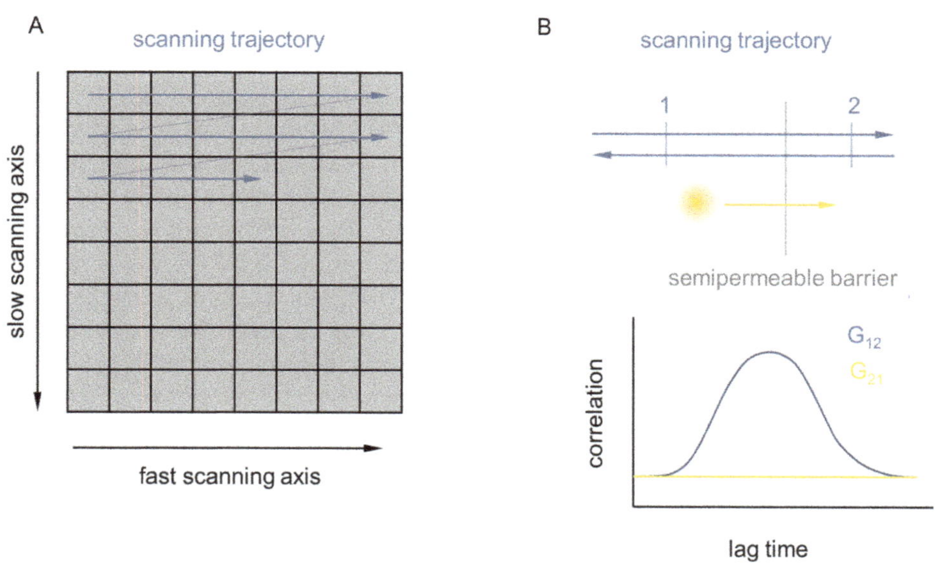

Figure 9.3. A schematic illustration of the fast and slow scanning axis in RICS (panel A) and of the identification of barriers to diffusion in scanning FCS by cross-correlating fluorescence intensity time traces obtained at different positions along the scan trajectory (panel B). The situation for a semipermeable barrier is illustrated here. The particles can cross the barrier only in one direction indicated by the yellow arrow. As a consequence, the CCF is asymmetric, in one direction it is zero, while in the opposite direction it is positive with a maximum at the lag time corresponding the average time needed for the particles to travel from point 1 to point 2.

direction, each of which contains temporal information on a different timescale. The temporal difference between neighbouring pixels in one of the directions, the direction parallel to the movement of the scanner along a line in the image (the fast scanning axis), is equal to the pixel dwell time; in the other direction (the slow scanning axis) it is equal to the time per line (see section 5.7.6 and figures 5.22 and 9.3). The spatial correlation can be again regarded as the PSF of the microscope broadened by the particle motion; however, the broadening would be different in the two directions because of the different timescales probed. If the scanning speed is too slow compared to the particle motion, the correlation function will extend only along the fast scanning axis; intensities in pixels neighbouring in the direction of the slow scanning axis will not be correlated, because the particles have moved too far before the observation volume has completed scanning a line. The correlation function is then effectively a one-dimensional function which cannot be reliably fitted with both D and w_0 as free parameters. On the other hand, if the scanning is too fast compared to the particle motion, the particles appear as nearly immobile and correlation function differs only very little from the PSF. The correlation function depends, therefore, only weakly on particle diffusion which would also compromise the reliability of fitting D.

An additional advantage of the scanning-based approaches is reduction in fluorophore photobleaching. This makes scanning FCS or RICS very well suited for live-cell FCS studies [26].

9.2.3 Imaging FCS and its calibration

The lateral extent of the effective observation volume in imaging FCS is determined by the convolution of the pixel area in the sample space (the area of a pixel on the camera chip divided by the total magnification of the imaging system) with the microscope PSF. The axial extent of the effective observation volume depends on the illumination scheme employed. In ITIR-FCS, it is defined by the penetration depth of the evanescent field in the TIRF illumination; in SPIM-FCS it is related to the thickness of the light sheet (see section 5.7.7). The thickness of the light sheet in SPIM can be characterised, for example, by reflecting the light sheet to the imaging objective and measuring the width of the imaged intensity profile [27]. Note that ITIR-FCS is typically used to study systems that are effectively two-dimensional such as cell membranes, in which case the ACFs are independent of the axial extent of the effective observation volume (see section 9.2.5).

If we look at the theoretical models for the imaging FCS ACFs derived in math box 5.7, we can notice, that unlike in standard confocal FCS, D and w_0 are here not fully interdependent. This is the result of the dependence on the camera pixel size, which represents an intrinsic length scale similarly to the inter-foci distance in 2-focus FCS or scan trajectory in scanning FCS. Unfortunately, the mutual dependence between D and w_0 in the model is still too strong to allow reliable determination of both w_0 and D as free fitting parameters, which is possible in the calibration-free modalities discussed in section 9.2.2. Therefore, an external calibration of w_0 is needed for quantitative diffusion coefficient measurements by imaging FCS.

However, it has been demonstrated experimentally that the calibration is in this case very robust and therefore introduces little uncertainty in the D measurements [28]. Since the applicable calibration strategies differ from those used in confocal FCS (see sections 9.1.2 and 9.2.1) we will present them briefly here.

In imaging FCS, the e^{-2} radius of the PSF, w_0, can be calibrated using the fact that the diffusion coefficient, D, of freely diffusing particles is independent of the observation volume. No matter what the size of the effective observation volume is, fitting the ACF with an appropriate model should return the same value of D, provided we input the correct w_0 value into the model. A single imaging FCS measurement contains information on diffusion in effective observation volumes ranging in size from the effective volume corresponding to a single pixel to the effective volume corresponding to the whole imaged area. Intensity time traces corresponding to different effective observation volumes are obtained by simply adding the intensity traces from several neighbouring pixels. The effective observation volume depends on the convolution of the area of the summed pixels with the microscope PSF; small effective volumes depend mainly on the PSF, while large volumes depend mainly on the area of the pixels which have been summed. As a result, the effective observation volume is not simply proportional to the area of the summed pixels. The following calibration procedure can be then employed to determine w_0:

(i) An imaging FCS measurement is taken in a sample of freely diffusing particles. Aqueous fluorophore solutions, used for confocal FCS calibration, are not suitable in this case, because the diffusion of small water-soluble fluorophores is too fast to be accurately characterised by imaging FCS, because of the limited temporal resolution of commonly available cameras. Instead, suspensions of fluorescent beads can be used for SPIM-FCS calibration and supported lipid bilayers (SLBs), in which the fluorescent lipid analogues are supposed to diffuse freely in the plane of the bilayer, for ITIR-FCS calibration [27, 28]. If the diffusion of the particles in the calibration sample is hindered, for example, by transient binding of the particles to immobile binding sites, the apparent diffusion coefficient changes with the effective observation volume, which invalidates the key assumption of the calibration procedure. This can be an issue with SLBs used for ITIR-FCS calibration if the fluorescently labelled lipids interact electrostatically with the surface of the solid support [29]. Another potential issue that can invalidate the calibration results is the influence of photobleaching. With increasing effective observation volume, the fraction of fluorescent particles that get bleached while transiting the observation volume increases. The residence time, τ_D, of those particles in the effective observation volume is shortened by photobleaching which results in a positive bias in the diffusion coefficient value. The bias in D increases with the effective observation volume, which again invalidates the key assumption of invariance of D.

(ii) ACFs for different effective observation volumes are calculated, starting with the ACFs for each individual pixel, then ACFs for 4 neighbouring pixels (a square of 2 × 2 pixels) and so on. In all cases, the diffusion time should be more than an order of magnitude shorter than the total measurement time [30]. If this condition is not satisfied, the diffusion time of those particles which stay the longest in the effective observation volume will exceed the accessible lag time range, introducing a bias towards shorter τ_D. Since τ_D increases with the effective observation volume, this effect will be more pronounced for larger observation volumes. The apparent diffusion coefficient will therefore increase with effective observation volume, similarly to the effect of bleaching mentioned in point (i).

(iii) Diffusion coefficient as a function of the effective observation volume is obtained by fitting the ACFs with an appropriate theoretical model using w_0 as a fixed parameter. If the correct value of w_0 is used, D is independent of the effective observation volume. Thus, by performing the fitting for a range of w_0 values, we can estimate the correct one. D is underestimated when w_0 is smaller than the actual value and overestimated in the opposite case. With increasing effective observation volume D converges to its correct value independently of the w_0 value used, because the larger effective observation volumes are determined mainly by the area of the pixels and less by the PSF.

Note that the calibration procedure does not require the knowledge of the diffusion coefficient of the fluorescent particles in the calibration sample; the only requirement is the invariance of D with respect to the effective observation volume.

Instead of ACFs for different effective observation volumes, CCFs between pixels at different distances apart could be used in the procedure outlined above. However, since the probability that the trajectory of a particle will connect two particular pixels decreases with the inter-pixel distance, the CCF become increasingly noisy for larger distances between pixels, which reduces the reliability of their fitting. Alternatively, w_0 can be also extracted from images of small fluorescent beads in analogy to the calibration procedure described in section 9.1.2.3 [28].

9.2.4 Separation of multiple processes contributing to the ACF temporal decay

We have so far in this chapter considered mainly the concentration fluctuations ACF, which contains information on fluctuations caused by translational motion of the fluorescent particles. In this section we will look at other processes that can contribute to the measured ACFs and discuss shortly how these different contributions can be separated experimentally.

The easiest way to separate two processes is using their functional dependence on time as this will lead to different shapes of the ACF. We have seen already that diffusion leads to a hyperbolic decay in the ACF while triplets, rotation, isomerization and other photophysical processes have typically an exponential decay with the lag time τ. In section 5.5, we saw that active transport and flow will also lead to an exponential term in the ACF. However, in cases that the time scales of the different processes are similar it can become difficult to fit and differentiate the distinct functional forms and their characteristic times. We therefore provide here some simple suggestions on how the different processes can be separated.

Concentration fluctuations caused by translational motion, that is by diffusion and flow, depend on the size of the observation volume, while all other processes, such as fluorophore blinking or transient binding to immobile binding sites, do not. Therefore, changing the observation volume is a simple way to differentiate translational motion from other processes. There exist several ways of changing the effective observation volume:

(i) By changing the size of the confocal pinhole [31]; this approach can be readily used with any commercial confocal microscope. The confocal pinhole controls mainly the axial extent of the effective observation volume; increasing the pinhole diameter will increase V_{eff} mainly by increasing z_0.

(ii) By under-filling the back aperture of the objective; only when the back aperture of the objective is fully filled with the excitation laser beam, the radius of the beam in the objective focus, w_0, reaches its smallest possible value determined by the numerical aperture of the objective. If the excitation laser beam entering the objective has a diameter smaller than the objective back aperture, the radius of the focused beam becomes larger; the narrower the beam entering the back objective back aperture, the larger the focus radius, w_0, and therefore also the effective observation volume (see section 6.2.2 where we mentioned objective under-filling as a possible strategy of compensating wavelength dependence of the effective observation volume). Only some commercial confocal systems, allow users to control the diameter of the laser beam entering the objective back aperture.

(iii) By reducing the effective observation volume using stimulated emission depletion (STED) microscopy; the PSF in STED is controlled by the power of the depletion laser, starting from a standard confocal PSF when the depletion laser is turned off and becoming smaller with increasing depletion laser power [32]. Microscopes with STED capability are commercially available from multiple manufacturers.

(iv) By confining the sample to volumes smaller than the microscope focal volume. In this case the effective observation volume is limited by the sample extent rather than by the PSF of the microscope. While not requiring specific microscope setup, this approach depends on specialised sample preparation, which limits the range of system that can be studied in this way [8, 33].

(v) In imaging FCS, we can generate larger effective observation volumes by simply binning pixels in the image. Compared to the previous four approaches applicable to confocal setups, this widefield approach has the advantage of not requiring separate measurements for different effective observation volumes. A single dataset contains information on a range of effective observation volumes starting with the smallest volumes corresponding to individual pixels and ending with the volume corresponding to the entire image. This straightforward approach is also used in imaging FCS for PSF calibration (see section 9.2.3).

Spatiotemporal ACFs obtained in techniques such as scanning FCS or RICS also allow distinguishing translational motion from other processes without the need for repeated measurements with different effective observation volumes. Processes other than translational motion of the particles influence only the temporal decay of the spatiotemporal ACF without broadening its spatial extent. This strategy has been used, for example, in RICS as a robust way of distinguishing translational diffusion from transient binding to immobile binding sites [34].

Diffusion changes also with viscosity of the solution and in some cases this can also help to identify a diffusive process. However, this is only applicable to *in vitro* systems, where we can control the viscosity of the medium and will not influence only translational but also rotational diffusion.

If we are, on the other hand interested in other processes than translational motion, we can in some cases eliminate the concentration fluctuations completely by immobilising the molecules of interest. In this way blinking of fluorophores used in single molecule localisation microscopy (SMLSM) has been characterised by FCS [35].

A more difficult task is to differentiate multiple processes that do not depend on the effective observation volume, such as blinking caused by triplet state dynamics, photo-induced isomerization, protonation dynamics or rotational diffusion. Dependence of the individual processes on experimental conditions can help to distinguish their contributions. Let us mention a few examples here. The characteristic blinking time corresponding to protonation–deprotonation dependent blinking will change with pH, unlike the characteristic blinking time caused by other processes. As we have already mentioned, rotational diffusion becomes slower with increased viscosity. Photo-induced processes such as photo-isomerisation or triplet blinking depend on the excitation intensity; however, how they depend on the excitation intensity is not identical and can be used to differentiate those two processes. An increase in laser power will accelerate photo-isomerisation [36] and at the same time increase the fraction of molecules in the triplet state. It has been also demonstrated that the build-up of triplet-state population can be efficiently suppressed by excitation intensity modulation with periods of the order of μs to 10 μs [37]. Under such excitation, the ACF will be also modulated with the same frequency as the excitation. The authors of the cited study used a simple and efficient way to obtain ACFs free from the periodic modulation; this strategy is quite generic with many other potential applications in FCS; we will therefore explain it here.

Under the modulated excitation, the measured intensity trace $F(t)$ is a product of the excitation intensity modulation $M(t)$ and an 'ideal' fluorescence intensity trace $\tilde{F}(t)$, which is the intensity trace that we would measure with constant excitation intensity and its fluctuations are caused purely by the processes of interest occurring in the sample (such as concentration fluctuations or fluorophore blinking).

$$F(t) = \tilde{F}(t)M(t) \tag{9.15}$$

According to equation (2.20), the ACF can be written as:

$$\begin{aligned} G(\tau) &= \frac{\langle F(t)F(t+\tau)\rangle}{\langle F(t)\rangle^2} = \frac{\langle \tilde{F}(t)M(t)\tilde{F}(t+\tau)M(t+\tau)\rangle}{\langle \tilde{F}(t)M(t)\rangle^2} \\ &= \frac{\langle \tilde{F}(t)\tilde{F}(t+\tau)\rangle\langle M(t)M(t+\tau)\rangle}{\langle \tilde{F}(t)\rangle^2\langle M(t)\rangle^2} = \tilde{G}(\tau)G_M(\tau) \end{aligned} \tag{9.16}$$

where we used the fact that $\tilde{F}(t)$ and $M(t)$ are processes independent of each other. We can see that the ACF is in this case a product of the 'ideal' ACF, $\tilde{G}(\tau)$, which contains information on the processes occurring in the sample, and the ACF of the excitation intensity modulation, $G_M(\tau)$. $\tilde{G}(\tau)$ is then obtained simply by dividing the overall ACF, $G(\tau)$, with $G_M(\tau)$. Note that the same strategy can be applied to any process that modulates the fluorescence intensity trace of interest, $\tilde{F}(t)$, as long as its time trace, $M(t)$, is known. It was used, for example, to eliminate the influence of instability in excitation laser power [5]. In both applications mentioned, $M(t)$ is obtained by continuously monitoring the laser power by a dedicated photodetector.

9.2.5 Diffusion in two-dimensional systems studied by FCS

If the diffusion of the particles of interest is confined to a surface or a thin layer of thickness negligible compared to the dimensions of the effective observation volume, the signal will be collected only from an effective observation area defined by the intersection of the effective observation volume with the surface or thin layer to which the particles are confined. Examples of such systems commonly studied by FCS are biological membranes and their artificial models, supported lipid bilayers (SLBs) or giant unilamellar vesicles (GUVs). The theoretical models for the ACF simplify in this case; considering that the membrane is oriented parallel to the focal plane of the microscope, the particle number and diffusion time will be independent of the axial extent of the microscope PSF. For example, the model for standard confocal FCS, expressed by equation (5.5) will be simplified to the form of equation (5.8).

However, the effectively two-dimensional sample geometry is connected with specific challenges in FCS, namely with what we will call here the positioning problem. The underlying effect is illustrated schematically in figure 9.4. The effective observation area is the smallest when the sample is positioned to the focal plane of the microscope objective, where the effective beam radius w_0 is the smallest. Moving the sample below or above the focal plane results in a larger effective observation area, which translates to a higher particle number and a longer diffusion time measured at such a position. The dependence of the diffusion time on the sample position means that uncertainty in the position will introduce additional uncertainty to the measured diffusion coefficient, D [38]. The signal-to noise ratio of the ACFs measured with the sample above or below the focal plane will be also compromised. The excitation intensity (which is the power of the laser beam divided by its cross-

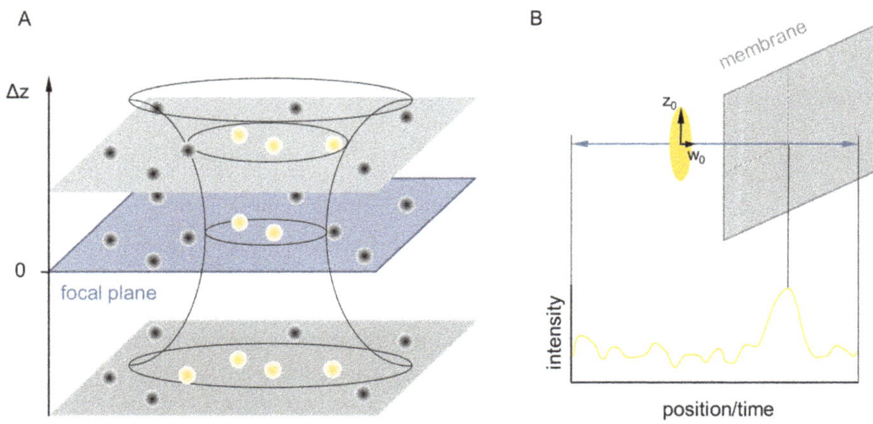

Figure 9.4. A schematic illustration of the positioning problem of FCS in two-dimensional systems (panel A) and of eliminating the influence of membrane undulations by scanning with the effective observation volume in a direction perpendicular to the membrane plane (panel B). The intensity trace recorded during each iteration of the scan trajectory will contain a peak corresponding to the point where the effective observation volume crosses the membrane.

section) and the fluorescence collection efficiency are both decreasing with the distance from the focal plane, Δz, leading to reduced particle brightness.

The correct positioning of the sample to the focal plane can be, therefore, identified by the maximum of the particle brightness measured by FCS. However, searching for the optimal sample position in this way is not very easy. We need to monitor the brightness continuously while focusing the microscope; this means that the brightness values need to be obtained from very short FCS measurements, to give us a real-time feedback on the brightness changes. ACFs obtained in such a way are likely to be very noisy, resulting in large uncertainty in their amplitudes and consequently in the brightness values calculated from those amplitudes. This then translates to large uncertainty in the sample positioning. The average fluorescence intensity also has a maximum at the focal plane; however, its dependence on Δz is less steep as the decrease in particle brightness is partially compensated by the increase in particle number. Furthermore, the intensity is expected to fluctuate a lot (which is what is optimal for FCS), making positioning the sample according to the intensity maximum equally challenging and likely less precise than when the brightness maximum is used. In live cells, the search for the optimal position may be further complicated by bleaching of the fluorescently labelled molecules, which requires that the correct position is found fast without exposing the cells to excitation light for too long prior to the actual FCS acquisition.

The results of calibration-free FCS modalities are robust against the positioning problem, because the information on the actual size of the effective observation area is extracted from the fit of each correlation function. The correct values of D and C can be then obtained from measurements in samples positioned above or below the focal plane as long as the brightness of the fluorescent particles is not reduced too much by the distance from the optimal position. This makes techniques such as scanning FCS or RICS well suited for FCS measurements in membranes. Additionally, the slow diffusion of molecules in membranes relaxes the requirements on the speed of scanning and confocal microscopes without resonant scanners can be used in these applications. In some cases, when measuring FCS in free-standing membranes, such as membranes of GUVs or cell membranes that do not adhere to a surface, the membranes can rapidly fluctuate, changing the effective observation area during the FCS acquisition. The fluorescence intensity trace is in such cases not stationary leading to distorted ACFs. Calibration-free FCS modalities are not robust against this effect which could be viewed as a 'positioning problem on a faster timescale'. One way of addressing this issue is by performing scanning FCS in a modified geometry. Instead of placing the membrane parallel to the focal plane and scanning the observation volume along a trajectory parallel to the membrane, we can place the membrane perpendicular to the focal plane and scan the observation volume along a line perpendicular to the membrane, as schematically shown in figure 9.4 [39]. The recorded florescence intensity trace contains peaks at the times when the effective observation volume crosses the membrane. As the membrane fluctuates, the membrane-crossing happens at a different point along the scan trajectory. If we then take the fluorescence intensities of the membrane-crossing peaks from all iterations of the scan trajectory, we can reconstruct a fluorescence intensity time trace corresponding to the membrane positioned correctly in focus. The temporal resolution of the intensity time trace obtained in this way is determined by the time per line of the scanning, which is not limiting for studying diffusion in membranes. Note that in this case the effective observation volume is not radially symmetric but stretched in the axial direction, therefore the theoretical model of equation (5.9) applies in this case.

An FCS modality that was developed particularly to address the positioning problem is z-scan FCS. This modality makes use of the Δz dependence of τ_D and N, which is the basis of the positioning problem [40]. FCS measurements are taken with the sample plane at different positions along the z-axis (the direction along the optical axis of the microscope) and τ_D and N are obtained for each position by fitting the respective ACF. Assuming that the effective observation area depends quadratically on the distance from the focal plane, the Δz dependence of τ_D and N can be written as:

$$\tau_D(\Delta z) = \frac{w_0^2}{4D}\left[1 + \left(\frac{\lambda\ \Delta z}{\pi w_0^2}\right)^2\right] \tag{9.17a}$$

$$N(\Delta z) = \langle C\rangle \pi w_0^2\left[1 + \left(\frac{\lambda\ \Delta z}{\pi w_0^2}\right)^2\right] \tag{9.17b}$$

where λ is the excitation wavelength. By fitting the experimentally obtained Δz dependence of τ_D and N with equations (9.17a) and (9.17b), we can extract the values of D, average particle surface concentration, $\langle C \rangle$, and of the effective observation area radius in the focal plane, w_0. The main disadvantage of z-scan FCS is the necessity to perform multiple FCS measurements in series, which prolongs the measurement time and in the case of cells can lead to increased photobleaching and phototoxicity. Another concern is the mechanical stability of the microscope over the prolonged times needed for z-scan measurements. Slow axial mechanical drift of the sample stage or the objective z-drive will compromise the accuracy of z-spacing of the planes at which the measurements are taken. This will then compromise the accuracy with which D and $\langle C \rangle$ are obtained from equations (9.17a) and (9.17b). For these reasons z-scan FCS has been largely superseded by other calibration-free modalities such as scanning FCS or RICS.

Another FCS modality with a proven track-record in membrane studies is imaging FCS. The large membrane area probed by imaging FCS provides a more representative picture of membrane heterogeneity. Moreover, each imaging FCS measurement contains information about a broad range of effective observation areas, which can be utilised to investigate membrane organisation by analysing the FCS diffusion law (see section 5.7.8).

9.3 Molecular interactions studies by FCS

9.3.1 Using changes in the ACF amplitude

As discussed in section 9.1, the changes in the ACF amplitude reflect the changes in the average particle number N, that is in the concentration of the particles. Therefore, ACF amplitude changes can be used to study molecular interactions only when those interactions lead to changes in concentration of fluorescent particles. In simple 1:1 binding reactions, i.e. when one fluorescently labelled ligand binds to one non-fluorescent receptor, or when multiple (H) non-fluorescent molecules bind to one fluorescent probe in a 1: H binding process, the concentration of independently moving fluorescent particles will not change and the amplitude will stay constant. In these cases, only the diffusion time can be used to investigate whether binding has taken place (see section 9.3.2). However, if the stoichiometry is H:1, i.e. multiple fluorescent particles bind to a non-fluorescent receptor, the amplitude will change, as the number of independently moving fluorescent particles will decrease.

As an example, we will discuss the binding of biotin to avidin, a classical ligand receptor model with very strong binding affinity ($K_D \sim 10^{-15}$ M) [41]. In this system four biotin molecules bind to one avidin protein. Using solutions of fluorescently labelled biotin in the nM range one titrates avidin into the solution. Due to the very high affinity, all four binding sites of avidin are immediately occupied until the concentration of avidin is a quarter of the biotin concentration at which the binding saturates, and no more free ligands are available. Increasing avidin concentration beyond the saturation concentration will not alter the measured ACF as the excessive avidin molecules will not have any fluorescent ligands attached to them. As each avidin carries now four fluorescent biotins, the number of independently diffusing fluorescent particles reduces by a factor of four. It should be noted that for these experiments it is important to reach a receptor concentration that can bind all available ligands [42].

If the titration is performed the other way round by adding fluorescently labelled biotin to a fixed concentration of unlabelled avidin, the interpretation of the ACF amplitude in terms of binding stoichiometry is not as straightforward as in the former case. No saturation will be reached in this case as adding the labelled ligand in excess will increase the number of independently diffusing fluorescent particles and, thus, decrease the ACF amplitude. There will be a phase during the titration when all avidin molecules are bound to at least one fluorescent ligand but at the same time the available binding sites are not yet saturated. Within this phase, increasing the biotin concentration does not change the number of independently moving fluorescent particles, nevertheless, that does not mean that the ACF amplitude is constant throughout this phase. The reason behind that is the changing distribution of particle brightness resulting from growing fraction of binding sites occupied by fluorescent ligands. According to equation (5.28), the ACF amplitude obtained in a sample containing multiple populations of particles differing in brightness is given by

$$G_0^{(\mathrm{conc})} = \frac{\sum_i \kappa_i^2 N_i}{\left(\sum_i \kappa_i N_i\right)^2} \tag{9.18}$$

where κ_i and N_i are the brightness and number of particles of an individual population. The probability of a substrate with H binding sites carrying η ligands would be in such a situation described by a binomial probability distribution

$$P_B(\eta) = \binom{H}{\eta} p^{\eta}(1 - p)^{H-\eta} \tag{9.19}$$

where the probability, p, of a binding site being occupied is given by the ratio of the total ligand molecules number to the total binding site number. In terms of ligand and substrate concentrations, C_L and C_S, it can be expressed as $p = \frac{C_L}{HC_S}$. If we assume that the brightness of a complex containing η ligands is η times the brightness of a free fluorescently labelled ligand (which assumption is not necessarily always satisfied as we discuss in the next section), we can write for the ACF amplitude

$$G_0^{(\mathrm{conc})} = \frac{\sum_{\eta=1}^{H} (\kappa\eta)^2 P_B(\eta) N_S}{\left(\sum_{\eta=1}^{H} \kappa\eta P_B(\eta) N_S\right)^2} = \frac{\sum_{\eta=1}^{H} \eta^2 P_B(\eta)}{N_S \sum_{\eta=1}^{H} \eta P_B(\eta)} = \frac{Hp - p + 1}{N_S} \tag{9.20}$$

where N_S is the number of substrate molecules, $N_S = C_S V_{\mathrm{eff}}$. In deriving the equation above we used the fact that the sums in the numerator and denominator are by definition the expressions for the second and the first moment of binomial distribution, respectively [43] (see exercise 3). Note that this is true only under the assumption that the brightness of a complex is proportional to η; for a general dependence of brightness on η, $\kappa(\eta)$, the ACF amplitude could be written as

$$G_0^{(\text{conc})} = \frac{\sum_{\eta=1}^{H} \kappa^2(\eta) P_B(\eta)}{N_S \left(\sum_{\eta=1}^{H} \kappa(\eta) P_B(\eta) \right)^2} \tag{9.21}$$

In general, all terms in the sums would have to be evaluate using the appropriate form of $\kappa(\eta)$.

When the number of binding sites, H, available on each particle is very high and their average occupancy, determined by p, is low, the probability of finding a particle with η ligands is given by the Poisson distribution $P_P(\eta)$, which can be regarded as a limiting case of the binomial distribution (equation (9.19)) for H approaching infinity. This is, for example, the case of lipophilic fluorescent molecules binding to liposomes or detergent micelles [44]. The probability distribution is then

$$P_P(\eta) = \frac{\bar{\eta}^{\eta} e^{-\bar{\eta}}}{\eta!} \tag{9.22}$$

where the average number of ligands per particle, $\bar{\eta}$, is given by the ratio of ligand and substrate concentrations, $\bar{\eta} = \frac{C_L}{C_S}$. In analogy to equation (9.20), the ACF amplitude, under the assumption of the brightness being proportional to η, can be written as:

$$G_0^{(\text{conc})} = \frac{\sum_{\eta=1}^{\infty} \eta^2 P_P(\eta)}{N_S \sum_{\eta=1}^{\infty} \eta P_P(\eta)} = \frac{\bar{\eta} + 1}{N_S} = \frac{C_L + C_S}{C_S^2 V_{\text{eff}}} \tag{9.23}$$

where we used the fact that the second moment of Poisson distribution is $\bar{\eta} + \bar{\eta}^2$ and the first moment is the average value $\bar{\eta}$.

9.3.2 Using changes in the ACF temporal decay

In addition to the amplitude of the ACF, one can also use the width, i.e. the diffusion time τ_D, to measure interactions. As we have seen earlier, the diffusion time is proportional to the particle hydrodynamic radius (see equations (5.6) and (9.7)), which also means it is approximately proportional to the cubic root of its volume; assuming constant particle density, it is, thus, proportional to the cubic root of the particle mass. This is rather a week dependence on the mass. To obtain, for example, a factor 2 difference in diffusion time, the two particles need to differ by factor 8 in their mass. Discussing the implications of this weak mass dependence of the diffusion time requires the distinction of at least three cases:

(i) A quantitative measurement of the interaction of two binding partners is to be achieved and there is no information on the sample available.

(ii) A quantitative measurement of the interaction of two binding partners is to be achieved and we have the information of the diffusion time of at least one partner.
(iii) A qualitative answer is required whether binding takes place.

The first case is the most demanding one. It was shown that in order to retrieve reliably the unknown diffusion times of the two particle populations, the complex and the free fluorescently labelled binding partner, and the fraction of the bound population, the diffusion time ratio needs to be at least 1.6, corresponding to a minimum mass ratio of $1.6^3 \approx 4$ [45]. In this case, if binding is to be measured by FCS one has to label the smaller binding partner and the mass ratio between the non-labelled to the labelled binding partner needs to be at least 3. This also means that binding between equal sized particles cannot be measured by FCS under these conditions. This case is often found in complex and/or live samples, where it is not easily possible to isolate and measure the diffusion time of individual binding partners. For these cases, FCCS is a viable alternative that can measure interactions independent of the masses of the interacting partners but requires that both binding partners are fluorescently labelled by distinct fluorophores (see section 9.4).

In the second case, the diffusion coefficient of the labelled binding partner is known. This is typically the case when measuring *in vitro* and when one has full control of the composition of the sample. In this case, the known diffusion coefficient can be fixed and only the second diffusion coefficient and fraction of the complex are kept as fitting parameters. This improves the stability of the fits and allows distinguishing populations with diffusion coefficients differing by a factor smaller than 1.6. If the diffusion time of the bound population can be fitted reproducibly and with low variability, the fitted bound fraction gives quantitative information about molecular affinities [41, 46, 47]. The reliability of bound fraction determination further improves if the diffusion time of the bound particles is also known and can be fixed in the fits leaving the bound fraction as the only unknown parameter.

This would be, for example, the case in the example of biotin–avidin binding discussed in section 9.3.1. When a solution of fluorescently labelled biotin is titrated with unlabelled avidin, the final point of the titration corresponds to the situation when all biotin molecules are bound to avidin and any further addition of avidin does not change the measured ACF. Under this situation biotin–avidin complexes are the only fluorescent particles present in the sample and their diffusion time can be obtained from the measured ACF. The diffusion time of free biotin is obtained from the ACF measured in the absence of avidin and the knowledge of both diffusion times will allow reliable fitting of ACFs corresponding to intermediate stages of the titration. Note, however, that a quantitative interpretation of the bound fraction will, in this case, need to consider the fact that, because of the 4:1 binding stoichiometry, the brightness of the complex is different from the brightness of free biotin. Particles contribute to the measured ACF with the square of their brightness (see equation (5.27)); therefore, the contributions of particle populations differing in brightness are not proportional to the fractions of the respective populations. If the brightness of the fluorophore does not change upon the labelled ligand binding to its substrate and the binding stoichiometry, H, is known, the brightness of the complex would be H-times higher than the brightness of a free ligand. The knowledge of the brightness of the complex and the free ligand would allow us to use equation (5.27) to correctly quantify the actual bound fraction (see exercise 4). However, in general, the brightness of the complex may be larger or smaller than H-times the brightness of individual fluorescently labelled ligands, because of changes in fluorophore brightness upon the ligand binding to its substrate. The changes can be caused by: (i) changes in the local environment of the fluorophore or (ii) interactions between multiple fluorophores bound to the same complex. While the former effect can lead to higher or lower brightness depending on the actual environment, the latter effect will be most likely mutual quenching of the fluorophores leading to reduced brightness.

In the last case, when one looks only for a qualitative answer whether binding takes place, then one can just monitor whether the average diffusion time shifts in a simple one-component model. When using good photostable fluorophores, e.g. most organic dyes, diffusion time shifts as little as 10% can be detected and interpreted as evidence for binding [41].

9.4 Molecular interactions studies by FCCS

FCCS in detecting molecular interactions does not rely on changes in diffusion time; instead it extracts the information from the cross-correlation of concentration fluctuations of two populations of fluorescent particles differing in their fluorescence properties, fluorescence emission spectra in most cases. This allows FCCS to quantify interactions between binding partners of comparable mass, in cases where the difference in diffusion time between the complex and the individual binding partners is too small for reliable quantification by FCS. The requirement in FCCS is that both binding partners are fluorescently labelled by fluorophores with distinct emission spectra or distinct lifetimes in the case of FLCS.

The theory of quantitative FCCS and the sources of artefacts that need to be addressed are presented in chapter 6. Here we will, therefore, give only a brief summary of the main steps needed in a quantitative FCCS experiment (table 9.1). Besides that, let us highlight a few points relevant to FCCS experiments in general.

9.4.1 The importance of relating the CCF amplitude to ACF amplitudes

The CCF amplitude should be always related to the ACF amplitudes. This is important in all FCCS experiments, including those seeking only qualitative answers whether the binding partners interact or not. To understand the reason for this, let us look at the theoretical models for the CCF amplitude. For simplicity, we can take the model for the CCF amplitude in an ideal FCCS experiment expressed by equation (6.6c), which we repeat here for convenience:

$$G_{0_X} = \frac{N_{ab}}{(N_a + N_{ab})(N_b + N_{ab})} = \frac{\langle C_{ab}\rangle}{V(\langle C_a\rangle + \langle C_{ab}\rangle)(\langle C_b\rangle + \langle C_{ab}\rangle)} \tag{6.6c}$$

Table 9.1. Summary of factors that need to be accounted for in a quantitative FCCS experiment.

Effect to address	More details in section:	Control experiment	Remarks
Imperfect overlap of observation volumes	6.2.2	(1) Standard FCS calibration of the observation volume for at least one detection channel (2) CCF effective observation volume calibration by FCCS measurement in a positive control sample (only double-labelled particles)	Not present in FLCS with two spectrally similar fluorophores
Uncorrelated background	6.2.4	Average intensity measurement in a blank sample (e.g. non-labelled cells)	Can be removed by statistical filtering
Spectral crosstalk	6.2.3 6.3.2 6.4	FCCS measurements in single-labelled samples (labelled by fluorophores of a single species) to determine the brightness of each fluorophore in each detection channel	Not present in PIE- or ALEX-FCCS and FCCS with statistical filtering
FRET and quenching	6.2.6	Comparing fluorophore brightness measured in positive control sample with the brightness measured in single-labelled samples	PIE-FCCS is free from the influence of FRET
Non-fluorescent binding partners (imperfect labelling and fluorophore dark states)	6.2.5	FCCS measurements in samples containing fluorophore tandems (particles labelled by two fluorophores of the same species) to determine the probability of the fluorophore being fluorescent	
Non-fluorescent binding partners (endogenous proteins)	6.2.5	Endogenous protein concentration cannot be directly estimated by fluorescent techniques	Preferably avoided by genome editing

Subscript *ab* refers to the complex, while *a* and *b* refer to the individual binding partners. We can see that the CCF amplitude depends on the concentrations of the complex and the individual binding partners in a way which prevents its unambiguous interpretation either in the terms of the absolute concentration or relative abundance of the complex. The latter is usually the more relevant information as it reflects the affinity of the binding partners to form a complex. It can be, for example, measured by the ratio of the complex concentration to the total concentration of the less abundant of the binding partners. In section 6.2.1, we have shown that for an ideal FCCS experiment this ratio is equal to a parameter called cross-correlation amount Q, defined by equation (6.11) (repeated here for convenience):

$$Q = \max\left[\frac{G_{0_X}}{G_{0_B}}; \frac{G_{0_X}}{G_{0_A}}\right] \tag{6.11}$$

Here, the function $\max[x; y]$ returns the higher of the two values x and y and the subscripts A and B refer to the two detection channels. The importance of relating the CCF amplitude to the ACF amplitudes can be illustrated by the following two FCCS experiment scenarios:

(i) The two binding partners are present in the sample at constant concentration and they slowly react to form a complex. FCCS can be used to monitor the increasing concentration of the complex in the system as the reaction progresses, assuming that the reaction is slow enough that the system can be considered stationary within the timeframe of an FCCS experiment (which is typically of the order of 10 s). An example of correlation functions corresponding to such a scenario is shown in figure 9.5. The ACF amplitudes remain nearly constant throughout the experiment reflecting the constant overall concentrations of the binding partners. The CCF amplitude, on the other hand, keeps growing as a result of increasing relative abundance of the complex (increasing cross-correlation amount).

(ii) In the second scenario, we are gradually increasing the concentration of one or both binding partners. This will lead also to an increase in the concentration of the complex. We assume that the complex formation is a fast reaction in this case and the FCCS measurements are therefore performed after the equilibrium between the complex and the free binding partners has been established for the increased concentrations of the binding partners. Figure 9.6 illustrates an example of such an experiment. We can see that in this case the CCF amplitude keeps decreasing with the increasing concentration of the complex as a result of the growing concentrations of the binding partners.

The examples in figures 9.5 and 9.6 demonstrate that the change in the CCF amplitude alone cannot be interpreted in terms of changes in the complex concentration. An increase in the complex concentration can be accompanied by either an increase or a decrease in the CCF amplitude depending on the changes in the binding partner concentrations which accompany the increase in the concentration

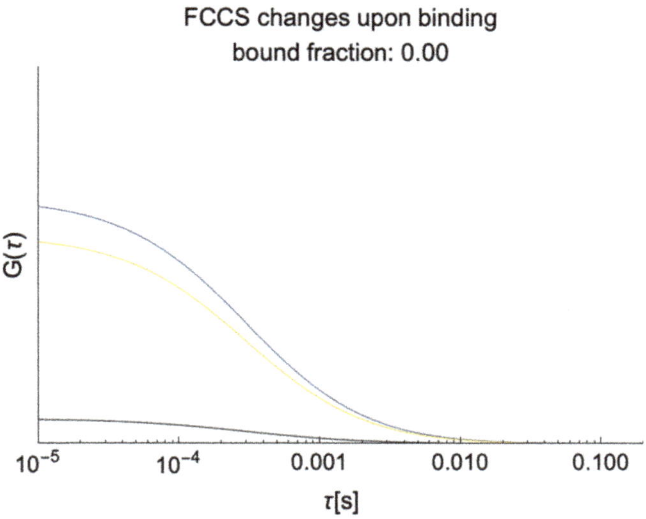

Figure 9.5. Dependence of the CCF on bound fraction, c_{ab}, assuming an ideal setup. The concentrations of the particles varied according to $c_a = c_b = 5$ nM $- c_{ab}$, $c_{ab} = 0 \ldots 5$ nM. Otherwise, we assumed identical detection volumes for both channels with the following geometrical parameters: $w_{0A} = w_{0B} = w_{0X} = 260$ nm, $K_A = K_B = 5$, $z_{0A} = z_{0B} = z_{0X} = 1.3$ μm, resulting in observation volumes of $V_A = V_B = V_X^{eff} = 0.5$ fl; the brightness of the two particles in the two detection channels was set to $\kappa_{aA} = 1500$, $\kappa_{aB} = 150$, $\kappa_{bA} = 15$, $\kappa_{bB} = 1500$; the concentration of the particles were $c_a = c_b = c_{ab} = 2.5$ nM; the diffusion times for the two particles were set to $\tau_{Da} = 3$ ms, $\tau_{Db} = 0.3$ ms. Animation available at https://iopscience.iop.org/book/978-0-7503-2080-1.

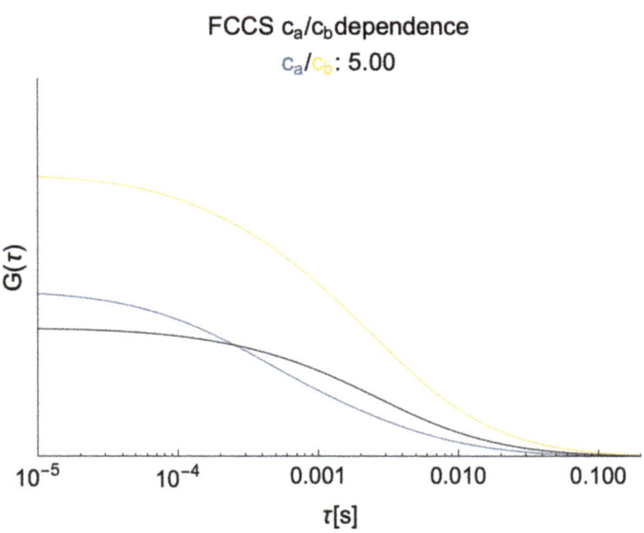

Figure 9.6. Dependence of the CCF on concentration ratio of the two binding partners for which the concentration of the particles were varied according to $c_a = 2.5$ nM, $c_b = 0.5 \ldots 10$ nM, $c_{ab} = 1.25$ nM$^{-1} \times (c_a \times c_b)$. Otherwise, we assumed identical detection volumes for both channels with the following geometrical parameters: $w_{0A} = w_{0B} = w_{0X} = 260$ nm, $K_A = K_B = 5$, $z_{0A} = z_{0B} = z_{0X} = 1.3$ μm, resulting in observation volumes of $V_A = V_B = V_X^{eff} = 0.5$ fl; the brightness of the two particles in the two detection channels was set to $\kappa_{aA} = 1500$, $\kappa_{aB} = 150$, $\kappa_{bA} = 15$, $\kappa_{bB} = 1500$; the concentration of the particles were $c_a = c_b = c_{ab} = 2.5$ nM; the diffusion times for the two particles were set to $\tau_{Da} = 3$ ms, $\tau_{Db} = 0.3$ ms. Animation available at https://iopscience.iop.org/book/978-0-7503-2080-1.

of the complex. Only a complete analysis of the correlation function amplitudes using an appropriate theoretical model, such as those derived in chapter 6, can provide a definite answer on the changes in the complex concentration.

9.4.2 Determination of the FCCS experiment dynamic range

It is a good practice to determine the range within which the measured values of cross-correlations are expected by taking negative and positive control measurements, which represent the limiting cases of no cross-correlation and complete cross-correlation, respectively.

Except for PIE- or ALEX-FCCS and FCCS with statistical filtering (see sections 6.3.2 and 6.4) cross-correlation functions in FCCS contain a positive contribution from spectral crosstalk (see section 6.2.3). The contribution of spectral crosstalk can be estimated by measurements in negative control samples containing non-interacting particles labelled by the same fluorophores as the particles of interest. Since the particles in such a sample do not interact, any positive cross-correlation amount can be attributed to spectral crosstalk. Cross-correlation amounts that are not larger (with sufficient statistical significance) than the values for negative control cannot be used as an indication of interaction between the binding partners.

On the other hand, the cross-correlation amounts are artificially decreased by several factors, mainly by the imperfect overlap of the effective observation volumes of the two detection channels (see section 6.2.2) and by the presence of non-fluorescent binding partners in the system (see section 6.2.5). The influence of these effects can be estimated by a measurement in a sample in which all particles are labelled by both fluorophores used in the FCCS experiment. The cross-correlation amount measured in such a sample represents then the upper limit on the values that can be expected for the actual sample. While any double labelled particles with fluorophores spectrally similar to those used in the actual FCCS experiment will give a good estimate of the influence of the differences in observation volumes, quantifying the influence of non-fluorescent binding partners is more difficult. Many factors which influence the fractions of non-fluorescent binding partners, such as the population of fluorophores in a dark state or the fraction of photobleached fluorophores (see section 6.2.5), are strongly dependent on the fluorophore local environment in the sample as well as conditions under which the FCCS experiments are conducted. For this reason, it is important to make sure that fluorophores identical to those in the sample of interest are used and all other conditions in the actual and the control experiment resemble each other as closely as possible. Low cross-correlation amounts measured for the positive control can be an indication of either imperfect alignment of the setup resulting in a large displacement of the two observation volumes or of problems with fluorescent labelling resulting in large fractions of non-fluorescent binding partners.

9.5 Synopsis

- The amplitude of concentration fluctuations ACF provides information on the average number of independently diffusing particles present in the effective observation volume. When the effective observation volume is known, this information can be translated to concentration of the fluorescent particles. In this way, FCS allows measurement of absolute concentrations in living cells or organisms.

- The effective observation volume can be calibrated by one of the following three approaches: (i) FCS measurement in a sample of known concentration of fluorescent particles, (ii) FCS measurement in a sample of fluorescent particles of known diffusion coefficient or (iii) imaging of a fluorescent bead significantly smaller than the effective observation volume. Beside allowing absolute concentration measurements, the effective observation volume calibration is useful as a benchmark of the experimental setup performance.
- The knowledge of fluorescent particle number in the effective observation volume allows us to determine the particle brightness as a ratio of the average fluorescence intensity and the particle number. This information can be used to estimate degrees of oligomerisation, quantitative FCCS analysis or FCS-calibrated quantitative imaging. The last approach yields information analogous to number and brightness analysis.
- The temporal decay of the concentration fluctuations ACF is related to the average time the diffusing particles spend in the effective observation volume. The knowledge of the effective observation volume allows quantitative interpretation of the ACF temporal decay in terms of particle diffusion coefficients.
- In some FCS modalities, such as two-focus or scanning FCS, the diffusion coefficient and the effective observation volume can be obtained as independent fit parameters by fitting the measured ACF. The reproducibility of the results is, thus, improved by avoiding the need for an external calibration.
- FCS in two-dimensional structures, such as biological membranes, is further complicated by the uncertainty in positioning the sample with respect to the focal plane of the microscope. The effective observation area is the smallest when the sample lies in the focal plane; moving the sample below or above the focal plane results in increased effective observation area, which translates to longer diffusion time and higher particle number. The results of calibration-free FCS modalities such as scanning FCS are to a large degree independent of these effects, making those modalities well suited for measurements in membranes.
- FCS is often applied to study molecular interactions. If the interaction leads to a change in the number of fluorescent particles (e.g. if multiple fluorescent molecules are involved in forming one complex particle), it can be characterised by changes in the ACF amplitude. If the interaction leads to changes in the florescent particles size, the changes in the ACF temporal decay can be used.
- Because of the relatively week dependence of the diffusion time on the particle mass, interactions resulting in small changes in mass are difficult to quantify by FCS. To maximise the change in particle mass upon interaction, the smaller of the binding partners need to be the fluorescently labelled one. The reliability of quantification of fractions of free and complexed particles can be greatly improved if the diffusion times of the individual particle populations (or at least one of them) are known *a priori*.

- FCCS detects interaction between binding partners labelled by distinct fluorophores via measuring cross-correlation of their respective concentration fluctuations. The information on complex formation is, thus, independent of changes in diffusion time. This makes FCCS suitable for studying interactions between binding partners of similar mass.

9.6 Exercises

1. Consider a typical confocal setup with the effective observation volume $V_{\text{eff}} = 0.5$ fL.
   (a) At what fluorescent particle concentration is the average particles number, N, equal to 1? Express the concentration both as number concentration C and as molar concentration c.
   (b) Under the condition that $N = 1$, what is the expected fraction of FCS acquisition time during which the effective observation volume is empty? What is the expected fraction of the acquisition time during which the effective observation volume contains more than one particle? *Hint*: use the fact that the probability of finding a particular number of particles, ξ, in the effective observation volume follows Poisson distribution:

$$P(\xi) = \frac{N^{\xi}e^{-N}}{\xi!}$$

2. Apply number and brightness analysis on the simulated intensity time traces shown in figure 9.2. The mean and standard deviation of the 100 simulated values are for the trace shown in panel A: $\langle F_A \rangle = 108$, $\sigma_A = 49$, and for the trace shown in panel B: $\langle F_B \rangle = 96$, $\sigma_B = 22$. Compare the results with the particle number and brightness values used in the simulations.

3. Derive equation (9.20) by deriving the expressions for the first and the second moment of binomial probability distribution $P_B(\eta)$. *Hint*: use the facts the sum of the binomial distribution is equal to 1:

$$\sum_{\eta=0}^{H} \binom{H}{\eta} p^{\eta}(1-p)^{H-\eta} = 1, \text{ for } 0 \leqslant p \leqslant 1$$

   and use the explicit expression for the binomial coefficients:

$$\binom{H}{\eta} = \frac{H!}{\eta!(H-\eta)!}$$

4. Consider the avidin–biotin binding experiment in which a solution of fluorescently labelled biotin is titrated by unlabelled avidin. By fitting the measured ACF with the following model:

$$G^{(\mathrm{conc})}(\tau) = \frac{1}{N_{app}}\left[f\left(1 + \frac{\tau}{\tau_{D_L}}\right)^{-1}\left(1 + \frac{\tau}{K^2\tau_{D_L}}\right)^{-1/2} + (1 - f)\left(1 + \frac{\tau}{\tau_{D_S}}\right)^{-1}\left(1 + \frac{\tau}{K^2\tau_{D_S}}\right)^{-1/2}\right] + G_\infty$$

where τ_{D_L} and τ_{D_S} are the diffusion times of free biotin and of the avidin–biotin complex, you find that the apparent fraction, f, of the faster component is ½.

(a) What is the actual fraction of free biotin molecules in the sample, assuming that the brightness of the complex is four times the brightness of a free fluorescently labelled biotin molecule?

(b) What is the actual particle number under such situation?

(c) What is the average particle brightness? Compare it with the apparent average brightness, $\langle F\rangle / N_{app}$.

References

[1] Pack C G *et al* 2019 Quantitative live-cell imaging reveals spatio-temporal dynamics and cytoplasmic assembly of the 26S proteasome *Nat. Commun.* 1–10

[2] Sasaki A, Yamamoto J, Kinjo M and Noda N 2018 Absolute quantification of RNA molecules using fluorescence correlation spectroscopy with certified reference materials *Anal. Chem.* **90** 10865–71

[3] Koppel D 1974 Statistical accuracy in fluorescence correlation spectroscopy *Phys. Rev.* A **10** 1938

[4] Qian H 1990 On the statistics of fluorescence correlation spectroscopy *Biophys. Chem.* **38** 49–57

[5] Laurence T A, Ly S, Bourguet F, Fischer N O and Coleman M A 2014 Fluorescence correlation spectroscopy at micromolar concentrations without optical nanoconfinement *J. Phys. Chem.* B **118** 9662–67

[6] Enderlein J, Gregor I, Patra D, Dertinger T and Kaupp U B 2005 Performance of fluorescence correlation spectroscopy for measuring diffusion and concentration *ChemPhysChem* **6** 2324–36

[7] Mütze J, Ohrt T and Schwille P 2010 Fluorescence correlation spectroscopy *in vivo Laser Photon. Rev.* **5** 52–67

[8] Blom H, Kastrup L and Eggeling C 2006 Fluorescence fluctuation spectroscopy in reduced detection volumes *Curr. Pharm. Biotechnol.* **7** 51–66

[9] Enderlein J, Gregor I, Patra D and Fitter J 2005 Statistical analysis of diffusion coefficient determination by fluorescence correlation spectroscopy *J Fluoresc.* **15** 415–22

[10] Sasaki A, Halter M and Elliot J T 2019 Fluorescence correlation methods for determining absolute numbers of molecules from microscopy images *Bioimages* **27** 13–22

[11] Schwille P, Haupts U, Maiti S and Webb W W 1999 Molecular dynamics in living cells observed by fluorescence correlation spectroscopy with one- and two-photon excitation *Biophys. J.* **77** 2251–65

[12] Rüttinger S, Buschmann V, Krämer B, Erdmann R, Macdonald R and Koberling F 2008 Comparison and accuracy of methods to determine the confocal volume for quantitative fluorescence correlation spectroscopy *J. Microsc.* **232** 343–52

[13] Pan X, Foo W, Lim W, Fok M H Y, Liu P, Yu H, Maruyama I and Wohland T 2007 Multifunctional fluorescence correlation microscope for intracellular and microfluidic measurements *Rev. Sci. Instrum.* **78** 053711

[14] Politi A Z, Cai Y, Walther N, Hossain M J, Koch B, Wachsmuth M and Ellenberg J 2018 Quantitative mapping of fluorescently tagged cellular proteins using FCS-calibrated four-dimensional imaging *Nat. Protoc.* **13** 1445–64

[15] Shivaraju M, Unruh J R, Slaughter B D, Mattingly M, Berman J and Gerton J L 2012 Cell-cycle-coupled structural oscillation of centromeric nucleosomes in yeast *Cell* **150** 304–16

[16] Digman M A, Dalal R, Horwitz A F and Gratton E 2008 Mapping the number of molecules and brightness in the laser scanning microscope *Biophys. J.* **94** 2320–32

[17] Unruh J R and Gratton E 2008 Analysis of molecular concentration and brightness from fluorescence fluctuation data with an electron multiplied CCD camera *Biophys. J.* **95** 5385–98

[18] Bancaud A, Huet S, Daigle N, Mozziconacci J, Beaudouin J and Ellenberg J 2009 Molecular crowding affects diffusion and binding of nuclear proteins in heterochromatin and reveals the fractal organization of chromatin *EMBO J.* **28** 3785–98

[19] Habimana O, Steenkeste K, Fontaine-Aupart M-P, Bellon-Fontaine M-N, Kulakauskas S and Briandet R 2010 Diffusion of nanoparticles in biofilms is altered by bacterial cell wall hydrophobicity *Appl. Environ. Microbiol.* **77** 367–68

[20] Sankaran J, Tan N J H J, But K P, Cohen Y, Rice S A and Wohland T 2019 Single microcolony diffusion analysis in Pseudomonas aeruginosa biofilms *npj Biofilms Microbiomes* **5** 95–37

[21] Dross N, Spriet C, Zwerger M, Müller G, Waldeck W and Langowski J 2009 Mapping eGFP oligomer mobility in living cell nuclei *PLoS One* **4** e5041

[22] Culbertson M J, Williams J T B, Cheng W W L, Stults D A, Wiebracht E R, Kasianowicz J J and Burden D L 2007 Numerical fluorescence correlation spectroscopy for the analysis of molecular dynamics under nonstandard conditions *Anal. Chem.* **79** 4031–39

[23] Dertinger T, Pacheco V, Von Der Hocht I, Hartmann R, Gregor I and Enderlein J 2007 Two-focus fluorescence correlation spectroscopy: a new tool for accurate and absolute diffusion measurements *ChemPhysChem* **8** 433–43

[24] Benda A, Ma Y and Gaus K 2015 Self-calibrated line-scan STED-FCS to quantify lipid dynamics in model and cell membranes *Biophys. J.* **108** 596–609

[25] Digman M A and Gratton E 2009 Imaging barriers to diffusion by pair correlation functions *Biophys. J.* **97** 665–73

[26] Satsoura D, Leber B, Andrews D W and Fradin C 2007 Circumvention of fluorophore photobleaching in fluorescence fluctuation experiments: a beam scanning approach *ChemPhysChem* **8** 834–48

[27] Krieger J W, Singh A P, Garbe C S, Wohland T and Langowski J 2014 Dual-color fluorescence cross-correlation spectroscopy on a single plane illumination microscope (SPIM-FCCS) *Opt. Express* **22** 2358

[28] Bag N, Sankaran J, Paul A, Kraut R S and Wohland T 2012 Calibration and limits of camera-based fluorescence correlation spectroscopy: a supported lipid bilayer study *ChemPhysChem* **13** 2784–94
[29] Otosu T and Yamaguchi S 2020 Effect of electrostatic interaction on the leaflet-specific diffusion in a supported lipid bilayer revealed by fluorescence lifetime correlation analysis *Phys. Chem. Chem. Phys.* **22** 1242–49
[30] Sankaran J, Bag N, Kraut R S and Wohland T 2013 Accuracy and precision in camera-based fluorescence correlation spectroscopy measurements *Anal. Chem.* **85** 3948–54
[31] Vukojevic V, Papadopoulos D K, Terenius L, Gehring W J and Rigler R 2010 Quantitative study of synthetic Hox transcription factor-DNA interactions in live cells *Proc. Natl Acad. Sci.* **107** 4093–98
[32] Eggeling C *et al* 2009 Direct observation of the nanoscale dynamics of membrane lipids in a living cell *Nature* **457** 1159–62
[33] Wenger J, Conchonaud F, Dintinger J, Wawrezinieck L, Ebbesen T W, Rigneault H, Marguet D and Lenne P-F 2007 Diffusion analysis within single nanometric apertures reveals the ultrafine cell membrane organization *Biophys. J.* **92** 913–19
[34] Digman M A and Gratton E 2009 Analysis of diffusion and binding in cells using the RICS approach *Microsc. Res. Tech.* **72** 323–32
[35] Vogelsang J, Kasper R, Steinhauer C, Person B, Heilemann M, Sauer M and Tinnefeld P 2008 A reducing and oxidizing system minimizes photobleaching and blinking of fluorescent dyes *Angew. Chem. Int. Ed.* **47** 5465–69
[36] Widengren J and Seidel C A M 2000 Manipulation and characterization of photo-induced transient states of Merocyanine 540 by fluorescence correlation spectroscopy *Phys. Chem. Chem. Phys.* **2** 3435–41
[37] Persson G, Thyberg P and Widengren J 2008 Modulated fluorescence correlation spectroscopy with complete time range information *Biophys. J.* **94** 977–85
[38] Milon S, Hovius R, Vogel H and Wohland T 2003 Factors influencing fluorescence correlation spectroscopy measurements on membranes: simulations and experiments *Chem. Phys.* **288** 171–86
[39] Ries J and Schwille P 2006 Studying slow membrane dynamics with continuous wave scanning fluorescence correlation spectroscopy *Biophys. J.* **91** 1915–24
[40] Benda A, Beneš M, Mareček V, Lhotský A, Hermens W T and Hof M 2003 How To determine diffusion coefficients in planar phospholipid systems by confocal fluorescence correlation spectroscopy *Langmuir* **19** 4120–6
[41] Wohland T, Friedrich K, Hovius R and Vogel H 1999 Study of ligand-receptor interactions by fluorescence correlation spectroscopy with different fluorophores: evidence that the homopentameric 5-hydroxytryptamine type 3As receptor binds only one ligand *Biochemistry* **38** 8671–81
[42] Hwang L C and Wohland T 2005 Single wavelength excitation fluorescence cross-correlation spectroscopy with spectrally similar fluorophores: Resolution for binding studies *J. Chem. Phys.* **122** 114708
[43] Ries J, Petrášek Z, García-Sáez A J and Schwille P 2010 A comprehensive framework for fluorescence cross-correlation spectroscopy *New J. Phys.* **12** 113009
[44] Yu L, Tan M, Ho B, Ding J L and Wohland T 2006 Determination of critical micelle concentrations and aggregation numbers by fluorescence correlation spectroscopy: aggregation of a lipopolysaccharide *Anal. Chim. Acta* **556** 216–25

[45] Meseth U, Wohland T, Rigler R and Vogel H 1999 Resolution of fluorescence correlation measurements *Biophys. J.* **76** 1619–31

[46] Rauer B, Neumann E, Widengren J and Rigler R 1996 Fluorescence correlation spectrometry of the interaction kinetics of tetramethylrhodamin alpha-bungarotoxin with *Torpedo californica* acetylcholine receptor *Biophys. Chem.* **58** 3–12

[47] Van Craenenbroeck E and Engelborghs Y 1999 Quantitative characterization of the binding of fluorescently labeled colchicine to tubulin *in vitro* using fluorescence correlation spectroscopy *Biochemistry* **38** 5082–88

[48] Kapusta P 2010 Absolute diffusion coefficients: compilation of reference data for FCS calibration (Berlin: PicoQuant) https://www.picoquant.com/images/uploads/page/files/ 7353/ appnote_diffusioncoefficients.pdf

IOP Publishing

An Introduction to Fluorescence Correlation Spectroscopy

Thorsten Wohland, Sudipta Maiti and Radek Macháň

Chapter 10

Where to go from here?

The task of this last chapter is to point interested readers to some advanced FCS techniques that were not discussed so far, give an idea about related techniques, provide some information about software resources, and discuss possible future directions of FCS.

10.1 Introduction

The first fluorescence detection of single molecules by Tomas Hirschfeld [1] and the advancement of the analysis of fluctuations with FCS in the 1970s opened up new possibilities of quantitative measurements in very small, femtoliter-sized volumes with single molecule sensitivity. With the advancement of technology and the theory of FCS over the next decades, FCS became commercially available and single molecule measurements became a routine tool in the life sciences.

In this book we have covered the basics of FCS and various modalities. There is no doubt that the selection of the topics to be covered and which derivations to include and which to leave out is based on our own, the authors', experience and interests. Many more modalities have been published over the years some of them used more often than others. The frequency of usage of a technique, though, does not always say something about its pedagogical value. For example, circular scanning FCS might not be as widely used as confocal FCS or RICS, however, it is a nice illustration of the properties of the ACF when a combination of a harmonic change (circular scanning) and diffusion influences the correlation at the same time. We hope, however, that the selection we made provides an easy introduction into this topic and inspires readers to delve deeper into the world of fluctuations and correlations. In this last chapter we want to point the reader to the wide range of different data evaluation tools, developed in recent decades, that extend the extraction of information from single molecule fluctuations and that seek to overcome limitations and artefacts of FCS but also to some direction in which we expect new developments in the future.

doi:10.1088/978-0-7503-2080-1ch10

10.2 What FCS can and cannot do

In principle FCS has excellent performance parameters, especially when one takes all the different variations and customizations into account that have been implemented over the years. FCS has a time resolution of nanoseconds and on the longer time scale is only limited by the stability of the system. Required acquisition times can be as low as seconds, or even milliseconds [2], and acquisitions can run as long as the system is stable. Accessible concentrations range from the picomolar range to the high micromolar range [3]. And FCS and its related techniques have been applied over wide ranges of samples under different conditions ranging from simple solution measurements, over measurements in complex solutions and gels and solid matrices, to cell and *in vivo* experiments. Furthermore, in all these cases FCS provides quantitative data with single molecule sensitivity that has allowed researchers to discover and measure molecular interactions, aggregations, folding and dynamics. So where are the limitations? The first limitation is that many of the extreme ranges of the parameter space just mentioned are only accessible with customized systems and thus are often not widely available. Commercial systems are of course catching up but lag behind and only implement those customizations that are expected to be widely used. The second limitation is the data evaluation. FCS is not as easy to understand as some other techniques and in particular its data evaluation is intricate and fraught with pitfalls. And this is of course one of the motivations to write this book. However, FCS data evaluation has made great strides and free available data treatment software, that includes many of the concepts discussed in chapter 8, is nowadays available and helps the user to effectively treat data (table 10.1).

Further, more practical and experimental limitations and means to correct or avoid them were already mentioned in previous chapters. Immobile or very slow moving particles cannot be easily measured, bleaching and sample movement artefacts bias data evaluation, if not corrected. In multi-colour experiments, cross-talk will lead to artificial cross-correlations, and artefacts of detectors and lasers can lead to unwanted correlations not related to the measurement under investigation. A constraint that is not that easily addressed is the inherent long measurement times, typically 10s of seconds, compared to the time resolution of nanoseconds. The long measurement times are necessary to achieve an adequate signal-to-noise ratio and collect sufficient statistics to be able to determine parameters with acceptable accuracy and precision. But this means that changes in the measured parameters cannot be measured faster than the measurement time. One way to address this, is to measure intensity traces for a long time and then calculate ACFs for subsections of the intensity traces in a sliding window fashion. This way, changes in the FCS parameters can be detected faster. However, the parameters, e.g. the diffusion coefficient, are of course averages of the instantaneous values and data taken this way will provide parameter values in time that are themselves strongly correlated. Thus one needs to be careful when interpreting the data.

Another limitation in FCS has been its nonlinear sensitivity to the brightness of probed molecules. In samples where one cannot assume that all particles are

Table 10.1. Examples of freely available FCS-related software. Each one of these programs has a wide range of applications but none of them is doing everything and some are more specialized. For example, SimFCS, PAM, PyCorrFit, and QuickFit are quite comprehensive packages but are all stand-alone programs. They do not connect directly back to a commonly used software that allows wider applications. The FCS plugin for ImageJ and Imaging FCS both run in ImageJ/Fiji and thus data can be treated with a wide range of other options and the user is not limited to FCS evaluations. On the other hand, FCS plugin for ImageJ, PyCorrFit, and FCS for IgorPro are treating only confocal FCS data, Imaging FCS is treating only images, while QuickFit and SimFCS can deal with both. The interested reader should try several versions and decide which one is the easiest to use or performs the tasks at hand.

Name	URL	Required Software
SimFCS	https://www.lfd.uci.edu/globals/	stand-alone
FCS plugin for ImageJ	https://research.stowers.org/imagejplugins/fcs_plugins.html	Fiji/ImageJ
FCS for IgorPro	http://www.dbs.nus.edu.sg/lab/BFL/confocal_FCS.html	IgorPro
PAM [4]	https://www.cup.uni-muenchen.de/pc/lamb/software/pam.html	MATLAB®
PyCorrFit [5]	https://pycorrfit.readthedocs.io/en/stable/sec_about.html#	Stand-alone
QuickFit [6]	https://github.com/jkriege2/QuickFit3	Stand-alone
Imaging FCS [6, 7]	http://www.dbs.nus.edu.sg/lab/BFL/imfcs_image_j_plugin.html	Fiji/ImageJ

similarly bright this can lead to errors in the estimation of concentrations and mole fractions. We will discuss this issue in section 10.5.

Finally, FCS, as any other fluorescence technique, requires labelling, with the rare exceptions in which autofluorescence or scattering can be used as the signal [8–11]. However, labelling always carries along the possibility of altering biomolecular functions and thus is an inherent limitation.

10.3 Data acquisition

The essential requirements for FCS data acquisition are a sufficiently high signal from single molecules for detection and a small observation volume so that fluctuations caused by single molecules can be distinguished from the variability of the overall signal. This has been achieved in a variety of ways. In this book we have discussed already the most common setups, including single or multi-spot confocal, total internal reflection, and light sheet microscopes. However, there have been variations on these methods that are instructive for the reader and which we at least want to mention here. The most important confocal method not treated so far in this book is multiphoton excitation FCS. Multi-photon excitation has multiple advantages. Because it operates at longer wavelength, and longer wavelengths are less scattered by tissues, they possess higher penetration depths. As the intensity for multiphoton excitation needs to be high to reach a sufficient probability of simultaneous absorption of multiple photons, the fluorescence emission is restricted to the focal volume, unlike one-photon excitation which excites fluorescence over the whole laser beam and thus requires a pinhole to filter the relevant signal from the focal spot. But multiphoton FCS has in general somewhat smaller photon counts per molecule, requires pulsed excitation with typically higher intensity compared to one-photon excitation, and has potentially increased photobleaching in the focal region.

The labelling problem mentioned in the previous section was circumvented in inverse FCS [12]. In inverse FCS the sample contains a high concentration of fluorophores. But it is not the intensity fluctuation caused by the fluorophores that is the source of information but rather the fluctuations caused by a non-labeled probe that displaces fluorophores in the probe volume.

The concentration that can be measured in FCS is closely related to the size of the observation volume. The smaller the observation volume, the larger the concentrations that can be used. This is the result of the $1/N$ dependence of the ACF (equations (2.18) and (5.5)). We have already seen that STED-FCS can reduce the size of the observation volume. Another way to achieve an attoliter sized observation volume are zero-mode waveguides [13], which also have been used for FCS [14, 15].

Other illumination schemes are closely related to the three common ones discussed in the book, namely confocal, TIRF, and light sheet microscopy. Patterned illumination FCS was developed using the interference fringes of two focused laser beams for excitation [16]. By measuring with different fringe spacings one is able to verify the transport mode of the probe. Using a single spot in TIRF, the intensity can be spatially filtered by a pinhole and the intensity from the small observation volume, created by the evanescent field and the pinhole, can be correlated. As detection in TIR-FCS is performed by the much faster single-photon avalanche diodes, commonly used in confocal FCS, TIR-FCS has excellent time resolution and has been used, e.g. to determine receptor ligand binding events [17]. By coupling the laser to the sample under an angle slightly smaller than the critical angle, the laser is no longer totally internally reflected but penetrates into the sample at a steep angle [18]. This provides essentially a light sheet that can be used for imaging FCS [7].

Very instructive acquisition modalities in FCS stem from the combination of multiple schemes. For example, scanning FCCS [19], scanning inverse FCS [20], near-field inverse FCS [21], STED-FCS [22], STED-RICS [23], dual-colour dual-focus line scanning FCS [24] and others. In principle there are no limits on which combinations can be used and each of the combinations is typically customized to solve a particular problem. One can expect more combinations in the future, driven by experimental needs. But there have also been a number of combinations of FCS with other techniques. FRET-FCS, for instance, monitors the fluctuations caused by FRET efficiency changes due to molecular dynamics [25]. The combination of FCS and atomic force microscopy allowed investigation of lipid domains, with dimensions below the diffraction limit, in lipid bilayers [26].

Single spot confocal FCS setups still have the best time resolution and using single-photon counting, they will likely remain the gold standard for some time for

FCS with a high signal-to-noise ratio, best achievable time resolution, and thus best parameter estimation. However, they are limited in the number of measurements that can be acquired and do not allow simultaneous measurements in a sample. We therefore expect that imaging FCS modalities, whether by confocal scanning, multifocal spot combined with array detectors [27], or camera based imaging FCS, will play an increasingly important role. In particular, the latter modalities using fast and sensitive cameras as detectors allow easy combination with computational super-resolution and other fluorescence spectroscopy techniques, potentially providing much more information from single measurements.

10.4 Data analysis

Fluorescence correlation spectroscopy had its origin in the analysis of fluorescence fluctuations. Once this concept was introduced the number of statistical analysis techniques to treat signals and their fluctuations from single molecules has multiplied. This development was to some extent parallel to the development in super-resolution microscopy. While the latter improved the determination of spatial details down to the size of molecules on the nanometer range, the former increased time resolution to be able to see single molecule dynamics and even photophysics down to the nanosecond range. The combination of the ultimate molecular resolution in space and time required and still requires considerable theoretical work that allows us to extract meaningful parameters from these measurements and has to be paralleled by computational advances that provide the means to treat the huge amount of data these novel techniques produce.

FCS data analysis has at least three problems, the determination of theoretical fit models, the analysis of the signal statistics, and the development of computational means to treat the vast amount of data that can be produced when we acquire more than a 1000 frames per second on a full camera chip of a million pixels or more.

The derivation of theoretical fit models can be complicated by two issues. First, one does not necessarily know the underlying process and thus the required fit model. In that case either one does not have a fit model to use and interpretation of data has to be cautious. Or the underlying process is known but no analytic solution can be found for the ACF. In that case, numerical integration or fit by simulation can be used. Or there are multiple candidate models and one tries to select which model is the right one. In that case, one can use Bayesian model selection to determine the model that fits the data best given the uncertainties, which leads us to the second issue.

The second issue is the analysis of the experimental uncertainties of the ACF. These are required for proper model selection and quality of fit measures. As pointed out in chapter 8 the variance of the ACF requires higher-order correlation functions that are difficult to compute. Therefore, various approximations to the error in FCS have been proposed, which already improve data analysis and their combination with Bayesian model selection has put FCS data evaluation on a more robust basis. Nevertheless, there are prospects that parameter estimation and the resolution of multiple components within an ACF can still be improved. Recently, Presse *et al* have pioneered the analysis of the fluorescence signal by Bayesian non-parametrics that can shorten the required data acquisition time and thus can help analyse short events [2]. Another possibility to improve data evaluation could lie in machine/deep learning (ML) approaches. ML uses neural networks that can be trained to discover patterns in experimental data. The advantage of ML is that it does not require explicit instructions of how to discover patterns but is trained with existing data sets on the desired patterns. However, training requires large amounts of training data which might not be readily available. In imaging FCS this could be collected due to its parallel acquisition of thousands of correlation functions. But this is almost impossible to do for single point FCS methods. Therefore, simulated data are an alternative and are explored as possible training sets.

ML can be used in at least three different ways. First, ML can be used to classify FCS data and thus determine, similar to Bayesian model selection, which model should be used for fitting. Second ML can be used to fit ACFs directly and determine the parameters that fit the data best. In that case the question is whether ML performs at least as well as NLS and GLS algorithms or even outperforms them. Or ML can be directly applied to the intensity traces and extract information of diffusion coefficients directly without calculating the correlation functions [28]. Combined with automated decision making these approaches promise to make applications of FCS much simpler and more effective in the future.

Third, the problem of data size can be addressed by parallelization of computation. With GPU-based data evaluation imaging FCS has already improved data evaluation times by at least a factor 30 [29]. This can be probably improved by optimizing algorithms to GPU architectures in the future. Nevertheless, this is sufficient to bring data evaluation times to the same order as the acquisition time, which is a requirement for real-time evaluation of experiments. Combined with direct camera-readout and data evaluation during acquisition, this will minimize the impact of computation and allow real-time feedback to the researcher during experimentation.

10.5 Related techniques

Over time many related techniques have been developed due to limitations inherent in FCS. One set was aiming at the determination of samples with particles of different brightness, as happens during aggregation or oligomerization. The FCS signal is strongly dependent on the brightness of the molecules under investigation. However, it is not linearly dependent but rather depends on the square of the brightness (equation (5.26)). In addition, the diffusion time, τ_D, is only weakly dependent on the mass (equation (5.21)). This makes it difficult to determine the oligomerization state in a mixture of monomers and oligomers. Therefore, techniques that were more sensitive to the brightness and/or size of the molecules under investigation were required. This problem was addressed by photon counting histogram (PCH), fluorescence intensity distribution analysis (FIDA), and fluorescence cumulant analysis (FCA).

In PCH and FIDA one determines the brightness of the probes by recording the number of photons emitted from the observation volume per time interval and then fits the resulting histogram with the theoretical histogram describing the probability to observe a certain number of photons in dependence on the number of particles and brightness [30–32]. In FCA one calculates the cumulants of the fluorescence intensities. Cumulants contain the same information as moments, which are calculated in correlation techniques [33]. And although they are more abstract they are often easier to treat mathematically. Both subjects are highly interesting and are good advanced topics to improve one's understanding of the statistics of fluorescence intensities. A conceptually simpler possibility to measure molecular brightness is so-called number and brightness analysis (N&B), which essentially analyses the amplitude of the ACF and the measured intensity, or in other words compares the variance with the mean of the fluorescence signal [34, 35].

Other extensions to FCS include alternative implementations to the FCS diffusion law (section 5.7.8). The so-called image mean squared displacement (iMSD) is the ICS based equivalent of the FCS diffusion law [36], which has been implemented to test diffusive modes of molecules over whole cell membranes [37]. And an extension to two-focus FCS (section 5.7.8) is the pair correlation function (pCF) analysis that instead of being used to determine absolute diffusion coefficients was implemented to determine whether there were diffusive barriers between the two points that were correlated [38]. This is a very interesting use of the technique as it provides direct evidence of the connectivity within a sample.

Finally, the analysis of fluctuations is not limited to fluorescence intensity fluctuations and, in principle, any observable that shows fluctuations around its equilibrium value can be evaluated by correlation functions. An interesting application is Förster Resonance Energy Transfer FCS (FRET-FCS). If two fluorophores have overlapping emission (donor) and excitation (acceptor) spectra and are sufficiently close (typically < 10 nm), then an excited donor can transfer its excitation energy to an acceptor (for a more thorough discussion of FRET see, e.g. the textbook by Lakowicz [39]). FRET reduces the emission of the donor and increases the emission of the acceptor. As it is very sensitive to distance, the FRET efficiency scales with R^{-6}, distance fluctuations will lead to significant donor and acceptor intensity fluctuation. By analysing changes of the acceptor compared to the total, i.e. donor and acceptor, intensity, structural fluctuations between the labelled parts of a molecule can be measured and analysed by FCS, thus providing information on structural dynamics and stability if the system is measured under different conditions [25]. In addition, it should be noted that the ACFs carry different information from the CCFs. While the ACFs contain mainly the diffusional contribution, the CCFs carry both diffusional and kinetics information. Comparing the ACFs for acceptor and donor and their CCF allows thus separating diffusional from kinetic information [40].

10.6 Some final remarks

FCS is a powerful quantitative technique that has been used in many investigations from solutions to cells, tissues and organisms. And although its use is not as widely distributed as would be expected for such a technique, we hope that with this book we make a small contribution to its wider usage and make the start for new students of FCS a bit easier. This book has been possible because of a large community of researchers that contributed to the theoretical development and the practical applications of FCS over the last 50 years and we hope we have been able to summarize some of the most salient aspects of FCS in this book to provide students with a good foundation from which to start to apply FCS in their own research, delve deeper into advanced fluctuation techniques or develop FCS further in new directions. Although the material in this book is built on the work of the FCS community and has been drawn from many sources, any mistakes are of course our own.

References

[1] Hirschfeld T 1976 Optical microscopic observation of single small molecules *Appl. Opt.* **15** 2965–6

[2] Jazani S, Sgouralis I, Shafraz O M, Levitus M, Sivasankar S and Pressé S 2019 An alternative framework for fluorescence correlation spectroscopy *Nat. Commun.* **10** 3662

[3] Laurence T A, Ly S, Bourguet F, Fischer N O and Coleman M A 2014 Fluorescence correlation spectroscopy at micromolar concentrations without optical nanoconfinement *J. Phys. Chem.* B **118** 9662–7

[4] Schrimpf W, Barth A, Hendrix J and Lamb D C 2018 PAM: a framework for integrated analysis of imaging, single-molecule, and ensemble fluorescence data *Biophys. J.* **114** 1518–28

[5] Muller P, Schwille P and Weidemann T 2014 PyCorrFit—generic data evaluation for fluorescence correlation spectroscopy *Bioinformatics* **30** 2532–3

[6] Krieger J W, Singh A P, Bag N, Garbe C S, Saunders T E, Langowski J and Wohland T 2015 Imaging fluorescence (cross-) correlation spectroscopy in live cells and organisms *Nat. Protoc.* **10** 1948–74

[7] Sankaran J, Shi X, Ho L Y, Stelzer E H K and Wohland T 2010 ImFCS: a software for imaging FCS data analysis and visualization *Opt. Express* **18** 25468–81

[8] van den Berg P A, Widengren J, Hink M A, Rigler R and Visser A J 2001 Fluorescence correlation spectroscopy of flavins and flavoenzymes: photochemical and photophysical aspects *Spectrochim. Acta A Mol. Biomol. Spectrosc.* **57** 2135–44

[9] Wennmalm S, Blom H, Wallerman L and Rigler R 2001 UV-Fluorescence correlation spectroscopy of 2-aminopurine *Biol. Chem.* **382** 393–7

[10] Lippitz M, Erker W, Decker H, van Holde K E and Basché T 2002 Two-photon excitation microscopy of tryptophan-containing proteins *Proc. Natl. Acad. Sci. USA* **99** 2772–7

[11] Rabasovic M D, Sisamakis E, Wennmalm S and Widengren J 2016 Label-free fluctuation spectroscopy based on coherent anti-Stokes Raman scattering from bulk water molecules *ChemPhysChem* **17** 1025–33

[12] Wennmalm S and Widengren J 2010 Inverse-fluorescence cross-correlation spectroscopy *Anal. Chem.* **82** 5646–51

[13] Levene M J 2003 Zero-mode waveguides for single-molecule analysis at high concentrations *Science* **299** 682–6

[14] Samiee K, Moran-Mirabal J, Cheung Y and Craighead H 2006 Zero mode waveguides for single-molecule spectroscopy on lipid membranes *Biophys. J.* **90** 3288–99

[15] Liao D, Galajda P, Riehn R, Ilic R, Puchalla J L, Yu H G, Craighead H G and Austin R H 2008 Single molecule correlation spectroscopy in continuous flow mixers with zero-mode waveguides *Opt. Express* **16** 10077–90
[16] Hansen R L, Zhu X R and Harris J M 1998 Fluorescence correlation spectroscopy with patterned photoexcitation for measuring solution diffusion coefficients of robust fluorophores *Anal. Chem.* **70** 1281–7
[17] Thompson N L and Steele B L 2007 Total internal reflection with fluorescence correlation spectroscopy *Nat. Protoc.* **2** 878–90
[18] Konopka C A and Bednarek S Y 2008 Variable-angle epifluorescence microscopy: a new way to look at protein dynamics in the plant cell cortex *Plant J.* **53** 186–96
[19] Unsay J D, Murad F, Hermann E, Ries J and García-Sáez A J 2018 Scanning fluorescence correlation spectroscopy for quantification of the dynamics and interactions in tube organelles of living cells *ChemPhysChem* **19** 3273–8
[20] Bergstrand J, Rönnlund D, Widengren J and Wennmalm S 2014 Scanning inverse fluorescence correlation spectroscopy *Opt. Express* **22** 13073
[21] Sandén T, Wyss R, Santschi C, Hassaïne G, Deluz C, Martin O J F, Wennmalm S and Vogel H 2012 A zeptoliter volume meter for analysis of single protein molecules *Nano Lett.* **12** 370–5
[22] Eggeling C *et al* 2009 Direct observation of the nanoscale dynamics of membrane lipids in a living cell *Nature* **457** 1159–62
[23] Hedde P N, rlich R E M D O, Blomley R, Gradl D, Oppong E, Cato A C B and Nienhaus G U 2013 Stimulated emission depletion-based rasterimage correlation spectroscopy reveals biomolecular dynamics in live cells *Nat. Commun.* **4** 1–8
[24] Dörlich R M, Chen Q, Niklas Hedde P, Schuster V, Hippler M, Wesslowski J, Davidson G and Nienhaus G U 2015 Dual-color dual-focus line-scanning FCS for quantitative analysis of receptor-ligand interactions in living specimens *Sci. Rep.* **5** 1–11
[25] Wallace M I, Ying L, Balasubramanian S and Klenerman D 2000 FRET Fluctuation spectroscopy: exploring the conformational dynamics of a DNA hairpin loop *J. Phys. Chem.* B **104** 11551–5
[26] Chiantia S, Ries J, Chwastek G, Carrer D, Li Z, Bittman R and Schwille P 2008 Role of ceramide in membrane protein organization investigated by combined AFM and FCS *Biochim. Biophys. Acta (BBA)-Biomembranes* **1778** 1356–64
[27] Krmpot A J *et al* 2019 Functional fluorescence microscopy imaging: quantitative scanning-free confocal fluorescence microscopy for the characterization of fast dynamic processes in live cells *Anal. Chem.* **91** 11129–37
[28] Guan X 2018 Machine learning approach to fluorescence correlation spectroscopy *BSc Thesis* National University of Singapore
[29] Sankaran J, Balasubramanian H, Tang W H, Ng X W, Röllin A and Wohland T 2020 SRRF 'n' TIRF – FCS: Simultaneous spatiotemporal super-resolution microscopy. bioRxiv.: 2020.02.26.965905
[30] Kask P, Palo K, Ullmann D and Gall K 1999 Fluorescence-intensity distribution analysis and its application in biomolecular detection technology *Proc. Natl. Acad. Sci. USA* **96** 13756–61
[31] Müller J, Chen Y and Gratton E 2000 Resolving heterogeneity on the single molecular level with the photon-counting histogram *Biophys. J.* **78** 474–86
[32] Meng F and Ma H 2006 A comparison between photon counting histogram and fluorescence intensity distribution analysis *J. Phys. Chem.* B **110** 25716–20

[33] Müller J D 2004 Cumulant analysis in fluorescence fluctuation spectroscopy *Biophys J.* **86** 3981–92
[34] Digman M A, Dalal R, Horwitz A F and Gratton E 2008 Mapping the number of molecules and brightness in the laser scanning microscope *Biophys. J.* **94** 2320–32
[35] Unruh J R and Gratton E 2008 Analysis of molecular concentration and brightness from fluorescence fluctuation data with an electron multiplied CCD camera *Biophys. J.* **95** 5385–98
[36] Di Rienzo C, Gratton E, Beltram F and Cardarelli F 2013 Fast spatiotemporal correlation spectroscopy to determine protein lateral diffusion laws in live cell membranes *Proc. Natl Acad. Sci.* **110** 12307–12
[37] Moens P D J, Digman M A and Gratton E 2015 Modes of diffusion of cholera toxin bound to GM1 on live cell membrane by image mean square displacement analysis *Biophys. J.* **108** 1448–58
[38] Digman M A and Gratton E 2009 Imaging barriers to diffusion by pair correlation functions *Biophys. J.* **97** 665–73
[39] Lakowicz J R 2006 *Principles of Fluorescence Spectroscopy* 3rd edn (Berlin: Springer)
[40] Torres T and Levitus M 2007 Measuring conformational dynamics: a new FCS-FRET approach *J. Phys. Chem.* **B 111** 7392–400

IOP Publishing

An Introduction to Fluorescence Correlation Spectroscopy

Thorsten Wohland, Sudipta Maiti and Radek Macháň

Appendix A

Solutions to exercises

Chapter 2

1. Transfer the argument of section 2.2 about the fluctuations to diffusion. For this you need to assume that you have unequal concentrations within at least two connected volumes. How do the concentrations in the two volumes change and what is the role of concentration fluctuations? Are the concentration fluctuations in the two volumes correlated or anticorrelated? Describe the processes in the two volumes until concentrations equilibrate.

   Answer: In the below scheme you see two compartments, A and B, at different concentrations of a molecule M at an initial time t_0. *At time* t_0 *there will be more molecules moving from compartment A to B. This happens until the concentration is the same on both sides at* t_∞. *At* t_∞ *there will be as many molecules M diffusing from compartment A to B as there are molecules M diffusing from B to A. Thus, this system is analogous to our example of reactions where first more molecules A combine with B to produce AB until there is an equilibrium reached at which there as many molecules AB formed as fall apart. So, in the case of diffusion the fluctuations in equilibrium will contain the information about the dynamics of the system.*

doi:10.1088/978-0-7503-2080-1ch11

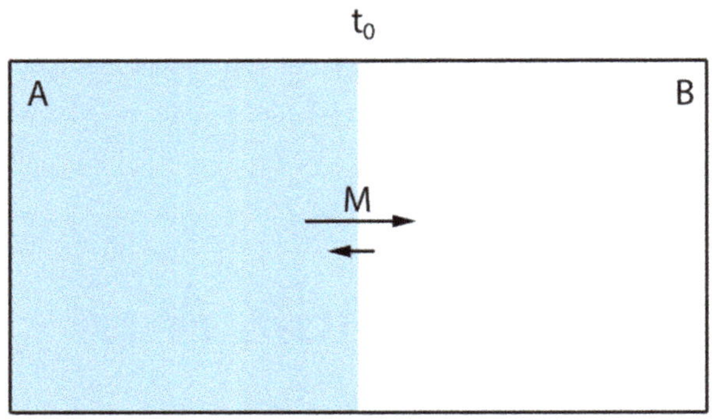

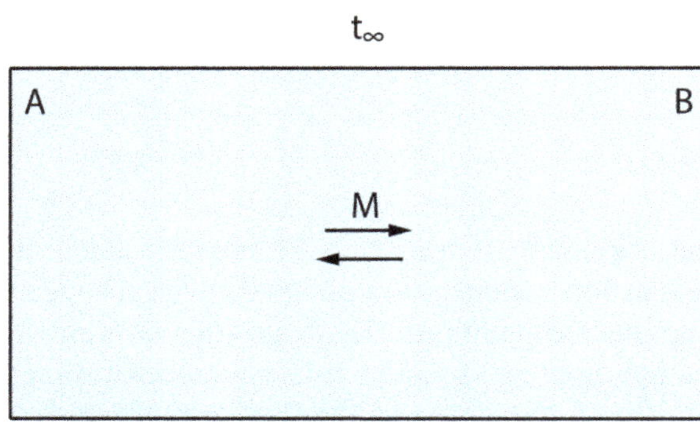

2. For the coin tossing examples in section 2.3, show that this is not only true when H and T are replaced by 1 and 0 but for any values for H and T as long as they are different.

Answer: Let's calculate the principle equations without putting in any concrete values:

$$\langle a(u)b(u)\rangle = \frac{1}{n}\sum_{u=1}^{n} a(u)b(u) = \sum_{m=1}^{4} p_{abm} a_m b_m$$
$$= p_{HH}HH + p_{HT}HT + p_{TH}TH + p_{TT}TT$$

$$\langle a(u)\rangle = \frac{1}{n}\sum_{u=1}^{n} a(u) = H \times p_H + T \times p_T$$

$$\langle b(u)\rangle = \frac{1}{n}\sum_{u=1}^{n} b(u) = H \times p_H + T \times p_T$$

$$\langle a(u)\rangle\langle b(u)\rangle = (H \times p_H + T \times p_T)(H \times p_H + T \times p_T)$$
$$= p_H p_H HH + p_H p_T HT + p_T p_H TH + p_T p_T TT$$

Thus, the question, whether the two variables are correlated is not dependent on the values of H and T but rather on whether the probabilities p_{HH}, p_{HT}, p_{TH}, p_{TT} are the same as the probabilities $p_H p_H$, $p_H p_T$, $p_T p_H$, $p_T p_T$. But this is of course just the question of correlations, namely whether the probabilities of a common event are just the product of the probabilities of the two individual events or whether a common event is more or less likely that what one would expect from the single events.
However, if $H = T$, then we obtain

$$\langle a(u)b(u)\rangle = (p_{HH} + p_{HT} + p_{TH} + p_{TT})HH$$
$$\langle a(u)\rangle\langle b(u)\rangle = (p_H p_H + p_H p_T + p_T p_H + p_T p_T)HH$$

As the sum of probabilities must be 1, then in this case we will have $\langle a(u)b(u)\rangle = \langle a(u)\rangle\langle b(u)\rangle$, due to the fact that $H = T$, and we can't make any statements about correlations.

3. Find examples of fully, partially and anti-correlated variables.

   Answer: There can be many examples, we just give some examples as a starting point to think more about correlations.
   Fully correlated: Particle size and diffusion time (assuming constant shape).
   Partially correlated: Ligand concentration and receptor activity in the case of receptors having a desensitized state. Initially, higher ligand concentration will lead to more activity but as receptors desensitize signalling will not rise but rather will be sustained and will fall when all receptors are in a desensitized state.
   Anti-correlated: Competitive ligand binding to a receptor. The more one ligand binds the less the other can bind.

4. How does the addition of a constant a_0 to $a(t)$ influence the correlation function as outlined in section 2.5? First calculate this for the correlation coefficient assuming that $a(t)$ is Poisson distributed. Then calculate it for $g(\tau)$.

   Answer: From the equation

$$G = \frac{\sigma^2}{\langle a(t)\rangle^2} + 1$$

We can replace $a(t)$ with $a(t) + a_0$ to obtain

$$G = \frac{\sigma^2}{\langle a(t) + a_0\rangle^2} + 1 = \frac{\mu}{\langle \mu + a_0\rangle^2} + 1$$

As we see, the amplitude is now not anymore proportional to μ^{-1}. One can also see that the effect of the constant a_0 will be the larger the smaller μ is.
For the correlation function $g(\tau)$ we have:

$$g(\tau) = \frac{\langle \delta a(t)\delta a(t+\tau)\rangle}{\langle a(t)\rangle^2}$$

Remember that $\delta a(t) = a(t) - \langle a(t)\rangle$. *Thus, adding* a_0 *to* $a(t)$ *leads to*

$$\begin{aligned}\delta a(t) &= (a(t) + a_0) - \langle a(t) + a_0\rangle = a(t) + a_0 - \langle a(t)\rangle - a_0 \\ &= a(t) - \langle a(t)\rangle\end{aligned}$$

So, the numerator doesn't change. Thus we get

$$g(\tau) = \frac{\langle \delta a(t)\delta a(t+\tau)\rangle}{\langle a(t) + a_0\rangle^2}$$

Meaning we change only the amplitude but not the time course of the correlation function. This is similar to the problem of uncorrelated background.

5. Following the same arguments as in section 2.5 (equations (2.16)–(2.18)), show the following relations:

$$g(0) = \frac{\langle \delta a(t)\delta a(t)\rangle}{\langle a(t)\rangle^2} = \frac{\sigma^2}{\langle a(t)\rangle^2} = \frac{\sigma^2}{\mu^2}$$

What is $g(0)$ for a Poisson distributed variable?

Answer: We start with the definition of the fluctuations

$$\delta a(t) = a(t) - \langle a(t)\rangle$$

and substitute it into the correlation function

$$\begin{aligned}g(0) &= \frac{\langle \delta a(t)\delta a(t)\rangle}{\langle a(t)\rangle^2} = \frac{\langle (a(t) - \langle a(t)\rangle)(a(t) - \langle a(t)\rangle)\rangle}{\langle a(t)\rangle^2} \\ &= \frac{\langle a(t)^2 - 2a(t)\langle a(t)\rangle + \langle a(t)\rangle^2\rangle}{\langle a(t)\rangle^2} \\ &= \frac{\langle a(t)^2\rangle - 2\langle a(t)\rangle^2 + \langle a(t)\rangle^2}{\langle a(t)\rangle^2} = \frac{\langle a(t)^2\rangle - \langle a(t)\rangle^2}{\langle a(t)\rangle^2}\end{aligned}$$

With the definition of the variance $\sigma^2 = \langle a(t)^2\rangle - \langle a(t)\rangle^2$, *we obtain*

$$g(0) = \frac{\sigma^2}{\langle a(t)\rangle^2} = \frac{\sigma^2}{\mu^2}$$

And for a Poisson process where the variance is equal to the mean,

$$g(0) = \frac{\sigma^2}{\mu^2} = \frac{\mu}{\mu^2} = \frac{1}{\mu}$$

Chapter 3

1. Suppose you are building an FCS spectrometer yourself. You are planning to use an optical fiber as the confocal aperture. Your objective is 40×, 1.2 NA, and the excitation laser is a 543 nm helium–neon. The microscope for the

objective has a 16 cm tube lens, but in your home built set-up you are using a 25 cm one. What fiber diameter should you choose?

Answer: The Airy disk size (diameter) is $1.22\lambda/NA$, *therefore the optimal diameter of the fiber (for magnification M) is* $M * 1.22\lambda/NA$. *Now, nominally the magnification is* 40×, *but the tube lens has a focal length of* 25 *cm instead of the standard* 16 *cm, so actual* $M = 40 \times 25/16 = 62.5\times$. *So in this case, the preferred diameter would be* 34.5 *μm.*

2. Suppose in a TIRF FCS experiment using 543 nm excitation and 1.49 NA objective lens, fluorescent beads of 1 μm diameter in aqueous solution are used for testing. The coverglass has a refractive index of 1.51, and the beam is incident at an angle of 70°. The viscosity of water is 1 mPa·s, at the laboratory temperature (20 °C). What would be the approximate diffusion time observed in the experiment?

   [We note that in the TIRF geometry, the characteristic diffusion time in the direction away from the surface is given by ~3.3 d^2/D, where d = characteristic decay length of TIRF illumination, and D = the diffusion constant].

   Answer: There are two timescales of diffusion in the experiment. The first one is the lateral diffusion, parallel to the surface. The resolution for that is given by the optical resolution, which would be of the order of 250 *nm. However, the second one would in the vertical direction, where the thickness of the probe volume would be given by*

$$d = \frac{\lambda_0}{4\pi\sqrt{\left(n_1^2 \sin^2 \phi_i - n_2^2\right)}}$$

   For the given parameters, d = 76 nm. *Since this is shorter, this would be the dominant distance scale along which bulk of the diffusion would occur.*

   Now, $D = \frac{kT}{6\pi\eta R} = (1.38 \times 10^{-23} \times 293) / (6 \times 3.14 \times 0.001 \times 0.5 \times 10^{-6}) = 4.3 \times 10^{-13}\ \mathrm{m^2\ s^{-1}}$

   So, the characteristic diffusion time is $3.3 \times (76 \times 10^{-9})^2/4.3 \times 10^{-13}$ *s* = 44 ms.

Chapter 4

1. Show that equation (4.4) for the number of products needed to calculate an ACF can be easily derived using the formula for triangular numbers. Triangular numbers can be written as sums of the form $\sum_{i=1}^{n} i$. For such sums it holds:

$$\sum_{i=1}^{n} i = \frac{1}{2} n(n+1).$$

Answer:

$$\sum_{m=1}^{m_{\max}}(M - m) = Mm_{\max} - \sum_{m=1}^{m_{\max}} m = Mm_{\max} - \frac{1}{2}m_{\max}(m_{\max} + 1)$$

2. Prove the formula for triangular numbers used in the previous exercise. *Hint*: use mathematical induction; show that the equation is valid for $n = 1$ and then that its validity for n implies its validity for $n + 1$. Alternatively use the fact that for an arithmetic series, the sum is equal to the number of terms times the average of the series.

   Answer:

$$\sum_{i=1}^{n+1} i = \left(\sum_{i=1}^{n} i\right) + n + 1 = \frac{1}{2}n(n + 1) + n + 1 = \frac{1}{2}(n + 1)(n + 2)$$

3. Consider an FCS time trace recorded for 30 s with 1 μs temporal resolution (typical values in confocal FCS experiments). How many products need to be calculated to obtain the ACF for lag times up to 2 s? How does the number change if the measurement time is increased twice?

   Answer: In this case $M = 30 \times 10^6$ *and* $m_{\max} = 2 \cdot 10^6$*; according to equation* (4.4), *the number of sums is approximately* 59×10^{12}. *Doubling the measurement time leads to approximately* 119×10^{12} *sums.*

4. Show that equation (4.5) can be rewritten in a non-recursive manner as:

$$\tau_m = \begin{cases} m\Delta\tau & 0 < m \leqslant \frac{p}{2} \\ \Delta\tau\left(l + \frac{p}{2}\right)2^{k-1} & \frac{p}{2} < m \leqslant (q + 1)\frac{p}{2} \end{cases}$$

   where $k = 1, 2, \ldots, q$ is the number of the group of lag times and $l = 1, 2, \ldots, \frac{p}{2}$ is the number of the lag time point within its group and $m = k\frac{p}{2} + l$.

   Answer: We want to show that:

$$\tau_m = \tau_{m-1} + \Delta\tau\, 2^{\left\lfloor \frac{m-1-p/2}{p/2} \right\rfloor} \qquad \frac{p}{2} < m \leqslant (q + 1)\frac{p}{2}$$

   can be written as:

$$\tau_m = \Delta\tau\left(l + \frac{p}{2}\right)2^{k-1} \qquad \frac{p}{2} < m \leqslant (q + 1)\frac{p}{2}$$

where $k = 1, 2, \ldots, q$ *is the number of the group of lag times and* $l = 1, 2, \ldots, \frac{p}{2}$ *is the number of the lag time point within its group and* $m = k\frac{p}{2} + l$. *Let us use the recursive formula to write the expression for* τ_{m+1} *and compare it with the non-recursive formula for the same lag time. For* $l < \frac{p}{2}$, *that is for all members of each group except for the last one we get:*

$$\begin{aligned}\tau_{m+1} &= \Delta\tau\left(l + \frac{p}{2}\right)2^{k-1} + \Delta\tau\, 2^{\left\lfloor \frac{m-1-p/2}{p/2} \right\rfloor} \\ &= \Delta\tau\left(l + \frac{p}{2}\right)2^{k-1} + \Delta\tau\, 2^{k-1} \\ &= \Delta\tau\left(l + 1 + \frac{p}{2}\right)2^{k-1}\end{aligned}$$

For the last member of each group we get:

$$\tau_{m+1} = \Delta\tau\left(2\frac{p}{2}\right)2^{k-1} + \Delta\tau\, 2^{k} = \Delta\tau\left(1 + \frac{p}{2}\right)2^{k}$$

which is the non-recursive expression for the first lag time of the following group.

5. Show that the maximum correlation lag time of a (p, q) correlator with $p = 16$ is $2^{q+3}\Delta\tau$.

   Answer: According to the non-recursive formula $\tau_{m_{\max}} = \Delta\tau\ (\frac{p}{2} + \frac{p}{2})2^{q-1} = \Delta\tau\ 2^4 2^{q-1}$, *where we used the fact that* $16 = 2^4$.

6. Use the result of the previous exercise to calculate the maximum correlation lag time (in units of $\Delta\tau$) achieved with a (16, 8) multiple-tau correlator. For how many lag time points τ_{m} is the value of $G(\tau_{\mathrm{m}})$ calculated? Compare the number of points with that needed for a linear correlator to reach the same maximum lag time.

   Answer: $\tau_{m_{\max}} = \Delta\tau\ 2^{q+3} = \Delta\tau\ 2^{11} = 2048$. *The number of lag time points* $m_{\max} = q\,p = 128$. *For a linear correlator* $m_{\max} = \tau_{m_{\max}}/\Delta\tau = 2048$.

7. Prove equation (4.6). *Hint*: use mathematical induction to prove that:

$$\sum_{i=1}^{n} 2^{-k} = 1 - \frac{1}{2^n}$$

Answer:

$$\sum_{i=1}^{n+1} 2^{-k} = \left(\sum_{i=1}^{n} 2^{-k}\right) + 2^{-(n+1)} = 1 - \frac{1}{2^n} + \frac{1}{2}\frac{1}{2^n} = 1 - \frac{1}{2^{n+1}}$$

Chapter 5

Note that for exercises 4–11, you will need to have the computable document format player which is freely available at https://www.wolfram.com/cdf/. With the CDF Player you can open the 'Introduction to FCS—interactive exercises.cdf' file available at https://iopscience.iop.org/book/978-0-7503-2080-1.

1. Verify the transition from 5.3 to 5.5.

Answer:

$$G(\tau) = \frac{1}{\langle C\rangle \pi^{3/2} w_0^2 z_0}\left(1 + \frac{4D\tau}{w_0^2}\right)^{-1/2}\left(1 + \frac{4D\tau}{w_0^2}\right)^{-1/2}\left(1 + \frac{4D\tau}{z_0^2}\right)^{-1/2} + 1$$

Using that $V_{\mathrm{eff}} = \pi^{3/2} w_0^2 z_0$ *(for a confocal system) and the average number of particles is* $N = \langle C\rangle V_{\mathrm{eff}}$, *we arrive at*

$$G(\tau) = \frac{1}{N}\left(1 + \frac{4D\tau}{w_0^2}\right)^{-1/2}\left(1 + \frac{4D\tau}{w_0^2}\right)^{-1/2}\left(1 + \frac{4D\tau}{z_0^2}\right)^{-1/2} + 1$$

The second and third term can be combined as they are the same for a rotational symmetric confocal volume:

$$G(\tau) = \frac{1}{N}\left(1 + \frac{4D\tau}{w_0^2}\right)^{-1}\left(1 + \frac{4D\tau}{z_0^2}\right)^{-1/2} + 1$$

Next we define the diffusion time over the following equation $4D\tau_D = w_0^2$, *which is reminiscent of the mean squared displacement:*

$$G(\tau) = \frac{1}{N}\left(1 + \frac{\tau}{\tau_D}\right)^{-1}\left(1 + \frac{\tau}{\frac{z_0^2}{w_0^2}\tau_D}\right)^{-1/2} + 1$$

And finally, we define the ration of the axial to the lateral extent of the confocal volume as $K = \frac{z_0}{w_0}$. *Thus, we arrive at*

$$G(\tau) = \frac{1}{N}\left(1 + \frac{\tau}{\tau_D}\right)^{-1}\left(1 + \frac{\tau}{K^2\tau_D}\right)^{-1/2} + G_\infty$$

2. Extend the multi-particle ACF to n particles

Answer: For two particles we obtained

$$G(\tau) = \frac{1}{N(\kappa_1 f + \kappa_2(1-f))^2}\left[\kappa_1^2 f\left(1 + \frac{\tau}{\tau_{D1}}\right)^{-1}\left(1 + \frac{\tau}{K^2\tau_{D1}}\right)^{-1/2} + \kappa_2^2(1-f)\left(1 + \frac{\tau}{\tau_{D2}}\right)^{-1}\left(1 + \frac{\tau}{K^2\tau_{D2}}\right)^{-1/2}\right] + G_\infty$$

For the extension to n particles we need to define $n-1$ fractions f_x and n brightness values κ_x, as well as n diffusion times τ_{Dx}. Then

$$G(\tau) = \frac{1}{N(\kappa_1 f_1 + \kappa_2 f_2 + \cdots + \kappa_n(1 - f_1 - f_2 \ldots f_{n-1}))^2}$$
$$\times \left[\kappa_1^2 f_1 \left(1 + \frac{\tau}{\tau_{D1}}\right)^{-1} \left(1 + \frac{\tau}{K^2 \tau_{D1}}\right)^{-1/2} \right.$$
$$+ \kappa_2^2 f_2 \left(1 + \frac{\tau}{\tau_{D2}}\right)^{-1} \left(1 + \frac{\tau}{K^2 \tau_{D2}}\right)^{-1/2} + \cdots$$
$$\left. + \kappa_n^2 (1 - f_1 - f_2 \ldots f_{n-1}) \left(1 + \frac{\tau}{\tau_{Dn}}\right)^{-1} \left(1 + \frac{\tau}{K^2 \tau_{Dn}}\right)^{-1/2} \right] + G_\infty$$

which is just

$$G(\tau) = \frac{\sum_{i=1}^{n} \kappa_i^2 f_i \left(1 + \frac{\tau}{\tau_{Di}}\right)^{-1} \left(1 + \frac{\tau}{K^2 \tau_{Di}}\right)^{-1/2}}{N\left(\sum_{i=1}^{n} \kappa_i f_i\right)^2} + G_\infty$$

with the condition that $\sum_i f_i = 1$.

3. Equation (5.21) describes the dependence of the diffusion coefficient on the mass and hydrodynamic radius of the particle under investigation.
   a. What is the expected value of D if the particle changes its hydrodynamic radius by a factor 2?

   Answer:

$$D \sim \frac{1}{r_h}$$

   Therefore, if the particle increases its hydrodynamic radius by a factor 2, *the diffusion coefficient will be smaller by a factor* 2.

   b. What is the expected value of D if the particle changes mass by a factor 2, assuming the particle remains spherical?

   Answer:

$$D \sim \frac{1}{r_h} \approx \frac{1}{\sqrt[3]{M}}$$

   If the particle increases its mass by a factor 2, *then the diffusion coefficient decreases only by a factor of* $\sqrt[3]{2} \approx 1.26$. *However, this is only true if the particle is spherical and if the density is constant.*

c. If the particle changes mass, e.g. as would happen in dimerization, do you expect a change in D smaller or greater than what you calculated in *b*?

Answer: As the condition that a particle is spherical is probably not fulfilled, one would expect a decrease of the diffusion coefficient by a factor larger than 1.26. *Of shapes with the same volume (assuming that the particles have comparable density), the spherical shape will have the fastest diffusion coefficient.*

4. In the 'Confocal FCS' section of the 'Introduction to FCS—interactive exercises.cdf' (available at https://iopscience.iop.org/book/978-0-7503-2080-1), choose G2D1P and G3D1P for model and reference function, respectively for comparison.
   a. Keeping N, τ_D and G_∞ constant, change the structure factor K and observe the differences. At what value for K do the functions become visibly indistinguishable?

   Answer: For $K \to \infty$ *the 2D and 3D functions become indistinguishable as*

$$\lim_{K\to\infty}\left(1+\frac{\tau}{K^2\tau_D}\right)=1$$

   b. What happens for $K = 1$.

   Answer: In this case

$$G(\tau)=\frac{1}{N}\left(1+\frac{\tau}{\tau_D}\right)^{-1}\left(1+\frac{\tau}{K^2\tau_D}\right)^{-1/2}+G_\infty=\frac{1}{N}\left(1+\frac{\tau}{\tau_D}\right)^{-3/2}+G_\infty$$

   c. Using equation (5.16), explain how the differences between the 2D and 3D model develop for small and large K.

   Answer:

$$G(\tau_D)=\frac{1}{2N\sqrt{1+K^{-2}}}+G_\infty$$

   We can plot this function for K from 1 *to* 20, *assuming for simplicity a value for* $G_\infty = 0$:

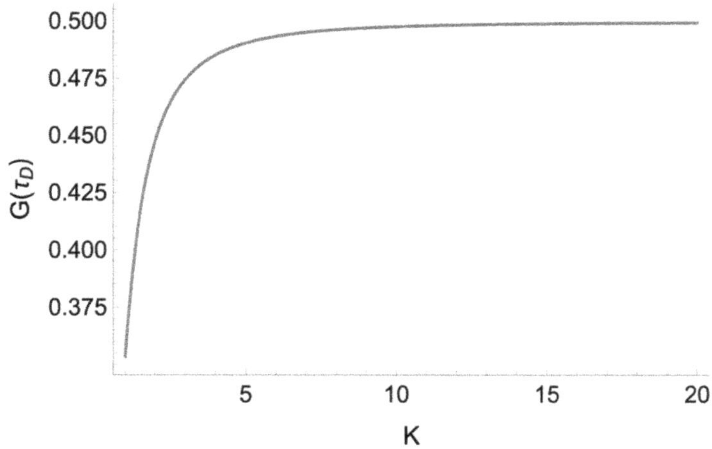

This shows how the correlation function at τ_D converges to the value to the half-amplitude $G(\tau_D) = G(0)/2 = 1/2N$ with increasing K. In this case it is easy to estimate the diffusion time τ_D from the half-amplitude. But it also points out that especially at low K values between 1 and 5 the strongest deviations from this rule are expected.

5. In the 'Confocal FCS' section of the 'Introduction to FCS—interactive exercises.cdf' (available at https://iopscience.iop.org/book/978-0-7503-2080-1) choose G3D2PQ for both, model and reference, functions. Keep the default values of the parameters, but set $\tau_{D2} = 0.004$ (s) for both functions.
   a. Change the q value (relative brightness, see equation (5.28)) to 2 in the model function. Which other parameter(s) do you need to change in the model function so that the reference and model function overlap again?

   Answer: You need to change both, the total number of particles N and the fraction F_2. With values of 1.4 for N, and 0.15 for F_2, model and reference are again about the same.

   b. Can you explain why these two parameters need to be changed but the other parameters stay constant? Verify this by changing the q value to other values (e.g. 3 or 4).

   Answer: From the equation

$$G(0) = \frac{\sum_{i=1}^{n} q_i^2 f_i}{N\left(\sum_{i=1}^{n} q_i f_i\right)^2} + G_\infty$$

   one can see that it is really q, f, and N, which are related and determine the amplitudes.

6. Choose G3D1P and G3D1P1T models for comparison. Keep common parameters equal for both models. And keep the triplet fraction at $F_{\text{trip}} = 0.5$.
   a. At what triplet time τ_{trip} is there no obvious transition anymore between the triplet and diffusive contribution (no two 'humps')?

   Answer: When $\frac{1}{10} \gtrsim \frac{\tau_D}{\tau_{\text{trip}}} \gtrsim \frac{1}{20}$ *then it is difficult to make a difference bwteen the two processes. This will get worse when the curves are noisy.*

   b. Does this change with the value of F_{trip}?

   Answer: This changes slightly with F_{trip}

7. Investigate the influence of the different parameters on the correlation function. For this purpose, go to the section 'Noise and Background' of the 'Introduction to FCS—interactive exercises.cdf' (available at https://iopscience.iop.org/book/978-0-7503-2080-1). Start with $N = 1$, cps = 1000, a background of 100 counts per second, and 20 s measurement time.
   a. Change the measurement time to see the increase in noise. Does the amplitude change?

   Answer: No, the amplitude does not change, only the noise on the ACF.

   b. Change the background to see the changes in amplitude. Does the noise change?

   Answer: Changing background changes only the amplitude but not the noise, because we assume here a constant background that is itself noise free.

   c. If you start with the default values as stated above and change the number of particles, how does the difference between the ACFs with and without background change? Explain why.

   Answer: The larger N, the closer the ACF with and without background will be. The background increases the apparent number N (see also chapter 7). The larger N the smaller will be the influence of the background.

8. Start as in the previous question with the default values but choose cps = 100, and a background of 1000 counts per second. Now change N from 1 to 30. How does the amplitude change? Why do you see this deviation from the background free case, where the amplitude always decreases with N?

   Answer: If you take a close look, the amplitude first increases, as the signal increases, before it decreases with higher N. This is an effect seen at low S/N.

9. Using the 'STICS and RICS.cdf' (available at https://iopscience.iop.org/book/978-0-7503-2080-1) verify how the STICS function (first graph) varies with different parameters. In particular compare the influence of beam radius and diffusion coefficient on the peak shape, and observe how the peak position depends on the flow velocities in the x- and y-directions.

   Answer: This is a purely investigative task to see how the peak width and position change in STICS and RICS.

10. Select the 'Imaging FCS' section of the 'Introduction to FCS—interactive exercises.cdf' file (available at https://iopscience.iop.org/book/978-0-7503-2080-1).
    a. Compare ITIR-FCS and SPIM-FCS models for varying parameters; in particular, investigate how the z-resolution (σ_z) influences SPIM FCS and how SPIM-FCS and ITIR-FCS converge for large values of σ_z.

       Answer: This is similar to the investigation of 2D *and* 3D *confocal models and shows how the* 3D *model converges to the* 2D *model for large structure factors. Note, however, that the program does not take account of cross-talk. In reality a larger light sheet thickness will increase the cross talk between pixels and this will lead to deviation from the presented models. This can lead to inaccurate determinations of D and to extra autocorrelation components in spatial cross-correlations.*

    b. Observe how spatial cross-talk leads to a non-zero amplitude at distances up to about 3 pixels in ITIR-FCS.

       Answer: As just discussed under a.*, the finite light sheet thickness will lead to cross-talk between pixels. The thicker the light sheet, the further this cross-talk will extend.*

Chapter 6

1. Use equations (6.6) and (6.8) to derive an explicit formula for K_D as a function of the CCF and ACF amplitudes measured in an ideal DC-FCCS experiment.

   Answer: From equation (6.6):

$$\langle C_a \rangle = \frac{1}{VG_{0_A}} - \langle C_{ab} \rangle$$

$$\langle C_b \rangle = \frac{1}{VG_{0_B}} - \langle C_{ab} \rangle$$

$$\langle C_{ab} \rangle = \frac{G_{0X}}{VG_{0_A}G_{0_B}}$$

then:

$$K_D = \frac{\langle C_a\rangle\langle C_b\rangle}{N_A\langle C_{ab}\rangle} = \frac{(G_{0_A} - G_{0_X})(G_{0_B} - G_{0_X})}{N_A V G_{0_A} G_{0_B} G_{0_X}}$$

2. Use equations (6.24) to show that, when p_a and/or p_b are smaller than 1, the apparent dissociation constant defined by equation (6.25) is larger than the actual dissociation constant calculated from the equilibrium concentrations $[M_a]$, $[M_b]$ and $[M_aM_b]$ using equation (6.8).

   Answer:

$$\begin{aligned} K_D^{app} &= \frac{\langle C_a^{app}\rangle\langle C_b^{app}\rangle}{N_A\langle C_{ab}^{app}\rangle} \\ &= \frac{(p_a[M_a] + p_a(1 - p_b)[M_aM_b])(p_b[M_b] + (1 - p_a)p_b[M_aM_b])}{p_a p_b[M_aM_b]} \\ &= \frac{[M_a][M_b]}{[M_aM_b]} + (1 - p_a)[M_a] + (1 - p_b)[M_b] \end{aligned}$$

3. Consider a DC-FCCS experiment studying complex formation between proteins M_a and M_b in the presence of endogenous forms of the proteins at concentrations $[M_a^{(e)}]$ and $[M_b^{(e)}]$, respectively:
   a. Write a model for the correlation function amplitudes in the case the fluorescent moiety a (which forms a part of $M_a^{(f)}$, the fluorescent analogue of protein M_a) is quenched by protein M_b. Use equations (6.27) to write the concentrations of the different particle types appearing in the model.

      Answer:

$$G_{0_A} = \frac{(q_a\kappa_{aA} + q_b\kappa_{bA})^2\langle C_{ab}^{app}\rangle + \kappa_{aA}^2\left(\langle C_a^f\rangle + q_a\langle C_a^c\rangle\right) + \kappa_{bA}^2\langle C_b^{app}\rangle}{V_A\left[\kappa_{aA}\left(\langle C_a^f\rangle + q_a\langle C_a^c\rangle\right) + \kappa_{bA}\langle C_b^{app}\rangle + (q_a\kappa_{aA} + q_b\kappa_{bA})\langle C_{ab}^{app}\rangle + \frac{\langle B_A\rangle}{V_A}\right]^2}$$

$$G_{0_B} = \frac{(q_a\kappa_{aB} + q_b\kappa_{bB})^2\langle C_{ab}^{app}\rangle + \kappa_{aB}^2\langle C_a^{app}\rangle + \kappa_{bB}^2\langle C_b^{app}\rangle}{V_B\left[\kappa_{aB}\left(\langle C_a^f\rangle + q_a\langle C_a^c\rangle\right) + \kappa_{bB}\langle C_b^{app}\rangle + (q_a\kappa_{aB} + q_b\kappa_{bB})\langle C_{ab}^{app}\rangle + \frac{\langle B_B\rangle}{V_B}\right]^2}$$

$$\begin{aligned} G_{0_X} = &\frac{(q_a\kappa_{aA} + q_b\kappa_{aB})(q_a\kappa_{aB} + q_b\kappa_{bB})\langle C_{ab}^{app}\rangle + \kappa_{aA}\kappa_{aB}\left(\langle C_a^f\rangle + q_a\langle C_a^c\rangle\right) + \kappa_{bA}\kappa_{bB}\langle C_b^{app}\rangle}{V_X^{\text{eff}}\left[\kappa_{aA}\left(\langle C_a^f\rangle + q_a\langle C_a^c\rangle\right) + \kappa_{bA}\langle C_b^{app}\rangle + (q_a\kappa_{aA} + q_b\kappa_{bA})\langle C_{ab}^{app}\rangle + \frac{\langle B_A\rangle}{V_X^{\text{eff}}}\right]} \\ &\times\left[\kappa_{aB}\left(\langle C_a^f\rangle + q_a\langle C_a^c\rangle\right) + \kappa_{bB}\langle C_b^{app}\rangle + (q_a\kappa_{aB} + q_b\kappa_{bB})\langle C_{ab}^{app}\rangle + \frac{\langle B_B\rangle}{V_X^{\text{eff}}}\right]^{-1} \end{aligned}$$

where the concentrations $\langle C_a^f \rangle$ *and* $\langle C_a^c \rangle$ *are:*

$$\langle C_a^f \rangle = [M_a^{(f)}]N_A$$
$$\langle C_a^c \rangle = [M_a^{(f)}M_b^{(e)}]N_A$$

Assuming that the concentrations of the endogenous proteins are known and that the dissociation constant of the complex does not depend on the fluorescent labelling (it is the same for fluorescently labelled as well as the endogenous proteins), $\langle C_a^f \rangle$ *and* $\langle C_a^c \rangle$ *can be expressed using the endogenous protein concentrations and the apparent concentrations,* $\langle C_{ab}^{app} \rangle$, $\langle C_a^{app} \rangle$ *and* $\langle C_b^{app} \rangle$, *similarly to equations* (6.32). *However, solving equations* (6.27) *is much more tedious than solving equations* (6.24) *(which we did to obtain equations* (6.32)*). We will, therefore, not attempt it here.*

b. How does the model change if, instead of being quenched by protein M_b, the fluorescent moiety a forms a FRET pair with the fluorescent moiety b and FRET between them occurs when the two fluorescently labelled protein analogues ($M_a^{(f)}$ and $M_b^{(f)}$) form a complex.

Answer: In this case FRET occurs only in the apparently double-labelled particles; therefore, equations (6.30) *apply.*

c. How does the model derived in a. change if the fluorescent moieties a and b undergo transitions to dark states? The probabilities of the moieties being fluorescent (residing in a bright state) are p_a and p_b, respectively

Answer: The equations for the correlation functions amplitudes remain unchanged, only the concentrations $\langle C_a^f \rangle$ *and* $\langle C_a^c \rangle$ *are in this case (see equations* (6.28)*):*

$$\langle C_a^f \rangle = p_a[M_a^{(f)}]N_A$$
$$\langle C_a^c \rangle = p_a(1 - p_b)[M_a^{(f)}M_b^{(f)}]N_A + p_a[M_a^{(f)}M_b^{(e)}]N_A$$

What we said in point a. *about solving equations* (6.27) *applies to equations* (6.28) *as well.*

4. Write explicitly the concentrations in equations (6.34) for the case of dimer formation ($\alpha = 2$). What is the concentration $\langle C_{0a0b}^{app} \rangle$ of particles that are non-fluorescent? Verify that summing the numbers of monomer units M present in each of the four sub-populations (double-labelled, single-labelled with a, single-labelled with b, non-fluorescent) gives the total number of monomer units in the sample: $N_A([M] + 2[M_2])$.

Answer:

$\langle C_{1a1b}^{app} \rangle = 2p_a p_b [M_2] N_A$

$\langle C_{2a0b}^{app} \rangle = p_a^2 [M_2] N_A$

$\langle C_{0a2b}^{app} \rangle = p_b^2 [M_2] N_A$

$\langle C_{1a0b}^{app} \rangle = p_a [M] N_A + 2p_a (1 - p_a - p_b)[M_2] N_A$

$\langle C_{0a1b}^{app} \rangle = p_b [M] N_A + 2p_b (1 - p_a - p_b)[M_2] N_A$

$\langle C_{0a0b}^{app} \rangle = (1 - p_a)[M] N_A + (1 - p_b)[M] N_A + (1 - p_a - p_b)^2 [M_2] N_A$

5. Consider an oligomerisation experiment under the assumptions of complete dimerisation, perfect labelling, no dark states, no background, no crosstalk and identical observation volumes.
   a. Verify the validity of equation (6.37) for the case of dimerisation ($\alpha = 2$). Show that the result is independent on p_a (the fraction of monomers labelled by fluorophore a) for $0 < p_a < 1$

      Answer: Under the given assumptions, the CCF for the dimer is given by:

$$G_{0X}^{(2)} = \frac{2p_a(1 - p_a)}{V\langle C_{ab}\rangle \left[2p_a^2 + 2p_a(1 - p_a)\right][2(1 - p_a)^2 + 2p_a(1 - p_a)]}$$

$$= \frac{2p_a(1 - p_a)}{V\langle C_{ab}\rangle 2p_a 2(1 - p_a)} = \frac{1}{2V\langle C_{ab}\rangle}$$

   b. Derive equation (6.37) for an arbitrary degree oligomerisation α. *Hint*: use the formula for the sum of the binomial distribution:

$$\sum_{i=1}^{n} \binom{n}{i} p^i (1 - p)^{n-i} = 1, \text{ for } 0 \leqslant p \leqslant 1$$

      Answer: In the general case we get:

$$G_{0X}^{(\alpha)} = \frac{\sum_{\rho=1}^{\alpha-1} \binom{\alpha}{\rho} p_a^{\rho} 1 (1 - p_a)^{\alpha-\rho} \rho(\alpha - \rho)}{V\langle C_{ab}\rangle \left[\sum_{\rho=1}^{\alpha} \binom{\alpha}{\rho} p_a^{\rho} (1 - p_a)^{\alpha-\rho} \rho\right]\left[\sum_{\rho=0}^{\alpha-1} \binom{\alpha}{\rho} p_a^{\rho} (1 - p_a)^{\alpha-\rho} (\alpha - \rho)\right]}$$

      The sum in the numerator can be evaluated in the following way:

$$\sum_{\rho=1}^{\alpha-1} \frac{\alpha!}{\rho!(\alpha-\rho)!} p_a^{\rho}(1-p_a)^{\alpha-\rho}\rho(\alpha-\rho)$$
$$= \sum_{\rho=1}^{\alpha-1} \frac{\alpha!}{(\rho-1)!(\alpha-1-\rho)!} p_a^{\rho}(1-p_a)^{\alpha-\rho}$$
$$= \sum_{\vartheta=0}^{\alpha-2} \frac{\alpha(\alpha-1)(\alpha-2)!}{\vartheta!(\alpha-2-\vartheta)!} p_a(1-p_a)p_a^{\vartheta}(1-p_a)^{\alpha-2-\vartheta}$$
$$= \alpha(\alpha-1)p_a(1-p_a)$$

where we substituted ϑ *for* $\rho - 1$ *and used the formula for the sum of the binomial distribution. The sums in the denominator can be evaluated analogously, leading to:*

$$G_{0X}^{(\alpha)} = \frac{\alpha(\alpha-1)p_a(1-p_a)}{V\langle C_{ab}\rangle \alpha^2 p_a(1-p_a)} = \left(1 - \frac{1}{\alpha}\right)\frac{1}{V\langle C_{ab}\rangle}$$

6. Consider a PIE-FCCS experiment using two pulsed laser lines with 80 MHz repetition rate (e.g. a Ti-sapphire laser commonly used for two-photon fluorescence excitation) shifted by half of their repetition period with respect to each other. Fluorophore *a* has fluorescence lifetime of 4 ns. By what factor will be the spectral crosstalk from channel *A* to channel *B* reduced in the PIE-FCCS experiment compared to the corresponding DC-FCCS experiment? Consider the excitation pulses to be very short and neglect the IRF. By what factor would the crosstalk be reduced if the laser repetition rate is 40 MHz?

   Answer: The relative residual crosstalk is given by the integral of the fluorescence intensity decay of fluorophore a *in the time window for the detection channel* B *divided by the integral of the decay from* 0 *to* ∞ *(DC-FCCS situation):*

$$\frac{\int_{\tau_s/2}^{\tau_s} e^{-\frac{t}{\tau}}dt}{\int_0^{\infty} e^{-\frac{t}{\tau}}dt} = \frac{\tau\left[e^{-\frac{t}{\tau}}\right]_{\tau_s}^{\tau_s/2}}{\tau} = e^{-\frac{\tau_s}{2\tau}}\left(1 + e^{-\frac{\tau_s}{2\tau}}\right)$$

   For 80 *MHz repetition frequency, the switching period,* τ_s, *is* 12.5 *ns, which gives a relative residual crosstalk of approximately* 17%. *By reducing the repetition rate to* 40 *MHz, the relative residual crosstalk is reduced to approximately* 4%.

7. In the 'Introduction to FCS—interactive exercises.cdf' (available at https://iopscience.iop.org/book/978-0-7503-2080-1) test the dependences of FCCS

on various parameters. First let us set all parameters to an ideal case so that ACFs and CCF match. For that purpose, set the following parameter values:

Concentration: 50 nM
Bound Fraction: 1
τ_{D1} and τ_{D2}: 400 μs
Molecular brightness η_{11} and η_{22}: 1500 cts
Crosstalk η_{12} and η_{21}: 0 cts
Background b_1 and b_2: 0 cts
Probability for molecules to be in a bright state p_1 and p_2: 1
FRET factor 1: 1 (corresponding to no FRET)
FRET factor 2: 1 (corresponding to no FRET)
All effective volumes to 5.0×10^{-16}

Now try varying each parameter individually to test how they affect the ACFs and the CCF. Take some time to systematically change the parameters and explain your observations. Here are some examples:

a. Note how the background in the two channels separates the three CFs. What happens if the background in both channels is the same?

   Answer: If the background is the same in both channels, all three correlation function amplitudes are equal (see equations (6.23) *in the case of identical observation volumes, identical brightness of both fluorophores, no crosstalk and a sample containing only double-labelled particles).*

b. Change the three effective volumes. The most common case is that $V_{\text{eff}_1} < V_{\text{eff}_2} < V_{\text{eff}_{12}}$.

   Answer: If one of the effective observation volumes is larger than the other, double-labelled particles present in the part of the larger volume that does not overlap with the smaller volume contribute only to a single detection channel, appearing effectively as single-labelled (see the discussion in section 6.2*). A similar effect is obtained by modifying the probabilities of the fluorophores being fluorescent. A double-labelled particle carrying only a single fluorophore that is fluorescent, appears as single-labelled. Therefore, by, for example, increasing the observation volume of one of the channels and at the same decreasing the probability of the corresponding fluorophore being fluorescent, the two effects can compensate each other.*

c. Vary p_1 and show that you can cancel the effect by changing V_{eff_1}. Why is that so?

   Answer: The CCF amplitude depends on the number of particles. As both the probability p_1 *to be fluorescent and the effective volume* V_{eff_1} *change the concentration of particles they can compensate for each other.*

d. With the parameters set as above, vary η_{12}, η_{21}, and the FRET factors. Why do you see no change? Now set the bound fraction to 0.5 and try again.

Answer: The correlation function amplitudes are in the case of a perfect positive control (double-labelled particles only) and no background entirely independent of brightness and, therefore, also of crosstalk and FRET or quenching. The independence of crosstalk and FRET is lost in the presence of single-labelled particles. Alternatively, you can also try introducing non-zero background.

Chapter 7

1. Suppose you are performing an FCS experiment with a relatively weak fluorophore, and the area of the spectrum under the Raman peak is 5% of the area under the fluorescence peak. If you chose emission filters such that the Raman peak is included in your detection signal, how much error would you have made in the estimation of the real concentration?

   Answer: $G(0) = a^2/N$, *where* $a = \langle F\rangle / (\langle F\rangle + \langle B\rangle)$, *where F is fluorescence, B is background, and N is the true number of particles in the volume. So here* $a^2 = (1/1.05)^2 = 0.907$. *Hence we would be overestimating the concentration by a factor of* $1/0.907 = 1.1025$, *or by* 10.25%.

2. Suppose you are studying a fluorescently labelled dimeric protein. With time, the protein slowly dissociates. Initially, at the laser power you were performing the FCS, there were some signs of photobleaching. With time, what changes do you expect to see in the ACFs?

   Answer: First, the number of particles will go up with dissociation, so G(0) will go down (by a factor 2 when complete dissociation occurs). Second, the diffusion time changes to a smaller value, as the smaller monomers are expected to diffuse faster. However, since the diffusion is faster, the effect of photobleaching should be lower after dissociation.

3. Suppose the molar absorptivity of your dye is 100 000 [l (mol^{-1} cm^{-1})], and its lifetime is 2.62 ns. At what laser power of your 543 nm laser (which is focused through an objective of 1.2NA) would you expect to start seeing substantial saturation?

   Answer: The molecular cross section is $3.82 \times 10^{-21} \times 100\,000 = 3.82 \times 10^{-16}\ \mathrm{cm}^2$. *Now, the effects of saturation are already observable when there is* ~10% *probability that a second excitation photon arrives when the molecule is still in the excited state. Then, approximately, intensity I is such that,* $I \times 3.82 \times 2.62 \times 10^{-25} = 0.1$. *This implies* $I = 10^{23}$ *photons/*($\mathrm{cm}^2{\cdot}\mathrm{s}$). *If we assume that the focus area is diffraction limited with a radius of about* 270 nm *(given $NA = 1.2$), this translates to about* 2.3×10^{14} *photons/s at the focus. So about* 83 μW *of laser power at the sample will cause noticeable saturation.*

Chapter 8

The accompanying CDF file for fitting (available at https://iopscience.iop.org/book/978-0-7503-2080-1) allows testing data fitting in three scenarios: (i) a one-component data set can be fitted to a one-component model; (ii) a two-component data set can be fitted to a one-component model; (iii) a two-component data set can be fitted to a two-component model. The following questions have no strict solutions but are exploratory exercises for the reader to investigate how residuals are influenced by the fitting models and how different parameters influence the outcome and residuals of fits.

1. Test how the structure parameter K influences the fits.
    a) Note that for high noise or wrong fit model it diverges and needs to be fixed to perform a proper fit.
    b) Investigate how the residuals change when a wrong structure factor is fixed in the fit.
2. Observe the difference in the residuals when a two-component data set is fitted with either a one- or two-component fit.
3. Observe how weighted fits improve on data fitting and how that is influenced by the structure factor and the signal-to-noise ratio (i.e. *CPS*).
4. Investigate how well two diffusive components can be solved …
    a) … in dependence of the ratio of D_2/D_1, keeping the fraction $F_2 = 0.5$ and changing the signal-to-noise ratio (i.e. the *CPS*).
    b) … in dependence on the fraction F_2, keeping $D_2/D_1 = 5$, and changing the signal-to-noise ratio (i.e. the *CPS*).

Chapter 9

1. Consider a typical confocal setup with the effective observation volume $V_{\text{eff}} = 0.5$ fL.
    a. At what fluorescent particle concentration is the average particles number, N, equal to 1? Express the concentration both as number concentration C and as molar concentration c.

       Answer: $C = 2 \times 10^{15}\ \text{L}^{-1}$, $c \doteq 3$ nM.

    b. Under the condition that $N = 1$, what is the expected fraction of FCS acquisition time during which the effective observation volume is empty? What is the expected fraction of the acquisition time during which the effective observation volume contains more than one particle? *Hint*: use the fact that the probability of finding a particular number of particles, ξ, in the effective observation volume follows Poisson distribution:

$$P(\xi) = \frac{N^{\xi} e^{-N}}{\xi!}$$

Answer: The fraction of time for which the effective observation volume is void of particles is equal to the probability

$$P(0) = e^{-1} \doteq 0.37$$

The probability of finding more than 1 particle in the effective observation volume is then:

$$P(\xi > 1) = 1 - P(0) - P(1) = 1 - 2e^{-1} \doteq 0.26$$

2. Apply number and brightness analysis on the simulated intensity time traces shown in figure 9.2. The mean and standard deviation of the 100 simulated values are for the trace shown in panel A: $\langle F_A \rangle = 108$, $\sigma_A = 49$, and for the trace shown in panel B: $\langle F_B \rangle = 96$, $\sigma_B = 22$. Compare the results with the particle number and brightness values used in the simulations

   Answer: For panel A we get: $\kappa \doteq 21$, $N \doteq 5.1$, *for panel B:* $\kappa \doteq 4.0$, $N \doteq 24$. *If we neglect the shot noise variance correction, we get for panel A:* $\kappa \doteq 22$, $N \doteq 4.9$ *and for pane B* $\kappa \doteq 5.0$, $N \doteq 19$.

3. Derive equation (9.20) by deriving the expressions for the first and the second moment of binomial probability distribution $P_B(\eta)$. *Hint*: use the facts the sum of the binomial distribution is equal to 1:

$$\sum_{\eta=0}^{H} \binom{H}{\eta} p^{\eta} (1 - p)^{H-\eta} = 1, \text{ for } 0 \leqslant p \leqslant 1$$

   and use the explicit expression for the binomial coefficients:

$$\binom{H}{\eta} = \frac{H!}{\eta!(H - \eta)!}$$

   Answer: The first moment of $P_B(\eta)$:

$$\begin{aligned} &\sum_{\eta=0}^{H} \frac{H!}{\eta!(H - \eta)!} \eta\, p^{\eta} (1 - p)^{H-\eta} \\ &\quad = \sum_{\eta=1}^{H} \frac{H!}{(\eta - 1)!(H - \eta)!} p^{\eta} (1 - p)^{H-\eta} \\ &\quad = \sum_{\alpha=0}^{H-1} \frac{H(H - 1)!}{\alpha!(H - 1 - \alpha)!} p p^{\alpha} (1 - p)^{H-1-\alpha} = Hp \end{aligned}$$

   First, we used the fact that the term for $\eta = 0$ *is equal to* 0 *to change the lower summation bound to* 1 *and then we substituted* α *for* $\eta - 1$, *which allowed us to use the formula for the sum of the binomial distribution. Analogously, we get the second moment of* $P_B(\eta)$:

$$\sum_{\eta=0}^{H}\frac{H!}{\eta!(H-\eta)!}\eta^2\,p^\eta(1-p)^{H-\eta}$$
$$=\sum_{\eta=0}^{H}\frac{H!}{\eta!(H-\eta)!}(\eta^2-\eta+\eta)p^\eta(1-p)^{H-\eta}$$
$$=\left[\sum_{\eta=2}^{H}\frac{H!}{\eta!(H-\eta)!}\eta(\eta-1)p^\eta(1-p)^{H-\eta}\right]+Hp$$
$$=\left[\sum_{\beta=2}^{H-2}\frac{H(H-1)(H-2)!}{\beta!(H-2-\beta)!}p^2p^\beta(1-p)^{H-2-\beta}\right]+Hp$$
$$=H(H-1)p^2+Hp$$

4. Consider the avidin–biotin binding experiment in which a solution of fluorescently labelled biotin is titrated by unlabelled avidin. By fitting the measured ACF with the following model:

$$G^{(\mathrm{conc})}(\tau)=\frac{1}{N_{app}}\left[f\left(1+\frac{\tau}{\tau_{D_L}}\right)^{-1}\left(1+\frac{\tau}{K^2\tau_{D_L}}\right)^{-1/2}\right.$$
$$\left.+(1-f)\left(1+\frac{\tau}{\tau_{D_S}}\right)^{-1}\left(1+\frac{\tau}{K^2\tau_{D_S}}\right)^{-1/2}\right]+G_\infty$$

where τ_{D_L} and τ_{D_S} are the diffusion times of free biotin and of the avidin–biotin complex, you find that the apparent fraction, f, of the faster component is ½.

   a. What is the actual fraction of free biotin molecules in the sample, assuming that the brightness of the complex is four times the brightness of a free fluorescently labelled biotin molecule?

   Answer: The correlation function can be written using the knowledge of the ligand-substrate complex stoichiometry, H, *and brightness,* $\kappa_S = H\kappa_L$, *as:*

$$G^{(\mathrm{conc})}(\tau)=\frac{1}{(N_L\kappa_L+N_SH\kappa_L)^2}\left[N_L\kappa_L^2\left(1+\frac{\tau}{\tau_{D_L}}\right)^{-1}\left(1+\frac{\tau}{K^2\tau_{D_L}}\right)^{-1/2}\right.$$
$$\left.+N_SH^2\kappa_L^2\left(1+\frac{\tau}{\tau_{D_S}}\right)^{-1}\left(1+\frac{\tau}{K^2\tau_{D_S}}\right)^{-1/2}\right]+G_\infty$$

where N_L and N_S are the particle numbers of the ligand and the ligand-substrate complex. If we introduce the total ligand particle number (comprising free ligand as well as the ligand molecules involved in the complexes), $N_L^{(tot)}$, and the fraction of free ligand, φ, the ACF becomes:

$$G^{(\mathrm{conc})}(\tau) = \frac{1}{N_L^{(tot)}}\left[\varphi\left(1 + \frac{\tau}{\tau_{D_L}}\right)^{-1}\left(1 + \frac{\tau}{K^2\tau_{D_L}}\right)^{-1/2} + H(1-\varphi)\left(1 + \frac{\tau}{\tau_{D_S}}\right)^{-1}\left(1 + \frac{\tau}{K^2\tau_{D_S}}\right)^{-1/2}\right] + G_\infty$$

Comparing this model with the one using N_{app} and f, we can express $N_L^{(tot)}$ and φ using N_{app} and f:

$$\varphi = \frac{fH}{1-f+fH}$$

$$N_L^{(tot)} = \frac{N_{app}H}{1-f+fH}$$

For $f = 0.5$ and $H = 4$, we get the fraction $\varphi = 0.8$. The apparent fraction f, therefore, underestimates the free ligand fraction by not considering the higher brightness of the complex.

b. What is the actual particle number under such situation?

Answer:

$$N = N_L^{(tot)}\left(\varphi + \frac{1-\varphi}{H}\right) = N_{app}\frac{1-f+fH^2}{(1-f+fH)^2}$$

which is in this case $1.36\ N_{app}$.

c. What is the average particle brightness? Compare it with the apparent average brightness, $\langle F\rangle / N_{app}$.

Answer: The average brightness:

$$\bar{\kappa} = \frac{N_L\kappa_L + N_S\kappa_S}{N} = \frac{\kappa_L}{\varphi + \frac{1-\varphi}{H}} = \kappa_L H\frac{1-f+fH}{1-f+fH^2}$$

The apparent brightness:

$$\kappa_{app} = \frac{N\bar{\kappa}}{N_{app}} = \frac{\kappa_L H}{1-f+fH}$$

In this case, $\bar{\kappa} \doteq 1.2\kappa_L$ *and* $\kappa_{app} = 1.6\kappa_L$.

www.ingramcontent.com/pod-product-compliance
Ingram Content Group UK Ltd.
Pitfield, Milton Keynes, MK11 3LW, UK
UKHW051905100926
8556IPUK00005B/773

9 780750 3207